"Impressive in its disciplinary, geographical, and topical breadth, Environmental Justice: Key Issues examines the many meanings of justice inherent to the field. Written as an introductory text, it manages to present scholarship that is simultaneously accessible and cutting-edge. A tremendous contribution for all teachers and students of environmental justice."

—Professor Laura Pulido, Indigenous, Race, and Ethnic Studies Department, University of Oregon

"This is an excellent collection of writing about the foundations, diversity and future of environmental justice, as both a site of activism and academic research. The range of authors is particularly impressive, spanning multiple disciplines, perspectives and contexts. For those studying environmental justice, there is much in here that will inform and inspire."

—Professor Gordon Walker, Lancaster Environment Centre, Lancaster University

"Over the last decades, the study of environmental justice has become one of the most prominent fields in global sustainability governance. This timely new volume, carefully edited by Brendan Coolsaet, now brings together the Who's Who of environmental justice scholarship for one of the first comprehensive reviews of all key issues at stake, covering the main concepts and issues and charting new research directions. The book is highly recommended for use in the classroom but also provides new insights and reflections for scholars and practitioners."

—Frank Biermann, Copernicus Institute of Sustainable Development, Utrecht University

Environmental Justice

Environmental Justice: Key Issues is the first textbook to offer a comprehensive and accessible overview of environmental justice, one of the most dynamic fields in environmental politics scholarship.

The rapidly growing body of research in this area has brought about a proliferation of approaches; as such, the breadth and depth of the field can sometimes be a barrier for aspiring environmental justice students and scholars. This book therefore is unique for its accessible style and innovative approach to exploring environmental justice. Written by leading international experts from a variety of professional, geographic, ethnic, and disciplinary backgrounds, its chapters combine authoritative commentary with real-life cases. Organised into four parts—approaches, issues, actors and future directions—the chapters help the reader to understand the foundations of the field, including the principal concepts, debates, and historical milestones. This volume also features sections with learning outcomes, follow-up questions, references for further reading and vivid photographs to make it a useful teaching and learning tool.

Environmental Justice: Key Issues is the ideal toolkit for junior researchers, graduate students, upper-level undergraduates, and anyone in need of a comprehensive introductory textbook on environmental justice.

Brendan Coolsaet is Associate Professor of Environmental Politics at ESPOL, the European School of Political and Social Sciences at Lille Catholic University, France.

Key Issues in Environment and Sustainability

This series provides comprehensive, original and accessible texts on the core topics in environment and sustainability. The texts take an interdisciplinary and international approach to the key issues in this field.

Low Carbon Development: Key Issues
Edited by Frauke Urban and Johan Nordensvärd

Sustainable Business: Key Issues
Helen Kopnina and John Blewitt

Sustainability: Key Issues
Helen Kopnina and Eleanor Shoreman-Ouimet

Ecomedia: Key Issues
Edited by Stephen Rust, Salma Monani and Sean Cubitt

Ecosystem Services: Key Issues
Mark Everard

Sustainability Science: Key Issues
Edited by Ariane König and Jerome Ravetz

Sustainable Business: Key Issues (Second Edition)
Helen Kopnina and John Blewitt

Sustainable Consumption: Key Issues
Lucie Middlemiss

Human Rights and the Environment: Key Issues
Sumudu Atapattu and Andrea Schaper

The Anthropocene: Key Issues for the Humanities
Eva Horn and Hannes Bergthaller

Environmental Justice: Key Issues
Edited by Brendan Coolsaet

Environmental Justice

Key Issues

Edited by Brendan Coolsaet

First published 2021
by Routledge
2 Park Square, Milton Park, Abingdon, Oxon OX14 4RN

and by Routledge
52 Vanderbilt Avenue, New York, NY 10017

Routledge is an imprint of the Taylor & Francis Group, an informa business

© 2021 selection and editorial matter, Brendan Coolsaet; individual chapters, the contributors

The right of Brendan Coolsaet to be identified as the author of the editorial material, and of the authors for their individual chapters, has been asserted in accordance with sections 77 and 78 of the Copyright, Designs and Patents Act 1988.

All rights reserved. No part of this book may be reprinted or reproduced or utilised in any form or by any electronic, mechanical, or other means, now known or hereafter invented, including photocopying and recording, or in any information storage or retrieval system, without permission in writing from the publishers.

Trademark notice: Product or corporate names may be trademarks or registered trademarks, and are used only for identification and explanation without intent to infringe.

British Library Cataloguing-in-Publication Data
A catalogue record for this book is available from the British Library

Library of Congress Cataloging-in-Publication Data
Names: Coolsaet, Brendan, editor.
Title: Environmental justice : key issues / edited by Brendan Coolsaet.
Other titles: Environmental justice (Routledge)
Description: Abingdon, Oxon ; New York, NY : Routledge, 2020. | Series: Key issues in environment and sustainability | Includes bibliographical references and index.
Identifiers: LCCN 2020006760 (print) | LCCN 2020006761 (ebook) | ISBN 9780367139926 (hardback) | ISBN 9780367139933 (paperback) | ISBN 9780429029585 (ebook)
Subjects: LCSH: Environmental justice. | Environmental justice—Research. | Environmental justice—Case studies.
Classification: LCC GE220 .E586 2020 (print) | LCC GE220 (ebook) | DDC 362.7—dc23
LC record available at https://lccn.loc.gov/2020006760
LC ebook record available at https://lccn.loc.gov/2020006761

ISBN: 978-0-367-13992-6 (hbk)
ISBN: 978-0-367-13993-3 (pbk)
ISBN: 978-0-429-02958-5 (ebk)

Typeset in Goudy
by Apex CoVantage, LLC

For Eleonora – may her generation inherit a more environmentally-just world.

Contents

List of figures	*xii*
List of boxes	*xiii*
List of tables	*xv*
List of plates	*xvi*
List of contributors	*xvii*
Acknowledgements	*xx*
Foreword Robert D. Bullard	*xxi*

1   Introduction   1
BRENDAN COOLSAET

2   A history of environmental justice: foundations, narratives, and perspectives   6
ESME G. MURDOCK

PART I
Defining and conceptualizing environmental justice   19

3   Distributive environmental justice   21
ALICE KASWAN

4   Procedural justice matters: power, representation, and participation in environmental governance   37
KIMBERLY R. MARION SUISEEYA

5   Recognition and environmental justice   52
BRENDAN COOLSAET AND PIERRE-YVES NÉRON

6   Capabilities, well-being, and environmental justice   64
BREENA HOLLAND

7   Latin American decolonial environmental justice   78
IOKIÑE RODRIGUEZ

8 Degrowth and environmental justice: an alliance between two movements? 94
JULIEN-FRANÇOIS GERBER, BENGI AKBULUT, FEDERICO DEMARIA, JOAN MARTÍNEZ-ALIER

9 Sustainability and environmental justice: parallel tracks or at the crossroads? 107
JULIE SZE

PART II
Issues of environmental justice 119

10 Toxic legacies and environmental justice 121
ALICE MAH

11 Biodiversity: crisis, conflict and justice 132
ADRIAN MARTIN

12 Climate justice 148
GARETH A.S. EDWARDS

13 Energy justice 161
ROSIE DAY

14 Food, agriculture, and environmental justice: perspectives on scholarship and activism in the field 176
KRISTIN REYNOLDS

15 Urbanisation: towns and cities as sites of environmental (in)justice 193
JASON BYRNE

16 Water justice: blatant grabbing practices, subtle recognition politics and the struggles for fair water worlds 207
RUTGERD BOELENS

PART III
Actors and subjects of environmental justice 223

17 Racial minorities in the United States: race, migration, and reimagining environmental justice 225
LISA SUN-HEE PARK AND STEVIE RUIZ

18 Gender matters in environmental justice 234
SHERILYN MACGREGOR

19 Labour unions and environmental justice: the trajectory and politics of
    just transition   249
    DIMITRIS STEVIS

20 Indigenous environmental justice: anti-colonial action through kinship   266
    KYLE WHYTE

21 Justice beyond humanity   279
    STEVE COOKE

PART IV
Future directions of environmental justice   291

22 Critical environmental justice studies   293
    DAVID N. PELLOW

23 Sustainable materialism and environmental justice   303
    DAVID SCHLOSBERG

24 Mobilizing 'intersectionality' in environmental justice research and
    action in a time of crisis   316
    GIOVANNA DI CHIRO

    Index   334

Figures

12.1 The three-dimensional climate justice pyramid 157

Boxes

3.1	Siting decisions and racial segregation in Richmond, California	26
3.2	The Flint water crisis	28
4.1	Key concepts in procedural justice	38
4.2	Tyranny of participation	44
4.3	Free prior informed consent	46
5.1	Hegel's master–slave dialectic	53
5.2	Recognition versus distribution: the Fraser–Honneth debate	57
5.3	Recognition and biodiversity conservation	60
6.1	What are capabilities and functionings?	65
6.2	Nussbaum's list of central human capabilities	67
7.1	Two important recent Latin American environmental justice mobilizations	81
7.2	Key features of the concept of *Buen Vivir*	83
8.1	The politics of the social metabolism	97
8.2	Do we need growth to fight poverty?	101
9.1	Sustainability from an ecological science perspective	110
9.2	Sustainable forestry and the Sustainable Development Institute on Menominee tribal lands	114
10.1	Love Canal	123
10.2	Cancer Alley	124
11.1	Food systems, biodiversity and recognition injustices	136
11.2	Debates about the future of conservation	141
12.1	Calling for climate justice	152
12.2	Conceptualising climate justice as a 'pyramid'	156
13.1	Justice concerns in solar panel production and disposal	165
13.2	Energy services and their relationship with everyday needs	167
14.1	Concept in action: solidarity activism for farmworker livelihoods	177
14.2	Black farmers in the US	180
14.3	The 'food desert' debate	182
14.4	Urban agriculture as activism	185
15.1	Environmental impacts of urbanisation	196
15.2	Drivers of urban environmental inequality	198
15.3	Rapid urbanisation in Asia and Africa	200
16.1	Agro-export or food security? Unequal water allocation and access patterns	210
16.2	Rural to urban water transfers: drinking water for the poor?	211
16.3	Integrated water resources management	213
17.1	Re-envisioning environmentalism	228

18.1	Principles of gender justice	236
18.2	The Green Belt Movement	240
21.1	Utilitarianism and non-human animals	280
21.2	Moral agents and moral patients	281
21.3	Including non-human animals: from theory to practice	287
22.1	Case study: US prisons and critical environmental justice	295

Tables

4.1	Examples of tools for participation	43
7.1	Environmental injustices generated by different forms of coloniality	86
8.1	Complementarities between environmental justice and degrowth	102

Plates

1 The government of the Maldives holds an underwater cabinet meeting to raise awareness about the threats of climate change for Small Island Developing States (SIDS). While being among the world's smallest emitters of greenhouse gases, the SIDS are the most affected by climate change.
2 An estimated 400,000 people took the streets demanding climate justice during the 2014 People's Climate March in New York City, USA. See Chapter 12 of this volume.
3 1500 anti-coal activists shutting down the Tagebau Garzweiler surface mine during the 2015 *Ende Gelände* ("here and no further") protests in North-Rhine Westphalia, Germany. In 2018, mining was found to be the deadliest environmental cause around the world, according the non-governmental organization Global Witness. See Chapter 8 of this volume.
4 Squatters erecting a barricade in a 'Zone-to-Defend' (ZAD) in Notre-Dame-des-Landes, France, against a boondoggle airport construction project. Starting in the 1970s, opposition to the project is said to have been France's longest running struggle for land. The project was canceled in 2018.
5 A resident in Flint, Michigan, USA collects water from a distribution center during the lead-contaminated water crisis, which Paul Mohai called 'the most egregious example of environmental injustice'. The crisis started in 2014. See Chapters 3, 10 and 16 of this volume.
6 A boy scavenging for metals at the world's largest e-dump, Agbogbloshie, Ghana, where most of the locals die from cancer while in their 20s. Unjust exposure to toxic hazards is a pivotal and long-standing concern within the environmental justice movement. See Chapter 10 of this volume.
7 A woman with a "Water is Life" headscarf protesting the construction of the Dakota Access Pipeline (DAPL) near the Standing Rock Indian Reservation, Northern USA. Beyond the ecological impacts, the #NoDAPL movement laid bare the common roots of carbon-intensive industries and settler colonialism in the USA. See Chapters 5, 7 and 20 of this volume.
8 In 2017, Mount Taranaki on North Island, New Zealand was given "legal personality, in its own right" by the government. Eight local Māori tribes and the government share guardianship of the sacred mountain. See Chapter 21 of this volume.

Contributors

Bengi Akbulut is Assistant Professor at the Department of Geography, Planning and Environment, Concordia University, Montreal, Canada.

Rutgerd Boelens is Professor of Water Governance and Social Justice at Wageningen University; Professor of Political Ecology of Water in Latin America with CEDLA, University of Amsterdam; and Visiting Professor at the Catholic University of Peru and the Central University of Ecuador. He directs the international Justicia Hídrica/Water Justice alliance (www.justiciahidrica.org). His research focuses on political ecology, water rights, legal pluralism, cultural politics, governmentality, hydrosocial territories, and social mobilization.

Jason Byrne is Professor of Human Geography and Planning at the University of Tasmania, where he teaches about—and researches—environmental justice issues related to city planning, greenspace provision, green infrastructure, urban ecology, and climate change adaptation.

Steve Cooke lectures in political philosophy at the University of Leicester. His research focuses on the ethical status of non-human animals and of their place within theories of justice. He is particularly interested in the ethics of direct action and the exercise of political authority in defence of non-humans and the environment.

Brendan Coolsaet is Associate Professor at the European School of Political and Social Sciences (ESPOL) at Lille Catholic University and Research Fellow with the Global Environmental Justice research group at the University of East Anglia. He works on environmental justice, biodiversity governance, and food politics.

Rosie Day is Senior Lecturer in Human Geography at the University of Birmingham, where her teaching and research addresses environmental knowledges and environmental justice in relation to energy and water, especially, in various global contexts.

Federico Demaria is Deputy Coordinator of the EnvJustice project at ICTA-UAB, Visiting Fellow at the ISS in The Hague, and a founding member of Research & Degrowth.

Giovanna Di Chiro is Professor of Environmental Studies at Swarthmore College, where she teaches courses on environmental justice theory, action research methods, ecofeminism(s), and community sustainability. She has published widely on the intersections of environmental science, policy, and grassroots activism, focusing on environmental and climate justice.

Gareth A.S. Edwards is Senior Lecturer in the School of International Development at the University of East Anglia. Positioned at the interface between political ecology and environmental justice, his research is concerned with understanding and theorising discourses and practices of justice in the context of environmental governance.

Julien-François Gerber is Assistant Professor of Environment and Development at the ISS in The Hague and a member of Research & Degrowth.

Breena Holland is Associate Professor at Lehigh University, where she holds a joint appointment in the Department of Political Science and the Environmental Initiative. Her research draws on capabilities theory to improve the design and evaluation of environmental policy in the USA in order to address various forms of environmental injustice.

Alice Kaswan has been a professor of law at University of San Francisco since 1999. Her writing focuses on environmental justice, climate justice, and climate governance. She received her BS in conservation and resource studies, with highest honours, from the University of California, Berkeley, and her JD, with honours, from Harvard Law School.

Sherilyn MacGregor is Reader in Environmental Politics, in the Politics Department and Sustainable Consumption Institute at the University of Manchester. Her research investigates the synergies and tensions between feminist and environmental politics. She is editor of *The Routledge Handbook of Gender and Environment* (2017).

Alice Mah is Professor of Sociology at the University of Warwick, and her most recent book is *Toxic Truths: Environmental Justice and Citizen Science in a Post-Truth Age* (with Thom Davies, Manchester University Press, 2020). Her research interests include global environmental injustice, legacies of industrial ruination, corporate environmentalism, and political ecology.

Adrian Martin is Professor of Environment and Development in the School of International Development, University of East Anglia. He has extensively researched and written about conservation in the Global South, exploring ways in which conservation practices can be better aligned with the needs and rights of local communities. This work has become strongly informed by ideas of environmental justice.

Joan Martínez-Alier is Emeritus Professor of Economic History and Institutions at the UAB, and a founding member and former president of the International Society for Ecological Economics.

Esme G. Murdock is Assistant Professor of Philosophy at San Diego State University. Her research explores the intersections of social/political relations and environmental health, integrity, and agency.

Pierre-Yves Néron is Associate Professor of Political Philosophy at the European School of Political and Social Sciences (ESPOL), Lille Catholic University. A specialist of contemporary political philosophy, his current research focuses on recent developments in egalitarian thought, especially the debates between distributive and relational accounts of it, contemporary studies of inequalities and the very idea of economic democracy.

Lisa Sun-Hee Park is Professor and Chair of Asian American Studies at the University of California, Santa Barbara, with affiliations in feminist studies and sociology. Her interdisciplinary research focuses on the politics of migration, race, and gender in health care and environmental justice.

David N. Pellow is the Dehlsen Chair and Professor of Environmental Studies and Director of the Global Environmental Justice Project at the University of California, Santa Barbara. His teaching and research focus on environmental and ecological justice in the USA and globally.

Kristin Reynolds is a critical food geographer and independent scholar in New York City. She teaches courses on food systems and the environment at The New School, and at Yale School of Forestry and Environmental Studies. She holds a PhD in geography and MS in international agricultural development from the University of California, Davis.

Iokiñe Rodriguez is a Venezuelan sociologist at the University of East Anglia, UK. Her work on environmental conflict transformation in Latin America focuses on issues of local history, local knowledge, power, environmental justice, equity, and intercultural dialogue, using participatory action research. She has worked in Venezuela, Guyana, Ecuador, Chile, Bolivia, and Colombia, building local and institutional capacity to transform environmental conflicts.

Stevie Ruiz is Assistant Professor of Chicana/o Studies at California State University, Northridge. His research focuses on chicana/o studies, environmental history, and land rights.

David Schlosberg is Professor of Environmental Politics in the Department of Government and International Relations, and Director of the Sydney Environment Institute, at the University of Sydney.

Dimitris Stevis is Professor of Politics at Colorado State University. His research focuses on the political economy of global labour and environmental politics, with particular attention to labour environmentalism and social and ecological justice. He is a founder of the Center for Environmental Justice at Colorado State University and most recently co-edited an issue of the journal *Globalizations* entitled 'Labour in the Web of Life'.

Kimberly R. Marion Suiseeya is Assistant Professor of Political Science and Environmental Policy and Culture at Northwestern University. Trained as an interdisciplinary environmental social scientist, her research examines the relationships between institutions, norms, and justice in global environmental governance and their impacts on communities in Southeast Asia.

Julie Sze is Professor and the Founding Chair of American Studies at the University of California, Davis. She was also the founding director of the Environmental Justice Project for the John Muir Institute of the Environment. She has published three books, numerous articles, and a collection on sustainability and social justice. Her research is on environmental inequalities, the relationship between social movements and policy implementation and in the areas of public and environmental humanities.

Kyle Whyte is Professor and Timnick Chair in the departments of Philosophy and Community Sustainability at Michigan State University. Kyle focuses on Indigenous environmental justice in advocacy, research and teaching.

Acknowledgements

The chapters in this book are the result of tireless and passionate scholarly work conducted by the wonderful colleagues included in this book. I am indebted to them, and I would like to thank them for trusting me with this project. By gathering their work in this book, I hope to encourage a new generation of environmental justice students and scholars to follow in their footsteps. Our collective work is inspired by the numerous communities across the world fighting for and contributing to a more environmentally just world. This book humbly tries to do justice to their stories.

I am grateful to my colleagues at the European School of Political and Social Sciences (ESPOL) for their support and friendship. This book was made possible with matching funds provided by Lille Catholic University's Research Commission and ESPOL's Research Centre for European and International Politics (ESPOL-Lab), for which I am very appreciative.

These funds allowed paying for copy-editing by Marie Selwood, who helped us make this book accessible for a broad audience and find the right balance between text coherence and complexity. The funds also allowed acquiring reproduction rights for the photographs featured in the book. I am deeply grateful to the authors of the photos for allowing me to use them in the context of this work.

Special thanks also go to David N. Pellow for his advice and encouragements when we discussed the initial idea of this book in New Orleans in 2018. At Routledge, I would like to thank my editor Rebecca Brennan and editorial assistants Rosie Anderson and Leila Walker for their availability and support, and I am grateful to the six anonymous reviewers for their time, their ideas, and their constructive criticism at the very early stages of the process.

Foreword

We have come a long way since the early days of the study of environmental injustice. In the United States, 25 years after the historic Environmental Justice Executive Order 12898 was signed by US President Bill Clinton, all states now have instituted some type of environmental justice law, executive order, or policy. Out of the small and seemingly isolated environmental struggles emerged a potent grassroots community-driven movement that is now a global movement. Thousands of environmental justice groups have emerged, and numerous environmental justice leaders have won recognition for their groundbreaking work.

And yet, after decades of hard work, struggle, and some victories along the way, we must recognise that much more work remains to be done. Race and poverty continue to be powerful predictors of the level of pollution communities have to endure. Too many residents in frontline communities still have the "wrong complexion for protection." They continue to fight for justice as if their lives and their children's health depended on it. Because it does. Last year, a study published in the Proceedings of the National Academy of Sciences (PNAS) found that Blacks and Hispanics in the US still experience 56% and 63%, respectively, more pollution than their consumption would generate.

This year marks the 30th anniversary of the publication of my book *Dumping in Dixie* (Bullard, R.D. 1990. *Dumping in Dixie: Race, Class, and Environmental Quality*. Boulder: Westview Press), which was one of the few environmental justice books in the 1990s. The environmental justice framework we developed grew out of grassroots community struggles committed to equal protection of environmental, energy, health, employment, education, housing, transportation, and civil and human rights for all people and communities.

Environmental justice research has since become one of the most dynamic fields in environmental politics scholarship. Hundreds of environmental justice books line the shelves of bookstores, and the subject is now widely taught at universities across the world, including by many of the contributors of this book. But this rapidly growing body of work has brought about a proliferation of terms, concepts and approaches, all of which draw on different community struggles, have distinctive philosophical foundations, and require adopting different policy measures.

Environmental Justice: Key Issues is an accessible handbook to help students and scholars navigate the environmental justice framework. The cast of distinguished colleagues provides the cutting edge of environmental justice understanding in their specific fields, from a wide variety of professional, geographic, racial, and disciplinary backgrounds. From the urban farming sites in New York City, the Menominee tribal lands in Wisconsin and the contaminated drinking water in Flint and Standing Rock, to the Pemon ancestral lands in Venezuela, the crop fields in Europe, and the shorelines of the Pacific island countries, the book draws on some of the voices from the many remaining sacrifice zones across the world, in both the Global North and the Global South. The accessibility of the book will make it an essential tool for training the next generation of environmental justice scholars.

Dr. Robert D. Bullard
Distinguished Professor of Urban Planning and Environmental Policy
Texas Southern University

1 Introduction

Brendan Coolsaet

As this book goes to press, we are leaving a decade in which the environment has (finally) become one of the main issues of global concern. From extreme bushfires in Australia to massive flooding in Indonesia, the turn of the decade coincided with a series of dramatic events, painfully reminding us of the urgency of environmental action.

While current greenhouse gas emission levels leave little room for optimism, the United States (US) and the European Union (EU), the two main historical contributors to greenhouse gas emissions, nevertheless experienced a major shift in public opinion over the last decade. In the US, 53% of the population now considers climate change to be "an urgent problem" warranting immediate action (ABC News/Stanford/University Resources for the Future, 2018). In the EU, 60% thinks climate change is one of the most serious problems facing the world, ahead of international terrorism (54%) or the economic situation (40%) (European Commission, 2019).

The most visible expression of this heightened awareness in recent years has undoubtedly been the Fridays for Future movement. The youth-led movement, led by Swedish schoolgirl Greta Thunberg, has inspired school students across the world to take part in so-called "school strikes" for the climate, in which an estimated 9.6 million strikers in 261 countries have participated in the last two years. Some of the readers of this book have likely been a part of this.

Beyond the sheer magnitude, and the fact that it was led by young school students, a striking feature of the movement was the language it was using. Climate justice, a theretofore little-used term in public debates on the environment, became the rallying cry for millions of likeminded kids. In a letter published by the British newspaper *The Guardian* in March 2019, the coordination group of Fridays for Future demanded "justice for all past, current and future victims of the climate crisis" and promised to "rise until [seeing] climate justice" (The Guardian, 2019). In August 2019, the movement laid out its official demands in the Lausanne Climate Declaration (2019, p. 3), including a call to "ensure climate justice and equity". While undoubtedly implicit in their demands, the group did not call for agenda items like "stopping dangerous climate change" or "halting biodiversity loss" but, instead, chose to focus on the idea of justice.

Although the injustices at hand were never clearly identified or expressed, the movement's word choice did provide visibility to the idea that social difference and environment intertwine. Wittingly or not, claims like the ones made by the Fridays for Future movement are drawing on the long-standing struggles of the numerous environmental justice movements around the world. The purpose of this book it to help the reader make sense of such justice-related claims in the context of the environment. It helps to understand their emergence and their political implications but also to spot the limits of certain types of discourses, including the one used in the previous example.

The book sets out to offer an introductory overview of the field of environmental justice, understood either as a political discourse, a grassroots movement, or a field of study. Written in an accessible style, it incorporates numerous features and elements to make it a useful teaching and learning tool. The chapters of the book include:

- **learning outcomes** introducing the main topics and allowing for a focused and structured reading of the chapters;
- **themed boxes** illustrating the main text with case studies and historical events, summarizing important concepts or diving deeper into critical issues related to the chapter;
- **bolded key terms** throughout the text help identify and learn new terms;
- **cross-chapter references** help the reader draw connections between different approaches, subjects, actors, and issues of environmental justice; and
- **end-of-chapter material,** including follow-up questions, references for further reading and chapter bibliographies, assists the reader to reflect and engage further with specific issues and find more detailed coverage on the subject of the chapter.

The book also contains a series of vivid photographs to illustrate the realities of some of the numerous ongoing environmental justice struggles across the world.

Structure of the book

The book starts with a broad historical overview of environmental justice by Esme G. Murdock. Going back to the origins of the usage of the term as it emerged in the United States in the 1980s, Murdock in Chapter 2 tells the story of a movement concerned with the disproportionate burdening of communities of colour and low-income populations with environmental ills. This history is grounded in the lived experiences and realities of the communities experiencing multiple forms of domination and oppression and, as such, examines various forms of resistance enacted by communities to realize environmental justice. In light of this narrative, and by way of introduction to the broad diversity of approaches in this volume, Murdock asks what environmental justice for all truly means in today's world.

The rest of the volume is structured into four parts, each with its own environmental justice theme on approaches; issues; actors and subjects; and future directions.

As this book does not aim to provide readers with a single, straightforward definition of environmental justice, Part I explores some of the most common theories and concepts used by scholars to define the "justice" of environmental justice. The first half of Part I focuses on what have arguably become the four dominant theories of environmental justice: distribution (Chapter 3), participation and procedure (Chapter 4), recognition (Chapter 5), and capabilities (Chapter 6).

Alice Kaswan starts by introducing us to the core distributive justice concepts (utilitarianism, equality, and well-being). **Distributive justice** theory helps us understand how the distribution of environmental goods and environmental ills follows predictable patterns of domination and oppression in our societies. One of the mechanisms identified for achieving distributive justice is a procedural one, which is the subject covered in Chapter 4 by Kimberly R. Marion Suiseeya. Drawing on key concepts such as power, representation, deliberation, and democracy, Suiseeya reminds us that meaningful involvement in decision-making bodies has long been essential to the environmental justice movement. But participation does not necessarily equate to **procedural justice**, which requires consideration of how power and influence work in environmental governance.

One way in which power is exerted is by recognizing the concerns of others, or by failing to do so. The next chapter, by Pierre-Yves Néron and myself, looks at how dynamics of **recognition** play out in the context of the environment. Working through key theories and approaches of recognition, we show how people express their differences through a multitude of relations to the world around them, and how this influences environmental action and relates to environmental injustices. In the following chapter, Breena Holland then introduces us to the **capabilities approach**, and how it helps us to conceptualize environmental justice concerns as broader questions about human well-being.

The second half of Part I explores three less well-established approaches to studying and enriching environmental justice. In Chapter 7, Iokiñe Rodriguez offers insight into a perspective emerging from the Global South: the **decolonial environmental justice** approach. While scholars have increasingly acknowledged the plurality of environmental justice, Rodriguez highlights that Indigenous peoples' movements in the Global South have long organized their struggles on the basis of non-Western conceptions of justice, nature, culture, and identity, which require understanding the persistence of colonial values (coloniality) and modernity as a cause of current injustices.

In the subsequent Chapter 8, Julien-François Gerber, Bengi Akbulut, Federico Demaria, and Joan Martínez-Alier identify parallels between environmental justice and **degrowth** movements. Drawing on the rich intellectual tradition of degrowth thought, the authors contend that injustice and growth form two sides of the same coin, and that attention to both dimensions is necessary for environmental justice to be achieved. In Chapter 9, Julie Sze discusses the dominant concept of **sustainability** and explores its relation to environmental justice. While both concepts have developed contemporaneously, they diverge ideologically. Drawing on environmental justice and interdisciplinarity, Sze's "situated sustainability" offers a justice-oriented and historically informed way of thinking about sustainability.

The second part of the book looks at different issues that have been the focus of environmental justice struggles. First-generation environmental justice studies were concerned primarily with injustices created by the unequal distribution of landfills and waste dumps. In the first chapter of Part II, Alice Mah reflects on the legacies of these **toxic disasters** and on the role of the anti-toxic movements within the broader environmental justice movement.

Subsequent environmental justice work has expanded the focus, using the environmental justice framework to explore a multitude of other environmental issues and leading to new concepts such as climate justice, energy justice, food justice, and water justice. As such, in Chapter 11 Adrian Martin focuses on **biodiversity** and nature conservation as a matter for justice. He shows how marginalized groups suffer, not only from the loss of biodiversity but often also from the way in which society chooses to respond to this loss. As existing conservation efforts cannot simply be abandoned, Martin suggests directions for a more just conservation.

Gareth A.S. Edwards follows in Chapter 12 with a discussion of the increasingly popular concept of **climate justice**. Comparing the ethical principles used by scholars to the political claims of the "climate justice movement", he explains why climate change should primarily be understood as a fundamental justice dilemma. Climate justice connects in important ways with **energy justice**, discussed by Rosie Day in Chapter 13. Day explores how justice problems arise in relation to both our production and consumption of energy, not only with regard to fossil fuels but also in relation to new developments in clean energy.

Next, Kristin Reynolds explores in Chapter 14 the intersection of food, **agriculture**, and environmental justice. She shows how the inherent injustices in our food system long pre-date the emergence of the environmental justice movement. Addressing the justice implications of food production and consumption compels us to consider the links between rural and urban areas which are often left unexplored in the wider environmental justice literature.

Urban areas are the focus of Chapter 15 by Jason Byrne. Byrne examines some of the environmental injustices that stem from the process of **urbanization**, including issues of waste management and green space accessibility, for example, in a context where more than half the world's population now lives in cities. Finally, in the last chapter of Part II, Rutgerd Boelens analyzes in Chapter 16 the consequences of the decreasing availability and the unequal distribution of **water** from a justice perspective. More than other resources perhaps, water has been at the heart of major environmental justice struggles in recent years.

Part III deals with some of the key actors and subjects of environmental justice. "Subjects of justice" refers to the stakeholders or users who are entitled to moral consideration, ought to be allowed a role in decision-making, or at least deserve recognition. This section was deliberately kept separate from the first two for reasons of clarity and readability, but its chapters also include important concepts, approaches, and issues which do not necessarily feature in Parts I and II. This illustrates what Robert D. Bullard notes in his foreword to this volume: that the environmental justice framework has grown out of a grassroots movement, and we, as (future) environmental justice scholars, should acknowledge this influence more.

The first chapter of Part III by Lisa Sun-Hee Park and Stevie Ruiz, Chapter 17, is devoted to racial minorities in the United States, the main historical actors of the environmental justice movement. Building on the concept of **nativist environmentalism**, the chapter introduces us to the ways in which race and racism have continuously shaped environmental inequality over time.

Women have been equally important actors in the environmental justice movement, and, yet, their role has long been ignored in the academic field of environmental justice, as Sherilyn MacGregor notes in Chapter 18. Writing from an **ecofeminist perspective**, MacGregor reminds us that women represent some of the most severely affected members of our communities and illustrates how this shapes environmental justice struggles.

In Chapter 19, Dimitris Stevis also focuses on actors who are often absent from the environmental justice narrative. He explores the idea of **just transition**, which he presents as the contribution of workers and labour unions to the practice and study of environmental justice. Just transition, Stevis argues, can help us challenge the dominant "jobs vs environment" discourse.

Next, Kyle White builds on his personal history as a member of the Citizen Potawatomi Nation in Chapter 20 to introduce us to a North American Indigenous perspective of environmental justice. Indigenous peoples have ancient traditions of conceptualizing and practicing environmental justice, which White examines by drawing on **kinship ethics**. A key idea of Indigenous kinship relationship is reciprocity between humans and non-humans. This implies challenging the implicit anthropocentric bias of most environmental justice theories. Steve Cook helps us do exactly that in Chapter 21, the last chapter of Part III, on justice beyond humanity. Focusing on two specific approaches—the **animal rights approach** and the **eco-centric approach**—Cook argues that there are moral reasons for protecting non-humans, including nature and the environment, and that this moral concern can give rise to justice claims, an idea which is increasingly making its way into the environmental justice literature.

Finally, the authors in Part IV reflect on some of the future directions the study of environmental justice may take. In Chapter 22, David N. Pellow puts forth his **critical environmental justice** perspective intended to address some limitations and tensions within environmental justice studies. Applying his ideas to the case of the US prison system, Pellow challenges us to expand our focus across a broader range of categories of difference, of temporal and spatial scales, and beyond forms of institutionalized power like that of the state.

David Schlosberg follows with Chapter 23 examining the justice claims of new social movements focused on sustainable food, energy, and fashion, which he calls **sustainable materialism**. Schlosberg asks us to consider how such movements, not generally considered to be a part of the traditional environmental justice landscape, still articulate a range of innovative conceptions of social and environmental justice.

In the closing chapter of the book, Giovanna Di Chiro draws on the concept of **intersectionality** to help us uncover the multiple and mutually reinforcing systems of oppression shaping environmental injustices. Di Chiro invites us to reflect on our role as scholars and to build diverse coalitions with frontline communities to make environmental justice scholarship truly transformative.

I hope these chapters will collectively provide the next generation of environmental justice students with the necessary tools to understand a wide range of environmental justice issues, and help with imagining and practicing a truly common future for all.

References

ABC News/Stanford University/Resources for The Future (2018). Public Attitudes on Global Warming Poll. Available at www.langerresearch.com/wp-content/uploads/1198a1Global-Warming.pdf.

European Commission (2019). Special Eurobarometer 490—Climate Change. Survey requested by the European Commission, Directorate-General for Climate Action and Coordinated by the Directorate-General for Communication. September 2019.

Lausanne Climate Declaration (2019). Available at http://www.climatestrike.ch.

The Guardian (2019). Climate Crisis and a Betrayed Generation. Available at https://www.theguardian.com/environment/2019/mar/01/youth-climate-change-strikers-open-letter-to-world-leaders.

2 A history of environmental justice

Foundations, narratives, and perspectives

Esme G. Murdock

> **Learning outcomes**
> - Learn the origin of the term environmental justice.
> - Learn important principles of environmental justice.
> - Contextualize environmental justice as a movement of the people.
> - Connect forms of resistance against environmental injustice to communities' histories.
> - Identify and apply environmental justice as a concept to various global realities.

Introduction

To begin, I would like to draw attention to the decision to entitle this chapter *a* history of environmental justice. What I mean by this is quite literal, in referencing the fact that every "history" is necessarily both a particular one and an incomplete one. Many histories include the same or similar events as part of their narrative structure, but, importantly, histories are themselves narratives or stories. This means that whoever writes the history imbues it with content shaped by their own experiences while also offering various interpretations of particular events. This matters deeply for the final product we read and interpret as history proper. History is not objective; it is not an apolitical accounting of events. Having said that, I am writing *a* history of environmental justice that will focus on particular salient events important to the formation of the movement and discipline we currently refer to as environmental justice. I will therefore focus on the origins of the usage of the term environmental justice as it emerged and came into popular parlance and use in the context of the United States (US) in response to the 1982 protest demonstrations against a decision to site a toxic waste landfill in a primarily African American community in **Warren County**, North Carolina. However, the history of environmental injustice is one that exceeds both chronologically and ideologically the timescape and temporal origin of the term itself. Consequently, I encourage readers to be expansive in their considerations of who, when, where, and what can be encompassed within the history and concept of environmental justice. The fact that I may not have included a particular event or reality does not mean that it does not qualify as falling under the category of environmental justice or that it is not a salient issue that should be mentioned in the annals of environmental justice history. I merely encourage scholars, students, and activists of environmental justice to use this history as a guideline and a call to action to go forward with their work in ways this history and I, the author, both can and cannot do or imagine. If histories are stories, they must continuously be told, added to, and retold. Dear reader, I hope you are and may become part of the telling.

US context

The term "environmental justice" is largely agreed to have come into wide use in the 1980s US context (Bullard, 1990; Shrader-Frechette, 2002; Gilio-Whitaker, 2019). In particular, it is associated with events and forms of resistance within the Black Belt region of the US Deep South (Alabama, Florida, Georgia, Louisiana, Mississippi, North Carolina, South Carolina, Texas) carried out by communities of colour, largely Black American communities, in the face of unfair siting practices of environmental externalities such as commercial noxious facilities (e.g. garbage dumps, chemical waste dumps, oil refineries, and the like) within their communities. Communities of colour and poor communities identified their neighbourhoods as being overburdened with particular environmental ills, especially those linked with toxicity and pollution related to the fossil fuel and petrochemical industries. Within this particular historical context of environmental justice, the 1982 protest demonstrations by Black community members in Warren County, North Carolina against the siting of a PCB (polychlorinated biphenyls) landfill is marked as a critical event in the development of the environmental justice movement. In 1982, Black community residents, civil rights activists, and political leaders gathered to demonstrate their opposition to the construction of the PCB landfill in Warren County. It is important to note that the community selected for this PCB landfill in Warren County was mostly Black and was located in one of the poorest counties in North Carolina at the time when it was selected as a "burial site for more than 32,000 cubic yards of soil contaminated with highly toxic PCB" (Bullard, 1990, p. 30). The protest demonstrations did not stop the soil from being dumped in the PCB landfill, and these acts of direct action and collective resistance carried out by the Black communities was punished with the arrest of 414 protestors. "The protest demonstrations in Warren County marked the first time anyone in the United States had been jailed trying to halt a toxic waste landfill" (Bullard, 1990, p. 31).

It is important to note that the protest demonstrations and collective resistance in Warren County were not the first forms of direct action or protest against issues of toxicity and pollution in residential areas carried out by Black or other communities in the US. A key event in the anti-toxics movement that preceded Warren County and gained mass media attention in the 1970s involved the incidents that occurred in **Love Canal**, New York, where community members responded to the poisoning of their neighbourhoods by chemicals buried by Hooker Chemical Company. Love Canal community members also participated in protest demonstrations and the face of the Love Canal movement, Lois Gibbs, even held hostage in her home officials from the **Environmental Protection Agency**, as a means to elicit action and response from the US Government (Gibbs, 1982; Kitchell, 2012). It is important to note the different demographic makeup of the Love Canal community in terms of **race**, where residents were majority white. Gibbs was never arrested or charged criminally for her actions that resulted in the taking of hostages, while hundreds of Black Americans peacefully and non-violently protesting a decade later in Warren County faced physical violence and criminal prosecution. Robert Bullard notes that there were and continue to be many **Black Love Canals**[1] that do not garner the same national outrage or media coverage as toxicity affecting white communities and neighbourhoods (Bullard, 1990). So while Black and white communities may face similar realities in terms of dealing with the siting and/or disposal of commercial noxious facilities in their neighbourhoods, the effects of these citing practices are experienced differently and handled differently on a racial and political level.

The events of Warren County, North Carolina in 1982 were not and are not unique or special in terms of the experiences many communities of colour and Black American communities were facing and continue to face in terms of the overburdening of their neighbourhoods and places of residence with environmental toxicity and inferior environmental quality. This trend was examined thoroughly in the publication of the **United Church of Christ (UCC) Commission for Racial Justice**'s report "Toxic Wastes and Race in the United States" published in 1987. The UCC's report is important for a variety of reasons. First, it collected and aggregated data that verified the experiences and knowledges of communities of colour living in environmentally degraded residential areas due to unfair siting practices. Second, the report's analysis of the data demonstrated that race was the most significant independent variable for commercial hazardous waste facilities and uncontrolled toxic waste sites being placed in particular communities (Commission for Racial Justice, 1987). The UCC's report laid the groundwork for a key organizing and foundational principle of the environmental justice movement, which is the reality of **environmental racism**.

Robert Bullard defines environmental racism as "any policy, practice, or directive that differentially affects or disadvantages (whether intended or unintended) individuals, groups, or communities based on race or colour" (Bullard, 1990, p. 98). Additionally, "[e]nvironmental racism combines with public policies and industry practices to provide *benefits* for whites while shifting industry *costs* to people of color" (Bullard, 1990, p. 98). Thus, environmental racism identifies the ways in which race and racial discrimination function in the selection of particular communities as the most probable sites of environmental toxicity. It also identifies the unfair distribution of benefits and costs associated with industrial development and production based on racial criteria, which compounds societal inequalities. Environmental racism goes hand in hand, for Bullard, with industries' identification of particular communities as those least likely to thwart or resist these predatory **siting practices**, or what he terms "the path of least resistance" (Bullard, 1990, p. xv). It is not coincidental for the UCC or Bullard that the most marginal and vulnerable communities socially, politically, and economically are overburdened with environmental ills. It is yet another area of existence that identifies particular communities as lesser than others and as disposable.[2]

Strategies of resistance against environmental injustice

The structural inequalities in distribution of environmental goods and environmental ills follow predictable patterns of domination and oppression, as demonstrated in some of the important research that has been conducted which shows that racialized communities are overburdened with environmental externalities. It is necessary to emphasize that the context of particular communities and how they experience historical and continuing injustices affects and connects to the forms of resistance and strategies these same communities employ to combat environmental injustice. Environmental injustice looks, feels, and is experienced differently by different communities. For example, Indigenous scholars discuss how **settler colonialism** is a major structural issue that deeply affects environmental realities for Indigenous nations and complicates the privileging of environmental racism as a foundational principle around which US environmental justice was structured from the 1980s onwards.[3] I think it is important to be careful when applying principles across situations and consider what is illuminated by noting the particularity of the experiences of different communities and how this contributes to our understanding of environmental justice (McGregor, 2009, 2018; Whyte, 2015, 2016; Gilio-Whitaker, 2019).

Environmental justice is, importantly, a movement, which means that it starts and lives with the people: in the communities that are experiencing the harms and in the challenges environmental injustices enact in the places where community members live, work, and play. What this means is that, while environmental justice has intellectual and academic dimensions, it is primarily a grassroots and people-driven movement. In fact, an important principle of environmental justice that emerges from activist communities is the maxim: "We speak for ourselves" (Cole and Foster, 2001). The imperative "We speak for ourselves" privileges the voices, experiences, and expertise of those directly confronted with issues of environmental injustice. It is related to the idea that the community members experiencing the realities of environmental injustice should be the leaders of movements to liberate and heal their own communities. This does not mean that solidarity and coalition-building are not an important part of and strategy for achieving environmental justice. However, it does mean there are often difficult conversations about privilege, as well as important differences in backgrounds, histories, and resources. For example, the Detroit Black Food Security Network (DBFSN) was established in 2006 to serve Detroit Black residents and communities who were/are experiencing high rates of food insecurity and limited access to fresh, nutritious, and affordable food.[4] This reality is the product of various intersecting processes including (but not limited to) economic recession related to industrial downturn, extraction of economic industrial capital from the city, legacies of slavery, racial apartheid and segregation, and government mismanagement at the city, municipal, state, and federal level (Downey and Reese, 2017; Ryan and Campo, 2013; Steinmetz, 2009). DBFSN outlines the stakes of community organizing and community service in its vision declaration, which states:

> We observed that many of the key players in the local urban agriculture movement were young whites, who while well-intentioned, never-the-less, exerted a degree of control inordinate to their numbers in Detroit's population. Many of those individuals moved to Detroit from other places specifically to engage in agricultural or other food security work. It was and is our view that the most effective movements grow organically from the people whom they are designed to serve. Representatives of Detroit's majority African-American population must be in the leadership of efforts to foster food justice and food security in Detroit. While our specific focus is on Detroit's African-American community, we realize that improved policy and an improved localized food system is a benefit to all Detroit residents.
> Detroit Black Community Food Security Network

This vision statement carefully grounds community work and service in community experiences. It draws our attention to the ways in which we need to honour and acknowledge the community members most affected by the realities they live and experience and align those people with the vision, leadership, and processes of community action. This is especially important when we consider that the realities of the communities experiencing environmental injustices are often rendered invisible and ignored by the dominant society and members of privileged communities. Additionally, the dominant society and more affluent communities largely silence the experiences of these communities, such that when community members speak about their lived experiences they are not believed or trusted as being qualified representatives of their own realities and interests (Dotson, 2011). This is another way of understanding the impetus behind the UCC's "Toxic Wastes and Race in the United States" report, as an initiative necessitated by the dominant society's ignoring and dismissal of disclosure by communities of colour of their own experiences of environmental injustice.

It is also important to track and honour the historical and continuous struggles of particular peoples and communities that contribute to and shape the forms of resistance apparent in the environmental justice movement. Strategies of resistance, such as direct action, non-violent demonstrations, and grassroots organizing, mark and emerge from various histories, especially those of global communities of colour (Bullard, 1990, 1997, 2005; Shiva, 2015). Often these communities, globally, were facing various and diverse struggles that called for and developed into particular forms of resistance, such as social movements for civil rights, labour rights, recognition of sovereignty, land rights, anti-colonial resistance, welfare rights, and farmworker movements (Di Chiro, 1992, p. 97). All of these social movements have in common resistance against systems of domination and oppression, whether those be colonialism, settler colonialism, imperialism, slavery, racial apartheid, segregation, capitalist exploitation, **class domination**, land expropriation, or the like. These forms of resistance importantly respond to the particular experiences of various communities globally, especially communities of colour. It is extremely important that we do not lose sight of the ways in which these forms of resistance belong to particular communities which are often constructed as powerless by the dominant society and its structures of domination. It is therefore important to honour and acknowledge the expert strategies of resistance and self-determination globally that communities have exercised and continue to enact in the face of overwhelming and powerful forms of domination and oppression. In this way, we can apply the concept of environmental justice to contexts that precede and exceed the term as it emerged in the 1980s.

Gender

So, who are the people on the **frontlines** of the environmental justice movement, and what strategies do they employ to combat the unfair systems which consistently mete out particular realities to similar communities? While environmental justice has paid attention to the distributive mechanisms through which environmental inequalities persist, especially by closely following along the axes of race and class, **gender** is also a very important vector or factor for considering environmental harm. Globally, women and children are some of the most vulnerable and most severely affected members of our global community, especially when it comes to incurring environmental risk and various environmental ills (Di Chiro, 1992; Shrader-Frechette, 2002; Mies and Shiva, 1993; Shiva, 2010; Warren, 2000; Plumwood, 1993).

In fact, many of the people on the frontlines of environmental activism and grassroots environmental change are women, specifically women of colour (Di Chiro, 1992; Shrader-Frechette, 2002). This makes sense if you consider that, worldwide, both women and children of colour are some of the most drastically and severely affected segments of our global community when it comes to forms of oppression such as poverty, gender violence, environmental degradation, and the like. Many women of colour take up the mantle of resisting environmental injustice and other related systems of oppression because no one else will fight this battle for them, and often, because of the sexual division of labour and patriarchal structuring of societies, women are charged with caring for their households and families (Hartsock, 1983; Sanchez, 1993; Stein, 2004). This means that women are frequently in more direct contact with the realities of the homestead and are monitoring the health and conditions of their immediate or even extended families. Given that, globally, gender is a critical axis of oppression and that the intersections of race, gender, and class oppression converge to create predictable recipients of environmental degradation and harm in the pronounced negative consequences for women and children of colour, the exclusion of explicit gender analysis from the traditional purview of environmental justice scholarship is particularly troubling.

Women of colour are often the behind-the-scenes leaders of resistance against environmental injustice in their communities, contributing the majority of the legwork to organizing at and on the ground level (as is true in many other social justice movements). However, they are the least visible, recognized, or acknowledged in the mainstream (hooks, 1991; Wright, 2015). Similar to the ways in which those with the most privilege tend to crowd the forefront of recognition in movements, even within communities of colour, the relegation of women to a subordinated category affects the recognition they are afforded for their tireless efforts to make their communities safer, healthier, and more liveable.[5]

Academic debates on environmental justice

While the environmental justice movement begins on the ground in the lives and experiences of those communities experiencing environmental injustices, academics have attempted to track and understand theoretically what environmental justice is all about. This has led to a series of conversations and debates that concern which paradigms of justice are found in the living forms of advocacy for clean, healthy, and safe environments for all communities, both human and **more-than-human**. Put another way, environmental justice scholars are involved in an ongoing debate about what kind of justice, theoretically, environmental justice requires. Theories about justice must contain a framework for assessing the nature and extent of harm, because one cannot make restitution for an injustice that cannot be measured in some generally agreed-upon way. Theories of environmental justice certainly have these components. Most examples of environmental justice theories offer accounts of what it means to suffer harm to environments where communities live, work, and play, from poor air and water quality to lack of green spaces. Environmental justice theories also offer parameters of what it means to repair communities that experience place-based harms, which range from increasing urban green spaces to projects that augment community members' involvement in monitoring air and water quality. In this way, accounts of harm are tied to accounts of repair, or what it takes to achieve justice.

Current frameworks of environmental justice are largely moving past a purely distributive model of justice when addressing the complex and intricate harms that result from environmental degradation. This is largely because the concept of harm within environmentally unjust contexts is considered too nuanced to be captured by a distributive solution. For example, people who live near polluting facilities are frequently treated as second-class citizens whom the rest of society is morally free to ignore. Thus, solutions to harms of this kind must involve not only the cleanup or abatement of the pollution's point source—that is, redistributing the environmental burdens—but creating or restoring a positive, more dynamic community image and concept and correcting the prejudices that outsiders may have about the community. Moreover, improving the community's image may also require deep involvement of community members in whatever actions are taken to address the pollution problems. "Deep involvement" means that community members themselves must have a direct role in the decision-making process, rather than just university experts or federal and state employees—a concept also referred to as participative and **procedural justice**.

Moving beyond distributive justice is difficult because in many policy contexts federal and state employees assume that distributive remedies are all that is required of them. Robert Figueroa's bivalent framework is an example of an environmental justice theory that owns up to the challenge of moving beyond **distributive justice** (Figueroa, 2011). I will describe it briefly.

Figueroa's conception of environmental justice is rich, and, in this chapter, I can only cover a few pertinent aspects of it, specifically Figueroa's definition of harm. Figueroa's theory of justice is "bivalent" in that it addresses both a community's need for a fair distribution of environmental resources and its symbolic need for culturally and historically appropriate paradigms of justice. Figueroa's bivalent conception is in dialogue with recognition-based theories of justice, such as those of Nancy Fraser and Axel Honneth.[6]

Figueroa claims that the damage and harm of injustice are not just about the deprivation of environmental goods, such as clean air. Rather, this harm also has deep physical, psychological, and existential dimensions and consequences that are far-reaching in terms of time, nature, and severity (Figueroa, 2011). For example, consider the declaration of ***terra nullius*** in Australia by white British settlers, which decreed that the land to be colonized was completely empty and devoid of human beings, and, in fact, attempted to construct Aboriginal peoples as part of the landscape itself. This allowed not only a certain instrumental attitude toward the physical land to be developed, but also and importantly a complete lack of existential **recognition** (also sometimes called **misrecognition**) of Aboriginal peoples' occupancy and deep relationship with the land. This is not a single harm or single event, but a network of harms and instance of structural, continuous violence with long-lasting consequences that go beyond how we conceive of and treat the physical environment and drastically alter the identity—and in this case, the very existence—of peoples who occupy a certain physical environment. Imagine what it might feel like to have your non-existence codified, which is something happening all over the planet as marginalized and vulnerable communities of colour pay the highest price of environmental toxicity for global "progress" elsewhere. Indigenous people are often construed as already having been driven into extinction, creating an inclination to feel they are beyond further harm. In the case of *terra nullius* in Australia, no harms can be perpetrated against Aboriginal peoples because they have already been deleted from the history and present context of the land. There is profound psychological and emotional harm in being told that you do not exist, and that, if you did, another group is morally free to visit all sorts of violence and injustice upon you. We can identify the similar logic of this tradition and apply this inheritance of structures of oppression and practices of violence to the same networks of injustice that prompted the Black residents of Warren County, North Carolina, to protest the siting of a PCB dump in their community.

Environmental injustice often solidifies and reinforces other axes of injustice and oppression that are already affecting vulnerable and marginalized communities of colour globally—a situation which requires theories of environmental justice that capture multiple dimensions of harm and repair. Figueroa's bivalent conception of environmental justice is just one example. Environmental justice scholarship has since expanded to trivalent theories and conceptions that attempt to address and combine frameworks of justice such as distributive justice, procedural justice, and recognition.[7] The debates and conversations in academic environmental justice scholarship are ongoing and will surely continue to change and expand as time goes on. One emerging area of environmental justice academic discourse is covered in the next section in the attention to diverse and positive **environmental histories**, heritages, and identities that communities experiencing environmental injustice also possess and maintain.

Global environmental justice: honouring diverse environmental histories, heritages, and identities

While the term environmental justice originated in the 1980s US context, what qualifies as environmental injustice is a broad and diverse set of issues, as we have seen with the identification of *terra nullius* in colonial Australia as a form of environmental injustice within environmental justice scholarship. To widen the lens of our thinking about the range of issues the term environmental justice as a movement and discipline can cover, it is necessary to examine how particular communities, such as poor communities of colour, have been globally manufactured as the probable, likely, and overwhelmingly actual recipients of environmental refuse, risk, and harm. When we talk about pervasive and severe forms of injustice that persist globally in our human community, it is necessary to think about the global regimes of power that overwhelmingly contribute both historically and continuously to the relegation of particular geopolitical places, geographies, and peoples to the bottom of our global hierarchy. However, it is also necessary to think about the ways in which these identifiable and predictable recipients of global environmental injustice have their own environmental histories, heritages, and identities that persist and modify to resist and survive environmental degradation. This section will focus on how movements globally for environmental justice manifest in ways that honour and acknowledge the power, resistance, and autonomy that global communities facing environmental injustice engage in to craft their own visions of environmental justice.

Environmental justice, as a theoretical lens and a global movement for safe, healthy, and fulfilling environments, expands the space of possibility for examining the intersectional sites of humans and environments while also multiplying the number of those whom we consider when we talk about environmental degradation. However, there are additional realms of concern related to the exclusion of people of colour when thinking about the natural world. The focus on environmental racism always limits the place and identity of people of colour in relation to the environment to already negative or degraded sites of interaction. It also has the tendency to oversimplify the struggles of diverse populations and communities of colour as the same or similar in content. This is not to dismiss or diminish the importance of environmental justice or the work being done by people of colour in relation to the environment, but the overwhelming focus on distributional issues and environmental racism does constitute a limited approach in terms of analyzing particular **environmental identities** and heritages.[8]

The environmental justice movement largely emerged as one intended to serve particular communities whose environmental realities were not acknowledged or considered to be "environmental" by the mainstream environmental movement, which historically was a grouping limited in terms of its founders by privilege in the form of race, class, and ability (Di Chiro, 1992; Bullard, 1990; Shrader-Frechette, 2002; Palamar, 2008). These experiences of privilege in another dimension of life and experience can lead to the development of what is termed "**environmental privilege**".[9] Thus, certain avenues of inquiry primarily represented the reactionary nature of the environmental justice movement, and action was dictated by the urgency of the specific contexts the communities of colour facing environmental injustice found themselves in. What this means is that most of the time, these same communities of colour, facing the most severe and drastic forms of environmental injustice, must primarily focus their efforts on surviving, resisting, and overcoming structural forms of injustice and distribution affecting the stability and health of their lives and homes. This often obscures other concerns, such as positive and historical sites of interaction by certain communities with the environment, which is an important aspect of environmental justice. This overdetermination of negative environmental realities also obscures the ways in which histories coexist and are simultaneously inscribed into the sites of conflict as they arise in the contemporary context.

This section will explore how another dimension of environmental injustice exists when the historical narratives, environmental heritages, and environmental identities of various communities of colour are overshadowed or obscured by experiences of environmental injustice. Thus, another important aspect of moving toward environmental justice for all involves the acknowledgement, honouring, and celebration of unique environmental histories, heritages, and identities that coexist and grow, even with the overwhelming experiences of environmental injustices visited upon particular human communities.

While environmental justice's focus on distribution and environmental racism draws our attention to the ways in which structural injustices manufacture already marginal and vulnerable communities into likely sites for society's refuse, it can overshadow positive and resilient relationships to environments that precede environmental degradation, persist within contexts of environmental injustice, emerge within the context of historical experiences, or all three. As discussed earlier, corporations and states relied on the marginality and vulnerability of communities of colour produced by structural inequalities to outwardly identify those same communities as the least likely or able to thwart predatory siting practices for environmental toxicity and other externalities. This "path of least resistance" strategy can also be described as the identification of communities of colour as powerless. However, the identification of communities of colour as powerless does not actually make them powerless. In fact, the fierce resistance enacted by these same communities in the face of the staggering odds against them is a testament to the falseness of that external identification and proclamation. Globally, in the case of many historically and continuously oppressed communities of colour, such communities often have hundreds of years of intergenerational knowledge and experience in combatting the global regimes of power that have attempted and continue to attempt to appropriate their lands, degrade their environments, extract their resources, and disenfranchise—as well as oppress—their peoples (Shiva, 2015; Gilio-Whitaker, 2019). These identifications of powerlessness fabricated by multi-national corporations, nation-states, and international governing bodies rest on the assumption that if you objectify a people, you also make them a literal object. However, these communities are not passive objects or recipients of the world's refuse; they are active subjects and agents whose histories precede and exceed the oppression and injustice that these regimes of power and their agents hope to make their perpetual state. Importantly, though, these histories and narratives of communities of colour globally are also marginalized within the dominant discourse.

The overdetermination of communities who experience environmental injustice as only having that particular history can lead to the overlooking of important **environmental heritages**, histories, and stories that are deep wells of knowledge and resilience directly applicable to our current planetary environmental crisis. For example, environmental justice scholar Kyle Whyte writes and discusses the ways in which climate change discourse predominantly reproduces the same environmental elitism and exclusionary tactics present in the mainstream environmental movement (Whyte, 2018). It does this through constructing climate change as an urgent and new problem, but, globally, Indigenous peoples and other minoritized peoples, such as enslaved Africans and their descendants and interred Japanese immigrants and their descendants, have been subjected to the violence of climate change for much longer through acts of spatial colonial violence, such as forced relocation, or confinement in reservations, plantations, and internment camps.[10] Similarly, these knowledges of intense and extreme climate adaptation are often excluded and entirely ignored in the scholarship on innovative research on climate change adaptation. This is just one example. Imagine what the possibility of achieving environmental justice (of which climate justice is a huge part) would look like or feel like if we actually opened our hearts and minds to honour the incredible wealth and diversity of knowledges, resistances, and adaptations of these powerful communities, who survive and thrive in the face of incredible obstacles.

This calls us as scholars and activists to decolonize and unsettle the dominant narrative of environmental history and environmental justice history that focuses on and reproduces the perspective of the conqueror, the colonizer, the settler, the polluter, and the West. What this means is that the majority of the world (the "rest" as defined by the West) not only incurs the burdens of a global community built on the backs of the most marginal and vulnerable, but that the West also understands itself as the only history worth telling, a history of achievements that are not natural successes but manufactured ones (Césaire, 2001). The "developing" world is developing because it was and is subject to the rapacious greed, **extractivism**, and corruption of the West (Shiva, 2015). It is not a story of natural inferiority that leads to underdevelopment, which is what the racist/racialized hierarchal order of the West would have us believe; rather, it is a structural system of land appropriation, brutalization, extraction, and outright theft that renders the balance so askew (Wynter, 2003). If we want to understand what moving closer to global environmental justice means, we have to reckon with the whole and truthful history of how and why our globe looks and feels the way it does. Environmental justice for all means a true and honourable place for all peoples that acknowledges, honours, and respects their environmental contexts, histories, heritages, and identities such that no community is ever labelled as powerless or disposable or forced to endure a reality whereby its members are made to experience themselves in that way.

> ### Follow-up questions
> - How is the context through which the term environmental justice emerged connected to particular communities' identities and histories?
> - What roles do race, class, and gender play in distribution of environmental injustices?
> - How might regimes of power, such as colonialism, slavery, and imperialism, matter for the large sections of our global community experiencing environmental injustices?
> - What is damaging about only being identified with injustices/oppressions?

Notes

1 For more on Black Love Canals, see also Chapter 10 of this volume.
2 On the role of anti-toxic struggles in the environmental justice movement, see also Chapter 10 of this volume.
3 On Indigenous environmental justice, see Chapter 20 of this volume
4 On food, agriculture, and environmental justice, see Chapter 14 of this volume
5 On the role of women in environmental justice movements, see also Chapter 18 of this volume.
6 For more on recognition, see Chapter 5 of this volume.
7 For more on the trivalent theories of environmental justice, see Chapters 3–5 of this volume.
8 On distributive (in)justice, see Chapter 3 of this volume; on environmental racism, see Chapter 17 of this volume.

9 For more on environmental privilege, see also Chapter 17 of this volume.
10 For more on the internment of Japanese and Japanese-descended peoples in the US, see also Chapter 17 of this volume.

References for further reading

Bullard, R.D., 1990. *Dumping in Dixie: Race, Class, and Environmental Quality*. Westview Press, Boulder.
Cole, L., Foster, S., 2001. We Speak for Ourselves: The Struggle of Kettleman City, in: *From the Ground Up: Environmental Racism and the Rise of the Environmental Justice Movement*. New York University Press, New York.
Gilio-Whitaker, D., 2019. *As Long as Grass Grows: The Indigenous Fight for Environmental Justice, from Colonization to Standing Rock*. Beacon Press, Boston.
Stein, R. (Ed.), 2004. *New Perspectives on Environmental Justice: Gender, Sexuality, and Activism*. Rutgers University Press, New Brunswick, NJ.

References used in this chapter

Bullard, R.D. (Ed.), 2005. *The Quest for Environmental Justice: Human Rights and the Politics of Pollution*. Sierra Club Books, San Francisco.
Bullard, R.D. (Ed.), 1997. *Unequal Protection: Environmental Justice and Communities of Color*. Sierra Club Books, San Francisco.
Bullard, R.D., 1990. *Dumping in Dixie: Race, Class, and Environmental Quality*. Westview Press, Boulder.
Césaire, A., 2001. *Discourse on Colonialism*. Monthly Review Press, New York.
Cole, L., Foster, S., 2001. We Speak for Ourselves: The Struggle of Kettleman City, in: *From the Ground Up: Environmental Racism and the Rise of the Environmental Justice Movement*. New York University Press, New York.
Commission for Racial Justice, U.C. of C., 1987. *Toxic Wastes and Race in the United States: A national Report on the Racial and Socio-Economic Characteristics of Communities with Hazardous Waste Sites*. Commission for Racial Justice, United Church of Christ, New York.
Detroit Black Community Food Security Network, n.d. About Us [WWW Document]. Dtownfarms-Copy. www.dbcfsn.org/about-us (accessed 1.9.19).
Di Chiro, G., 1992. Defining Environmental Justice: Women's Voices and Grassroots Politics. *Socialist Review* 22, 93–130.
Dotson, K., 2011. Tracking Epistemic Violence, Tracking Practices of Silencing. *Hypatia: A Journal of Feminist Philosophy* 26, 236–257.
Downey, D.C., Reese, L.A., 2017. Sudden Versus Slow Death of Cities: New Orleans and Detroit. *Du Bois Review: Social Science Ressearh on Race* 14, 219–243. https://doi.org/10.1017/S1742058X16000321
Figueroa, R., 2011. Indigenous Peoples and Cultural Losses, in: Dryzek, J.S., Norgaard, R.B., Schlosberg, D. (Eds.), *The Oxford Handbook of Climate Change and Society*. Oxford University Press, Oxford, pp. 232–247.
Gibbs, L.M., 1982. *Love Canal: My Story*. SUNY Press, Albany, NY.
Gilio-Whitaker, D., 2019. *As Long as Grass Grows: The Indigenous Fight for Environmental Justice, from Colonization to Standing Rock*. Beacon Press, Boston.
Hartsock, N., 1983. The Feminist Standpoint: Developing the Ground for a Specifically Feminist Historical Materialism, in: Harding, S., Hintikka, M.B. (Eds.), *Discovering Reality: Feminist Perspectives on Epistemology, Metaphysics, Methodology, and Philosophy of Science*, Synthese Library. Springer Netherlands, Dordrecht, pp. 283–310. https://doi.org/10.1007/0-306-48017-4_15
hooks, b., 1991. Theory as Liberatory Practice. *Yale Journal of Law & Feminism* 4, 1–12.
Kitchell, M., 2012. *A Fierce Green Fire: The Battle for a Living Planet*. First Run Features.
McGregor, D., 2009. Honouring Our Relations: An Anishnaabe Perspective on Environmental Justice, in: Agyeman, J., Cole, P., Haluza-Delay, R. (Eds.), *Speaking for Ourselves: Environmental Justice in Canada*. UBC Press, Vancouver.

McGregor, D., 2018. Mino-Mnaamodzawin: Achieving Indigenous Environmental Justice in Canada. *Environment and Society* 9 (1), 7–24. https://doi.org/10.3167/ares.2018.090102.

Mies, M., Shiva, V., 1993. *Ecofeminism*. Fernwood, Black Point, Nova Scotia.

Palamar, C.R., 2008. The Justice of Ecological Restoration: Environmental History, Health, Ecology, and Justice in the United States. *Human Ecology Review* 15, 82–94.

Plumwood, V., 1993. *Feminism and the Mastery of Nature*. Routledge, New York.

Ryan, B.D., Campo, D., 2013. Autopia's End: The Decline and Fall of Detroit's Automotive Manufacturing Landscape. *Journal of Planning History* 12, 95–132.

Sanchez, L., 1993. Women's Power and the Gendered Division of Domestic Labor in the Third World. *Gender & Society* 7, 434–459. https://doi.org/10.1177/089124393007003007

Shiva, V., 2015. *Earth Democracy: Justice, Sustainability, and Peace*, Reprint edition. North Atlantic Books, Berkeley, CA.

Shiva, V., 2010. *Staying Alive: Women, Ecology, and Development*, 10/31/10 edition. South End Press, Brooklyn, NY.

Shrader-Frechette, K., 2002. *Environmental Justice: Creating Equality, Reclaiming Democracy*. Oxford University Press, Oxford.

Stein, R. (Ed.), 2004. *New Perspectives on Environmental Justice: Gender, Sexuality, and Activism*. Rutgers University Press, New Brunswick, NJ.

Steinmetz, G., 2009. Detroit: A Tale of Two Crises. *Environment and Planning D: Society and Space* 27, 761–770.

Warren, K., 2000. *Ecofeminist Philosophy: A Western Perspective on What It Is and Why It Matters*. Rowman & Littlefield, Lanham, MD.

Whyte, K.P., 2015. Indigenous Food Systems, Environmental Justice, and Settler-Industrial States, in: Rawlinson, M., Ward, C. (Eds.), *Global Food, Global Justice: Essays on Eating under Globalization*. Cambridge Scholars Publishing, New Castle upon Tyne, UK, pp. 143–156.

Whyte, K.P., 2016. Indigenous Experience, Environmental Justice and Settler Colonialism, in: Bannon, B. (Ed.), *Nature and Experience: Phenomenology and the Environment*. Rowman & Littlefield, Lanham, MD.

Whyte, K.P., 2018. Indigenous Science (fiction) for the Anthropocene: Ancestral Dystopias and Fantasies of Climate Change Crises. *Environment and Planning E: Nature and Space* 1, 224–242. https://doi.org/10.1177/2514848618777621.

Wright, M.M., 2015. *Physics of Blackness: Beyond the Middle Passage Epistemology*, 1st edition. University of Minnesota Press, Minneapolis.

Wynter, S., 2003. Unsettling the Coloniality of Being/Power/Truth/Freedom: Towards the Human, After Man, Its Overrepresentation—An Argument. *CR: The New Centenniel Review* 3, 257–337. https://doi.org/10.1353/ncr.2004.0015

Part I
Defining and conceptualizing environmental justice

3 Distributive environmental justice

Alice Kaswan

Learning outcomes

- Gain an introductory understanding of several core distributive justice concepts.
- Gain an understanding of the different kinds of governmental decisions that have direct and indirect distributional consequences.
- Learn how policymakers and academics measure distributive justice.
- Consider possible remedies for distributional injustice.

Introduction

The drive down Interstate 880 in Oakland, California is not that different from the drive down many other industrial corridors in the United States. Industries, large and small, recycling facilities, waste storage facilities, tow yards, and other industrial and commercial uses are interspersed with patches of housing. Sometimes, there are just a few houses tucked in between industrial lots; other times, small neighbourhoods create a patchwork. Here in East Oakland, residents are invariably low-income and Latinx or African American.

Or take the plight of women in coastal Bangladesh, as climate change leads to increasingly severe cyclones and flooding. Due to its geographic location, Bangladesh is among the nation's most vulnerable to climate change and, within Bangladesh, women are disproportionately vulnerable to flooding events and post-disaster impacts. Women's low socio-economic status reduces their autonomy and options for responding to risk, leading to higher mortality rates, and their traditional roles as caretakers and food providers are particularly stressed when disaster strikes (Hossain and Punam, 2016).

These examples suggest the kinds of harms that trigger **distributive justice** concerns. But what is "distributive justice"? We start this chapter by exploring how distributive justice differs from other forms of justice and then identify several variants on what constitutes distributive justice. Moving beyond theory, we explore what distributive justice means on the ground. We identify the different decision-making contexts wherein issues of distributive justice arise, the kinds of evidence evaluating distributive justice, and the difficult question of what causes distributive injustice. Finally, drawing from examples in the United States, we consider mechanisms for achieving distributive justice.

Theories of distributive justice

What distinguishes "distributive justice" from other conceptions of justice is that it takes lived experience as its starting point. Distributive justice concentrates on how harms and benefits are distributed and experienced. For a distributive justice claim, it is not necessary to point to a specific actor causing injustice; disproportionate harms, or intolerably intense harms, trigger distributive justice concerns, even if they are not caused by direct discrimination. This is not to say that causation concerns are irrelevant; past or present **procedural injustice** can inform and amplify concerns about distributive injustice.[1] But concentrated misery or a dearth of basic necessities can constitute distributive injustice, regardless of causation.

Distributive justice is not a fixed and one-dimensional concept. Like the term "justice" itself, it has a wide variety of potential meanings (Walker, 2012, pp. 8–12). Different values and concerns inform differing conceptions, and these conceptions give us a range of lenses through which to view and assess current conditions. The following discussion provides a brief description of a number of the dominant theories: **utilitarianism**; **equality** and acceptable bases for diverging from strict equality; and a **"capabilities"** approach that focuses on each person having the minimum resources and opportunities to achieve a meaningful life.

Utilitarianism

The standard utilitarian creed is achieving the greatest good for the greatest number (Arthur and Shaw, 1978; Rescher, 1966). Under this approach, the goal is to maximize overall social welfare. Inequality in the distribution of goods is not per se unjust, if overall social welfare is high for many. In other words, poor conditions for a few would be justified if they allow much better conditions—superior preference satisfaction—for many.

Utilitarian theorists do not, however, ignore inequality (Singer, 2002, p. 42). Considering income inequality, more income is likely to be more meaningful to those who are poor than those who are rich—poor people will experience greater utility from a given sum relative to a rich person. Or people experiencing high levels of pollution would experience greater utility from a decrease in pollution than those who experience relatively modest levels of pollution. Because low-income and frontline communities would have more to gain from higher incomes and less pollution, overall social welfare would be enhanced by providing resources and improving conditions in these communities rather than in richer and less-polluted communities. This strand of utilitarianism could, therefore, support **frontline communities'** environmental justice claims.

Equality

The most common and intuitive starting point for distributive justice claims is equality (Berlin, 1961). Philosopher **John Rawls** anchors equality's moral appeal through his "veil of ignorance" exercise. He suggests that, if we were to put ourselves behind a veil that masked our identities as rich or poor, black or white, citizen or immigrant, and then asked to say what distribution would be fairest, not knowing where we would be on the spectrum of privilege, we would choose equality (Rawls, 1971, p. 150). No one would want to risk being worse off than others, and so the natural preference for distributive justice is an equal distribution, at least as a starting point. In environmental justice terms, that means that no one should be subject to more harms than anyone else—that disparate impacts are presumptively unjust.

Bases for deviating from strict equality: Rawls' difference principle

Though equality is the natural starting point, Rawls and other distributive justice theorists recognize that a range of factors could justify deviations from strict equality. Rawls' **"Difference Principle"** suggests that inequality is justified if it provides compensating benefits, particularly for "the least advantaged" (Rawls, 1971, p. 83). So, in a society that is already unjust, some degree of inequality would nonetheless be acceptable if, notwithstanding that inequality, the least advantaged accrued some benefits. Thus, income tax policies that allow the wealthy to become richer would be justified if the tax policies motivated investment and industry that created jobs and opportunities for the least advantaged, or some degree of concentrated pollution could be justified if it benefitted many other heavily exposed communities. So, to posit an example well outside of the milieu Rawls conceptualized, new gasoline-refining rules that reduced pollutants in gasoline, improving air quality for the many frontline communities living near highways, might be justifiable even if they increased pollution near some communities living near the refineries processing the gasoline.

Bases for deviating from strict equality: just deserts

Another basis for deviating from strict equality is the concept of **"just deserts"** (Arthur and Shaw, 1978; Rescher, 1966). Here, inequality could be justified if you deserve what you get. If you have worked harder to accumulate wealth than others, then you are entitled to the greater wealth; the inequality is not unjust. Or, if you live a resource-intensive lifestyle, then you may deserve the consequences of that use: exposure to the pollution or climate impacts that result from the lifestyle you lead. Or, if you make the foolish choice to live in an area that is vulnerable to climate change, like a floodplain or ocean cliff, then perhaps you deserve the consequences of the inevitable disasters you are likely to experience.

Conversely, if you work hard but earn little, or have been deprived of educational opportunities that would enable high earnings, you do not deserve poverty; or, if you use relatively few resources, but are exposed to pollution generated to create the products or energy that others consume, you do not deserve that disproportionate exposure; or, if you have made high-risk choices, like living in a flood-prone floodplain, but did so because your opportunities were highly constrained by poverty or racial discrimination, then arguably you did not have true free choice and would not deserve the higher risks resulting from your constrained choices.

Thus, the "just deserts" theory in distributive justice works in both directions. Sometimes, it provides a justification for existing inequality. At other times, it highlights the ways in which existing inequalities are not deserved.

Bases for deviating from strict equality: need

A **need-based theory** would consider underlying circumstances in determining how to distribute a good or a bad. If underlying conditions were unequal, then relative need could justify distributing more to those in need and less to those who are already well-off (Arthur and Shaw, 1978). In environmental terms, a need-based theory of environmental justice could justify refraining from further siting of undesirable facilities or prioritizing enforcement in areas already experiencing heavy pollution concentrations.

Bases for deviating from strict equality: differing preferences

The forgoing theories focused on the objective distribution of things and circumstances. Another strand of distributive justice focuses on preferences. Exposure to industrial or other facilities may differ because people have differing preferences for the kind of environments they choose to live in (Blais, 1996). For example, some may choose to live near industrial facilities due to the job opportunities and short commutes achieved by close proximity, and would object to government policies that impaired these opportunities in the name of an unwanted strict equality.

As in the context of the "just deserts" basis for inequality, however, one must again interrogate the degree to which preferences are unfairly constrained. People desperate for jobs, and who are faced with the choice of exposure to pollution or loss of livelihood, might choose to maintain their livelihood. But, lying behind this choice is underlying inequality: poverty and discrimination can lead people to choose exposure in order to have employment. Presumably, they would prefer not to face this trade-off. So, to the degree that differing "preferences" are used to justify distributional inequality, it is important to evaluate the extent to which choices are constrained by underlying and undeserved inequalities.

Why care about inequality? Underlying concerns about participatory and social injustice

Inequality is essentially a comparative test. Its focus is on how one community measures up against another—whether some people experience more harm than others. Arguably, if everyone lived in equally poor circumstances, that would not, under a strict equality approach to distributive justice, constitute injustice: everyone would suffer equally. It is the comparative gauge that marks this form of distributive justice.

Why do we care so much about equality? And why do environmental justice inquiries focus on disparities affecting marginalized groups, rather than all disparities, regardless of demographic features? Arguably, the environmental justice focus on comparisons involving disadvantaged groups reflects more than an abstract concern about equality. Instead, another concern may well lie behind these comparative inquiries: an underlying concern about the legacy of social injustice, and, in the United States, of racial discrimination, that has made these communities vulnerable. In other countries, underlying concerns about class, religion, or national origin may drive comparative concerns, and from a global perspective, concerns about distributional disparities are inevitably linked to the colonial roots of global inequality. As such, comparative inquiries, though focused on distribution, can indirectly reflect concerns about the current or historical social and political injustices that have led to existing conditions.

Absolute well-being: the minimum-capabilities approach

The "equality" approach to distributive justice focuses on comparative **well-being**—on defining injustice by whether some groups experience more harms or fewer benefits than others. In contrast, an absolute test focuses on overall well-being, regardless of what others experience. In other words, if everyone lived in equally poor conditions, that distribution would raise environmental justice concerns, despite the absence of distributional disparities.

A focus on absolute well-being is reflected in Amartya Sen's "capabilities" approach, which focuses on the degree to which people have the minimum capabilities necessary to lead a meaningful life (Sen, 2009).[2] Without delving into the controversy over what capabilities matter, it is clear that relevant capabilities include both physical well-being and the capabilities necessary to make free choices. A minimum level of wealth is not the only factor that counts, and minimum capabilities could vary by cultures and country (Sen, 2009, pp. 254–257). Defining what level of well-being would satisfy the capabilities approach is much disputed: should it be the bare minimum to survive, or something more?

For purposes of understanding distributive justice, one need not resolve the question of what capabilities count or how to measure injustice. What matters is recognizing that communities burdened by pollution are often concerned about absolute levels of pollution, not only comparative pollution levels. Thus, a community of colour subject to high levels of particulates might not suffer higher levels than a neighbouring white community, but would nonetheless argue that both their pollution levels and those of the neighbouring community should be reduced.

Human **rights-based approaches** to environmental justice resonate with a minimum-capabilities theory of distributive justice. Rights-based approaches suggest that governments should not allow environmental harms that interfere with basic rights to life and well-being. As such, they assume that people are entitled to a minimum level of well-being (Shelton, 2012).

Distributive justice and lived experience

As subsequent chapters will discuss in more detail, environmental justice issues arise in multiple contexts. Here, we provide an overview of the kinds of decisions that can affect distributive justice and then look at studies assessing distributive injustice to identify how academics and policymakers measure injustice.

Differing contexts

Siting decisions

In the public eye, the quintessential environmental justice dispute involves siting undesirable land uses, like waste sites, heavy industry, power plants, and other noxious uses. A 1982 land use siting controversy in North Carolina catalyzed the US environmental justice movement out of low-profile and isolated instances of community opposition. When national leaders and national groups, such as the National Association for the Advancement of Colored People, joined a low-income African American community in Warren County, North Carolina, to oppose a hazardous waste disposal site, leading to 500 arrests as they blocked truck traffic into the site, national attention at last focused on the plight of marginalized communities (McGurty, 2007).[3]

Land use patterns

In the United States, individual siting decisions are strongly shaped by underlying local land use regulations, like zoning. Areas zoned for purely residential uses preclude industrial uses, while areas allowing mixed uses, including industry, commercial uses, and residential uses, are more likely to host industrial facilities. Local governments have consistently permitted less desirable land uses in older low-income and of-colour neighbourhoods, while newly emerging, and exclusively white, suburbs have been zoned for residential and light commercial uses (Arnold, 1998). Industries seeking to locate in a new area are thus more likely to locate in the poorer of-colour areas that already permit industrial uses than in suburban areas zoned for residential use.

Current local government zoning and facility-siting decisions reflect the continuing legacy of historic racial segregation. From the 1920s through to the 1960s, developers and neighbourhood associations in newly emerging suburbs frequently included private land use contracts restricting occupancy to white residents, contracts designed to "maintain property values." These covenants were often a prerequisite to obtaining federal mortgage guarantees, guarantees that enabled many lower- and middle-income white citizens to purchase property (Rose and Brooks, 2013; Rothstein, 2017). Moreover, differing access to the wealth-accumulating benefits of home ownership have led to persistent gaps in wealth, access to education, and other opportunities that have perpetuated inequities that arose in an earlier time (Schermerhorn, 2019). Although restrictive covenants were finally outlawed by civil rights legislation in the late 1960s, differentiated zoning and differentiated uses persist and continue to affect the current distribution of existing land uses and siting decisions.

Box 3.1   Siting decisions and racial segregation in Richmond, California

As Richard Rothstein described in his book, *The Color of Law*, Richmond, California, initially built segregated public housing for African American workers who left the southern United States to build ships during World War II. Over time, white Richmond residents left for more suburban housing and to pursue new employment opportunities elsewhere in the Bay Area, opportunities more open to whites than blacks due to restrictive housing covenants (Rothstein, 2017, pp. 4–10). Richmond's residents experience on-going pollution from Richmond's heavy industry, including a Chevron oil refinery with a history of polluting accidents (Cagle, 2019).

Globally, natural resource exploitation, like oil development and mining, has often adversely impacted low-income and Indigenous communities (Gonzalez, 2015). For example, in Ecuador and Nigeria, oil company spills and dumping have significantly disrupted the environment and caused extensive health impacts (North, 2015; Kent, 2018), while large-scale mining in resource-rich countries like Indonesia has deprived Indigenous people of their land and community well-being (Yovanda, 2018).

In some cases, the distribution of undesirable land uses presents a tension among theories of justice. From a utilitarian perspective, there are potential societal benefits from placing industry where land is cheap and where there are existing infrastructure and resources worth extracting. If equality is the benchmark, however, the distribution is more problematic. While concentrated industry might meet the preferences of some residents and communities, environmental justice movement activism, alongside the dynamics of land use decision-making, suggests that such synergy is often lacking (Kaswan, 2003).

Environmental regulations

Environmental justice leaders have recognized that environmental regulation also raises distributive justice questions. For example, in 2005, the US General Accountability Office chastised the Environmental Protection Agency (EPA) for failing to discuss the environmental justice implications of a new regulation to reduce sulphur in fuels (US GAO, 2005). The new rule would reduce pollution along highways, improving air quality for the vulnerable communities who live, disproportionately, near major thoroughfares. But it would increase pollution near the refineries that would process the fuel, areas that are also disproportionately populated by people of colour. Thus, even broadly applicable rules can have significant distributive justice implications and present difficult trade-offs.

Choices among available regulatory policy instruments also raise distributive justice issues. Some environmental justice advocates have been deeply sceptical about cap-and-trade programmes for controlling pollution due, in part, to their potential distributive justice consequences. (Kaswan, 2013, pp. 160–161). Under a **cap-and-trade** approach, a government entity sets an overall pollution cap and then distributes pollution allowances to facilities. Facilities can respond in a variety of ways. Because they receive fewer allowances than existing emissions, they can simply reduce emissions to match their allowances. Or they could reduce by more than necessary and sell the remaining allowances. Finally, so long as they buy enough allowances, facilities could *increase* emissions, leading to pollution hot spots. In other words, cap-and-trade programmes are designed to achieve an overall cap on emissions, but, because they permit allowance-trading among facilities, they do not control the distribution of emissions.

In Europe, climate policies raise a different kind of distributional justice question: the distribution of the costs of control. For example, policymakers have highlighted the risk of increasing energy costs and their impact on low-income residents (Lohan, 2017). Transitioning away from inexpensive coal to more expensive renewables could increase electricity costs, particularly for those who cannot afford to invest in energy efficiency improvements that would reduce consumption and mitigate the effect of the rising costs. Similarly, shifting transportation from fossil fuels to electric vehicles could leave those reliant on older vehicles facing higher costs and unable to afford new electric vehicles.

From a utilitarian perspective, regulatory policies that reduce pollution for those most exposed or that provide transition resources to those with the least would maximize welfare by providing benefits to those who attach the most value to them. Because these policies recognize existing inequality, they also resonate with the "distribution based on need" perspective.

Infrastructure

The extent and quality of infrastructure has become an increasingly visible issue. Infrastructure disparities were evident in the United States when a low-income, largely African American community in Flint, Michigan, discovered that it had been drinking lead- and bacteria-laden water, a confluence of old lead-pipe infrastructure and mismanaged water supply and water quality control decisions (Flint Water Advisory Task Force, 2016). In California, immigrant farmworkers are drinking fertilizer-contaminated water because the communities in which they live lack the resources to provide adequate water supplies (Del Real, 2019).

In many parts of the world, a lack of access to drinking water and sanitation and the associated health consequences are among the most pressing distributive justice issues. As of 2015, 2.1 billion people, or almost one-third of the globe's population, half of them in Africa, did not have access to clean drinking water, and over half of the global population did not have proper sanitation facilities (UNESCO, 2019).

In addition to raising concerns about unjustified inequality, the absence of essential infrastructure resonates with the capabilities approach to distributive justice. Communities lacking basic infrastructure essential to a decent life, like drinking water and sanitation, have a strong claim, regardless of comparative differences.

Environmental enforcement

Government enforcement decisions have environmental justice implications. Environmental law violations far outnumber government resources to enforce, so government agencies often face tough choices about which violators to pursue. A critical question is whether state and federal environmental agencies pursue violations as vigorously in marginalized communities as they do in more privileged communities. For example, a US study of state enforcement of federal Clean Air Act violations found that, although the degree of risk did play a significant role in many enforcement priority decisions, political factors, which could privilege the privileged, appear to play a role as well, leading to relatively less vigorous enforcement in Latinx communities (Koninsky and Reenock, 2018).

Box 3.2   The Flint water crisis

Outrage about contaminated drinking water in Flint, Michigan, was not just about the poor infrastructure that led to the contamination, but about the failure of state and federal officials to respond to the complaints. State managers running the city's water system had failed to treat the water with an anti-corrosion additive, which caused lead and other sediment to leach from old pipes into the water supply. Community residents immediately alerted officials to the cloudy water. However, the state failed to address the contamination until over a year later. The federal EPA deferred to the state agency and likewise did not take prompt action (Flint Water Advisory Task Force, 2016).

In many parts of the world, citizens fear that global companies are operating with relative impunity. Victims of Union Carbide's deadly explosion at a pesticide plant in Bhopal, India, which killed thousands and left many blind and suffering from lifelong health ailments, were unable to obtain what they viewed as appropriate compensation for their losses from the US-based company (Withnall, 2019). In Latin America, Indigenous people have struggled for decades to hold international oil companies accountable for pollution (Kimerling, 2006; North, 2015).

A failure to enforce environmental laws in marginalized communities implicates the equality and capabilities approaches to distributive justice. It leads to disparities in pollution exposure and, in extreme cases, destroys the basic environmental and cultural conditions necessary for a decent life.

Evidence of distributional injustice: what gets measured, and why?

How do we know about the existence and distribution of environmental problems? After several decades of study, data on environmental justice is extensive—much more extensive than can be addressed in this overview chapter or even in this volume (Cole and Foster, 2001; Lester et al., 2001). We first note, without resolving, a few of the difficult methodological challenges likely to impact most distributional studies. One is the question of scale: at what level do we measure? By neighbourhood? And, if by neighbourhood, how is a "neighbourhood" to be defined? By census tract? ZIP code? Municipal boundaries? Metropolitan areas? Regions? States? (Zimmerman, 1994; Fahsbender, 1996). If we zoom in to too small a scale, we may miss important patterns. If we zoom out to too large a scale, we may dilute localized impacts. The "right" scale will depend upon the nature of the harm being analyzed and purpose for which information is being gathered. A second difficult question that arises in the context of studies of racial disparities is what counts as an "of-colour" community? Majority of-colour? A higher percentage relative to the average in the . . . City? Region? State? And a third challenging question concerns what constitutes a "disparate" impact. Is one of "x" enough, if other areas have nothing? More of "x" than others experience? A lot more than others?

Distributive justice inquiries usually measure one of two variables. One is "things," like the number of waste sites, power plants, jails, or other undesirable uses within the chosen unit of analysis (Mohai et al., 2009). The other is risk exposure, like the extent to which a facility poses a particular health risk, or, more broadly, the quality of the air or water in a particular area. (Morello-Frosch and Jesdale, 2006). Subsequent chapters will address actual data in numerous contexts. In this chapter, we focus on what gets measured, and why.

Measuring things

Measuring things serves a variety of purposes. One is to provide a proxy for exposure to harm. It is much simpler to count facilities than it is to determine emissions or discharges and the specific risks they present. Determining precise emission and hazard levels from stationary sources, like industry, and from mobile sources, like cars and trucks, is difficult and controversial. Counting the number of hazardous waste dumps, or the number of industrial facilities, or the number of other types of undesirable land uses, will provide at least a general sense that the communities with such land uses are likely to experience more hazardous risks, smells, noise, and other harms than a residential community of single-family homes.

Of course, counting things provides only a rough proxy for actual harm. For example, a hazardous waste landfill subject to extensive controls, like effective liners, prompt covering of contaminated material, and extensive boundary and groundwater monitoring, might present relatively little risk to a neighbouring community. In contrast, a poorly run facility subject to few limits, or one that is violating limits with impunity, might present a much greater risk. Moreover, actual risk is likely to depend upon proximity and a variety of environmental factors, including wind or water patterns.

In addition to providing a proxy for physical harm, assessing the distribution of things provides a proxy for stigmatic harm. Whatever the physical harm associated with an adverse land use, neighbouring residents may feel stigmatized by close proximity to undesirable land uses. The land uses could define their community as a "bad" neighbourhood, even if actual harms are low. Or, if a community is selected for a new noxious use, residents may believe that the decision reflects a lack of respect for their community. Dignity rights, as well as physical well-being, are often at stake. At times, these dignity impacts are reflected in something concrete: decreased property values.

Studying the distribution of things can also cast light on potential discriminatory decision-making, whether in the present or over time. While unequal distributions are not necessarily caused by discriminatory decisions, they raise a red flag. Disproportionate patterns can provide indirect evidence that some communities have endured a legacy of unfair treatment. Tracing causation is notoriously difficult, as discussed ahead, but demonstrated disparities in the distribution of burdens provides tangible evidence that decisions—often cumulative decisions by many actors—have concentrated societal burdens in certain communities.

Understanding the range of goals also illuminates a methodological issue noted previously: what is the appropriate scale for analysis? Arguably, highly localized assessments are best at establishing a proxy for physical risks, at least where the facilities in question could cause localized harms. To the degree that the inquiry is focused on assessing whether decision-makers are acting fairly, however, the appropriate unit of analysis could depend more on the scale of the jurisdiction responsible for making the decision (Kaswan, 2003). If the issue is whether local governments have sited municipal waste facilities fairly, then a city-wide inquiry would be appropriate because of what it reveals about patterns of local decision-making, not because of what it reveals about risk. And, if the issue is whether states have sited hazardous waste facilities fairly, then statewide decision-making patterns would be relevant to assessing potential discrimination, even if such studies provided little information about community risk.

Distribution of risk or harm

Moving beyond studies on the distribution of land uses, other studies assess the distribution of risk and harm. These studies arise in a variety of contexts. In the United States, assessments associated with permit decisions evaluate emissions and discharges from a specific facility. Environmental assessments accompanying new pollution control rules for an industrial category, like new air pollution standards for cement plants, would focus on expected impacts from that type of facility. Some broader environmental justice studies focus on levels of particular pollutants, like particulates, ozone (a contributor to smog), or one or more toxic pollutants (Bravo et al., 2016; Ash et al., 2009). Others focus on the incidence of demonstrated harms, like birth defects or deaths from heat waves (Ponce et al., 2005).

Continuing to broaden the lens, many environmental justice studies focus not just on a single pollutant or source of pollutants, but on cumulative exposures from multiple pollutants from multiple sources. For example, researchers from UC Berkeley analyzed the relationship between racial segregation and cancer risks presented by cumulative air toxics exposure across US metropolitan areas (Morello-Frosch and Jesdale, 2006). Even more broadly, one study assessed the distribution of environmental quality, considering a wide range of indicators, including both emissions and proximity to undesirable land uses (Hird and Reese, 1998).

Recognizing that the level of harm caused by a given level of risk depends upon underlying conditions, the newest wave of environmental justice analyses explores the larger question of vulnerability. Vulnerability assessments consider not only exposure levels, but conditions like access to health care, immigrant status, linguistic isolation, income, and potential workplace exposures to identify where a given risk could pose a particularly high burden (DeFur et al., 2007). The US EPA has developed "EJScreen," an "environmental justice screening and mapping tool" that incorporates numerous environmental and socio-economic factors and that could help inform environmental enforcement priorities and identify areas where greater pollution reduction measures would be most beneficial (Environmental Protection Agency n.d.).

The importance of underlying socio-economic conditions and their implications for vulnerability has received particular attention in the climate adaptation context. Building on evidence from disparities in disaster recovery, researchers and policymakers are increasingly recognizing how socio-economic disparities, including disparities in race and income, will affect communities' sensitivity to harm from climate impacts like heat waves and flooding, as well as their economic and social capacity to cope with climate disasters in the long-term. For example, a recent study in Latin America found that socio-economic factors played a major role in determining well-being after climate-sensitive disasters (Nagy et al., 2018). Globally, the historically developed countries have stronger infrastructure and a greater capacity to cope with climate disasters like hurricanes, flooding, and extreme heat (Kaswan, 2016).

What causes distributional injustice?

Distributive injustice, whether in comparative or absolute terms, is a concern no matter what the cause. From a comparative standpoint, one community experiencing greater harm than another is problematic. From an absolute perspective, a community experiencing intense harm is problematic even if there is no specific actor. Nonetheless, the reason behind a given distribution inevitably affects our perceptions of the degree of injustice. Particularly in the land use siting context, if environmental agency officials or industry decision-makers are purposefully concentrating undesirable land uses in of-colour and low-income communities, that heightens our concern. A key question, then, is whether current disparities reflect discriminatory decision-making, and, if so, whether they reflect on-going or past discrimination.

Numerous factors are likely to contribute to a given distribution of land uses or risks. Industry often seeks out lower land costs and access to transportation infrastructure, like highways and railways. Not surprisingly, lower-income people, with fewer choices, live in these areas. One could argue that "objective" factors, not discrimination, are one cause of disparities in land use siting and risk.

And, as noted previously, distributional outcomes are strongly influenced not just by individual industry or agency decisions, but by underlying land use policies. A given siting decision may reflect zoning-related opportunities, not the preferences of the siting or permitting officials. Moreover, past discrimination creating segregation and associated differences in land use patterns have strong staying power. Even if current housing policies and practices are not discriminatory, industrial areas cannot turn pristine overnight, and currently pristine areas will not relinquish that benefit lightly.

Moreover, some argue that current disparities may have arisen since the initial siting decision—that, at the time of siting, there might not have been any disparity at all (Lambert and Boerner, 1997; Samp, 1994). Instead, once industry or officials have situated an undesirable land use, like a hazardous waste facility, in a community, land values would drop, richer residents and white residents with more options for housing would leave, and lower-income and of-colour residents with fewer options would move in. Thus, existing disproportionate distributions might have been caused by post-siting housing market dynamics rather than by discriminatory decision-making (Been, 1993).

A major national study of hazardous waste sites found, however, that post-siting housing market dynamics did not match the hypothesized shifts. There was little evidence that siting a hazardous waste facility changed local demographics and, if it did, little evidence that the proportion of of-colour residents increased. The study also found evidence that waste sites had been disproportionately sited in Latinx neighbourhoods (Been and Gupta, 1997). This is not to say that the projected housing dynamics never occur; some localized studies have found evidence of increasing low-income and of-colour residents after siting (Lambert and Boerner, 1997). But the pattern is not universal, and the evidence suggests that siting itself, not just post-siting housing market dynamics, plays a role in existing disparities.

Whether current disparities are explained by "objective" factors, underlying past or present land use policies, or post-siting housing market dynamics, the justice question does not disappear. Underlying income inequalities and racial discrimination affect all of these dynamics. "Objective" factors, like cheaper land or proximity to transportation infrastructure, are associated with lower-income housing and land use policies that relegated people of colour to areas near polluting infrastructure. As noted previously, land use policies are rife with discriminatory and segregationist impulses.

To the degree that disparities arise after initial siting decisions, housing dynamics are themselves driven by disparities. Lower-income and of-colour residents have fewer options than people with means and people who do not fear social discrimination. Disadvantaged residents may locate close to undesirable land uses because they face a constricted range of choices. Although they are exercising choice, larger social inequities have constrained their choices in ways not experienced by more privileged residents.

Responding to distributive injustice

If a community believes it is experiencing injustice, what can it do? Achieving justice through the legal system is challenging. In the United States, antidiscrimination law does not provide an effective tool for addressing distributive justice. As elaborated in *Washington v Davis*, 426 US 229 (1976), the US Constitution's guarantee of equal protection under the law applies only to government decisions, not private decisions, and it prohibits only intentional discrimination, not disparate impacts. Thus, a case based solely on distributional inequities, and not traceable to a specific act of governmental discrimination, is not actionable.

Although the United States adopted civil rights legislation in the 1960s, it has not provided an effective mechanism for addressing distributional disparities. One provision had appeared promising: Title VI of the civil rights legislation prohibits state and local agencies receiving federal funding from discriminating (Civil Rights Act, 1964). Many of the federal agencies allocating federal funds to the states, including the EPA, adopted regulations that prohibited disparate impacts, as well as intentional discrimination (Environmental Protection Agency, 1996). But federal agencies have been reluctant to enforce the regulations for a variety of political and practical reasons, and so therefore, few Title VI complaints based on distributional inequities have succeeded (Kaswan, 2013).

Affirmative programmes are another vehicle for addressing injustice. For example, the US EPA's action plan for 2016–2020, adopted during the environment-friendly Obama administration, committed to directing enforcement resources to the nation's 100 most overburdened communities (Environmental Protection Agency, 2016, p. iii). On the local level, New York City has worked to increase the fairness of siting municipal facilities through its "fair share" rules (New York City Council, 2017). Beginning in 2018, California local governments with "disadvantaged communities" must, in their land use plans, include goals and develop policies to reduce health risks (Senate Bill 1000, 2016).

While not guaranteeing any substantive results, procedural mechanisms that generate information about disparate impacts can provide an important resource for impacted communities and increase accountability. A 1994 order by then-President Clinton established procedural requirements for agencies: when conducting environmental assessments under US law, which requires assessments for all major federal projects impacting the environment, federal agencies must document impacts on low-income and minority populations (EO 12, 898). The requirement delivers information that would be difficult for disadvantaged communities to assemble on their own, and provides a basis for political organizing.

Injustice resonates. When residents of Flint, Michigan, were finally able to draw national attention to their lead-contaminated water, political leaders responded. Although the local response remains slow and frustrating, the US Congress has now funded infrastructure improvements in Flint and elsewhere (Davenport, 2016). Not every environmental justice dispute will draw this kind of attention. But evidence of distributional injustice—in whatever form—can support political action and, ultimately, hold decision-makers more accountable for their decisions.

Conclusion

As with all claims for "justice," there is no single measure of distributive justice. Different conceptions highlight differing aspects, from utilitarian perspectives that focus on general well-being, through equality-based theories, to a minimum-capabilities perspective. The theories are not mutually exclusive; instead, each theory highlights differing measures of justice.

The theories of distributive justice provide lenses through which to evaluate justice in numerous contexts. Past and current decisions about siting, land use planning, regulatory strategies, and environmental enforcement all shape lived experience, and decisions made today are likely to have enduring impacts. Environmental justice scholars have devoted considerable attention to determining what is, in fact, occurring on the ground. Numerous studies, at a multiplicity of scales, analyze the distribution of a wide variety of land uses, as well as risk: what exposures, with what consequences, do people experience?

Although our inquiry into distributive justice does not require us to identify a specific discriminatory or unfair cause, we note that distributional disparities inevitably raise the causation question. In addition, we observe some of the complex factors that drive distributional disparities and their relationships to past and on-going social injustice.

Finally, we provide a sense of the substantive and procedural mechanisms that could address distributional disparities. Distributing the burdens and benefits of a modern society will always be highly contested and determined by deep power dynamics. Greater awareness and transparency about environmental harms and their consequences and a clear moral compass about what is fair could facilitate the political organizing that will ultimately be necessary to increase distributive justice at the local, state, national, and global levels.

Follow-up questions

- What conception or conceptions of distributive justice do you find most compelling, and why?
- When, if ever, should we require decision-makers to avoid distributional disparities, even if there is no direct evidence of discrimination? Explain why or why not in each instance.

> - What kinds of policies at the local, state, or federal level might lead to greater distributive justice? Should we adopt procedural mechanisms that require decision-makers to assess distributional impacts? Should we adopt substantive requirements that limit concentrated harms? What are the benefits and drawbacks to your suggested policies?

Notes

1 On procedural justice, see Chapter 4 of this volume.
2 On capabilities and well-being, see also Chapter 6 of this volume.
3 For more details on the case of Warren County, see Chapter 2 of this volume.

References for further reading

Been, Vicki. 1993. What's fairness got to do with it? Environmental justice and the siting of locally undesirable land uses. *Cornell Law Review* 78: 1001–1085.
Cole W. Luke and Sheila R. Foster. 2001. *From the ground up: Environmental racism and the rise of the environmental justice movement.* New York: New York University Press.
Kaswan, Alice. 2003. Distributive justice and the environment. *North Carolina Law Review* 81: 1031–1148.
Kuehn, Robert. 2000. A taxonomy of environmental justice. *Environmental Law Reporter* 30(10): 681–703.

References used in this chapter

Arnold, Craig Anthony. 1998. Planning Milagros: Environmental justice and land use regulation. *Denver Law Review* 76: 1–797.
Arthur, John and William H. Shaw. 1978. *Justice and economic distribution.* Upper Saddle River, NJ: Prentice-Hall.
Ash, Michael et al. 2009. *Justice in the Air: Tracking toxic pollution from American's industries and companies to our states, cities, and neighborhoods.* Amherst: Political Economy Research Institute. www.peri.umass.edu/publication/item/308-justice-in-the-air-tracking-toxic-pollution-from-america-s-industries-and-companies-to-our-states-cities-and-neighborhoods.
Been, Vicki. 1993. What's fairness got to do with it? Environmental justice and the siting of locally undesirable land uses. *Cornell Law Review* 78: 1001–1085.
Been, Vicki and Francis Gupta. 1997. Coming to the nuisance or going to the barrios? A longitudinal analysis of environmental justice claims. *Ecology Law Quarterly* 24: 1–56.
Berlin, Isaiah. 1961. Equality as an ideal. In Frederick A. Olafson, ed., *Justice and social policy*, 128. Upper Saddle River, NJ: Prentice-Hall, Inc.
Blais, Lynn E. 1996. Environmental racism reconsidered. *North Carolina Law Review* 75: 75–148.
Bravo, Mercedes A., Rebecca Anthopolos, Michelle L. Bell, and Marie Lynn Miranda. 2016. Racial isolation and exposure to airborne particulate matter and ozone in understudied US populations: Environmental justice applications of downscaled numerical model output. *Environment International* 92–93: 247–255.
Cagle, Susie. 2019. Richmond v Chevron: The California city taking on its most powerful polluter. *The Guardian*, Oct. 9. www.theguardian.com/environment/2019/oct/09/richmond-chevron-california-city-polluter-fossil-fuel
Civil Rights Act, Title VI, 42. 1964. U.S.C. §§ 2001e-1 to -17.

Cole, Luke W. and Sheila R. Foster. 2001. *From the ground up: Environmental racism and the rise of the environmental justice movement*. New York: New York University Press.

Davenport, Coral. 2016. Senate approves funding for flint water crisis. *The New York Times*, Sept. 15.

DeFur, P.L., G.W. Evans, E.A. Hubal, A.D. Kyle, R. Morello-Frosch, and D.A. Williams. 2007. Vulnerability as a function of individual and group resources in cumulative risk assessment. *Environmental Health Perspectives* 115: 817–824.

Del Real, Jose A. 2019. The grow the nation's food, but they can't drink the water. *The New York Times*, May 21.

Environmental Protection Agency. 1996. Nondiscrimination in programs receiving federal Assistance from the environmental protection agency. *Code of Federal Regulations* 40: §7.35(b)-(c).

Environmental Protection Agency. 2016. *EJ 2020 action agenda: The U.S. EPA's environmental justice strategic plan for 2016–2020*. www.epa.gov/sites/production/files/2016-05/documents/052216_ej_2020_strategic_plan_final_0.pdf

Environmental Protection Agency. n.d. *EJ screen: Environmental justice screening and mapping tool*. www.epa.gov/ejscreen.

Fahsbender, John J. 1996. An analytical approach to defining the affected neighborhood in the environmental justice context. *NYU Environmental Law Journal* 5: 120–180.

Flint Water Advisory Task Force. 2016. *Final report*. www.michigan.gov/documents/snyder/FWATF_FINAL REPORT_21March2016_517805_7.pdf

Gonzalez, Carmen G. 2015. Environmental justice, human rights, and the global South. *Santa Clara Journal of International Law* 13: 151–195.

Hird, John A. and Michael Reese. 1998. The distribution of environmental quality: An empirical analysis. *Social Science Quarterly* 79: 693–716.

Hossain, Kamrul and Noor Jahan Punam. 2016. Climate change impact and human rights of rural women in Bangladesh. In Randall S. Abate, ed., *Climate justice: Case studies in global and regional governance challenges*. Washington, DC: Environmental Law Institute.

Kaswan, Alice. 2003. Distributive justice and the environment. *North Carolina Law Review* 81:1031–1148.

Kaswan, Alice. 2013. Environmental justice and environmental law. *Fordham Environmental Law Review* XXIV: 149–179.

Kaswan, Alice. 2016. Climate adaptation and theories of justice. Published in *Philosophy, Law and Environmental Crisis* (Alain Papaux and Simone Zurbuchen, editors). *Archive für Rechts- und Sozialphilosophie Beihefte* 149: 97–118.

Kent, Sarah. 2018. Pollution worsens around shell oil spills in Nigeria. *Wall Street Journal*, May 25.

Kimerling, Judith. 2006. Indigenous peoples and the oil frontier in Amazonia: The case of Ecuador, Chevrontexaco, and *Aguida v. Texaco*. *International Law and Politics* 38: 413–664.

Koninsky, David M. and Christopher Reenock. 2018. Regulatory enforcement, riskscapes, and environmental justice. *Policy Studies Journal* 46: 7–36.

Lambert, Thomas A. and Christopher Boerner. 1997. Environmental inequity: Economic causes, economic solutions. *Yale Journal on Regulation* 14: 195–234.

Lester, James P., D.W. Allen, and K.M. Hill. 2001. *Environmental injustice in the United States: myths and realities*. Boulder, CO: Westview Press.

Lohan, Cillian (Rappoteur). 2017. *Climate Justice*. European Economic and Social Committee Opinion.

McGurty, Eileen M. 2007. *Transforming environmentalism: Warren county, PCBs, and the origins of environmental justice*. Piscataway, NJ: Rutgers University Press.

Mohai, Paul, Paula Lanz, Jeff Morenoff, James House, and Richard P. Mero. 2009. Racial and socioeconomic disparities in residential proximity to polluting industrial facilities: Evidence from the Americans' changing lives study. *American Journal of Public Health* 99: S649–S656.

Morello-Frosch, Rachel and Bill M. Jesdale. 2006. Separate and unequal: Residential segregation and estimated cancer risks associated with ambient air toxics in U.S. metropolitan areas. *Environmental Health Perspectives* 114: 386–393.

Nagy, Gustavo J., Walter Leal Filho, Ulisses M. Azeiteiro, Johanna Heimfarth, José E. Verocai, and Chunlan Li. 2018. An assessment of the relationships between extreme weather events, vulnerability, and the impacts on human wellbeing in Latin America. *International Journal of Environmental Research and Public Health* 15(9): 1802.

New York City Council. 2017. *Doing our fair share, getting our fair share: Reforming NYC's system for achieving fairness in siting municipal facilities.* https://council.nyc.gov/wp-content/uploads/2017/02/2017-Fair-Share-Report.pdf

North, James. 2015. Ecuador's battle for environmental justice against chevron. *The Nation*, June 22.

Ponce, Ninez A., Katherine J. Hoggatt, Michelle Wilhelm, and Beate Ritz. 2005. Preterm birth: The interaction of traffic-related air pollution with economic hardship in Los Angeles neighborhoods. *American Journal of Epidemiology* 162(2): 140–148.

Rawls, John. 1971. *A theory of justice*. Boston: Belknap Press.

Rescher, Nicholas. 1966. *Distributive Justice: A constructive critique of the utilitarian theory of distribution*. Indianapolis: Bobbs-Merrill.

Rose, Carol and Richard R.W. Brooks. 2013. *Saving the neighborhood: Racially restrictive covenants, law, and social norms*. Cambridge: Harvard University Press.

Rothstein, Richard. 2017. *The color of law: A forgotten history of how our government segregated America*. New York: Liveright Publishing Corporation.

Samp, Richard A. 1994. Fairness for sale in the marketplace. *St. John's Journal of Legal Commentary* 9: 503–507.

Senate Bill 1000. 2016. *Land use: General plans: Safety and environmental justice: 2015–16.* https://leginfo.legislature.ca.gov/faces/billTextClient.xhtml?bill_id=201520160SB1000

Schermerhorn, Calvin. 2019. Why the racial wealth gap persists, more than 150 years after emancipation. *The Washington Post*, June 19.

Sen, Amartya. 2009. *The idea of justice*. Cambridge: Harvard University Press.

Shelton, Dinah L. 2012. Using law and equity for poor and the environment. In Yves Le Bouthillier et al, eds., *Poverty Alleviation and Environmental Law*, 11–52. Cheltenham: Edward Elgar [confirm UK rather than US office].

Singer, Peter. 2002. *One world*. New Haven: Yale University Press.

UNESCO, United Nations World Water Development Report 2019—Leaving No One Behind. 2019. https://en.unesco.org/themes/water-security/wwap/wwdr/2019

U.S. General Accountability Office. 2005. *EPA should devote more attention to environmental justice when developing clean air rules*. Report to the Ranking Member, Subcommittee on Environment and Hazardous Materials, Committee on Energy and Commerce, House of Representatives.

Yovanda. 2018. Bornean villagers who fought off a mine prepare to do battle again. Trans. Aria Danaparamita. *Mongabay*, Aug. 14. https://news.mongabay.com/2018/08/bornean-villagers-who-fought-off-a-mine-prepare-to-do-battle-again/

Walker, Gordon. 2012. *Environmental justice: Concepts, evidence and politics*. Milton Park: Routledge.

Withnall, Adam. 2019. Bhopal gas leak: 30 years later and after nearly 600,000 were poisoned, victims still wait for justice. *Independent*, Feb. 14. www.independent.co.uk/news/world/asia/bhopal-gas-leak-anniversary-poison-deaths-compensation-union-carbide-dow-chemical-a8780126.html

Zimmerman, Rae. 1994. Issues of clarification in environmental equity: How we manage is how we measure. *Fordham Urban Law Journal* 21: 633.

4 Procedural justice matters

Power, representation, and participation in environmental governance

Kimberly R. Marion Suiseeya

Learning outcomes

- Identify and define key concepts related to procedural justice.
- Explain the relationships between participation, representation, and procedural justice.
- Evaluate how different tools, approaches, and principles for collective decision-making create opportunities for and constraints to procedural justice.

Introduction

> "If the underlying structures of process don't allow for equality, you won't get a good outcome."[1]

This statement by an Indigenous woman at the 2014 World Parks Congress highlights the importance of decision-making processes, especially the inclusion of **frontline communities**, such as Indigenous, traditional, minority, and poor communities, for governing the global environment. Despite the growing number of Indigenous representatives in global environmental governance, this participant was concerned about the ongoing lack of representation and sometimes mis-representation of Indigenous peoples. She was pointing out the structural inequalities in global environmental governance that participation alone could not overcome.

The Sixth World Parks Congress was significant for Indigenous Peoples for many reasons. It was the first Congress since the 2007 United Nations Declaration on the Rights of Indigenous Peoples was signed by all but three United Nations members.[2] The prior World Parks Congress held in 2003 in Durban, South Africa, had resulted in an ambitious agenda that, for the first time in the Congress's history, foregrounded Indigenous and local community rights (Brosius, 2004), and it also followed the signing of the 2010 Nagoya Protocol on Access and Benefits Sharing under the Convention on Biological Diversity. In selecting as the host country for this event Australia, a nation-state with a difficult history of violence and oppression against the continent's Indigenous Peoples, the IUCN[3] was also signalling the importance of addressing the ongoing tensions related to people and parks, especially with regards to Indigenous communities and lands, as part of its agenda for the Congress. Pressure from Indigenous leadership within and outside IUCN led to the development of a stream dedicated to engaging with and considering Indigenous experiences and ideas in biodiversity conservation in the Congress.[4] More than 6,000 delegates engaged in this stream, alongside seven other streams, to explore ways to reach conservation goals, respond to climate change, improve human health and well-being, support human life, reconcile different development challenges, enhance diversity and quality of governance, respect Indigenous and traditional knowledge and culture, and inspire a new generation to prioritize conservation (IUCN, 2014). Most sessions opened with a land acknowledgement, expressing recognition of the traditional stewards of the land upon which the meeting took place.

One unique and significant feature of the IUCN and its congresses, including the World Parks Congress, is the institutional design that grants nation-states and non-state delegates alike equal voting and engagement rights. Unlike other multilateral environmental institutions, such as the UN Framework Convention on Climate Change (UNFCCC) and the Convention on Biological Diversity, where nation-state parties are the only delegates with voting power and non-state actors may engage (with restrictions) at the pleasure of the parties, at IUCN congresses non-state actors often wield as much, and sometimes more, power than nation-states. And, although IUCN congresses do not result in international treaties or protocols (i.e. international law), historically they have provided strong foundations for advancing negotiations in the conventions of the biodiversity regime.[5]

Because of its unique structure in the landscape of global environmental governance that integrates state and non-state actors as equals, the IUCN's World Parks Congress can be seen as a reflection of contemporary and evolving understandings of procedural justice. At the most fundamental level, **procedural justice** is *the ability to participate in and influence decision-making processes* (see Box 4.1) (Clayton, 1998, Bell and Carrick, 2017). The IUCN, through its congresses, including the World Parks Congress and World Conservation Congress, is often at the forefront of developing and disseminating new innovations and concepts to improve environmental governance, including practices for engaging communities. Yet, even as participation in environmental governance has increased in breadth and depth across multiple scales of governance, concerns about procedural justice remain.

Box 4.1 Key concepts in procedural justice

Participation: processes, stakeholder analyses, and activities to support involvement in project and programme design, implementation, and operation

Procedural justice: the ability to participate in and influence environmental decision-making processes (Clayton, 1998)

Environmental democracy: institutions that ensure a "fair voice for all groups and promote fair outcomes by requiring participation to justify their proposals to other participants" (Carrick and Bell, 2017, p. 103)

Environmental citizenship: societal membership that requires consideration of public environmental goods, i.e. accounting for your individual interactions with the environment and subsequent impacts on the public good (Dobson, 2007)

Ecological citizenship: advances considerations of an individual's ecological footprint (Dobson, 2007)

Deliberative democracy: decision-making processes centred on free, fair, and authentic deliberation; e.g. free from political distortions that may create power imbalances (Dryzek and Stevenson, 2011)

Representation: the act of making something visible (Pitkin, 1967)

This chapter approaches procedural justice from the perspective of frontline communities that include groups like Indigenous Peoples, Traditional Peoples, those facing extreme poverty, and minority communities—the actors who disproportionately experience environmental injustice.[6] To illuminate the principles, drivers, and dynamics of procedural justice, the chapter draws on examples from multiple governance scales—from global to local—to examine *why* procedural justice is important for environmental governance, *what* procedural justice looks like in practice, *who* the central actors for procedural justice are, and *how* participation, representation, and **power** interact to shape the opportunities for and constraints to procedural justice. The chapter grounds its approach in several empirical examples that illuminate the dynamics of procedural justice in environmental governance at global, national, and local levels.

Procedural justice matters in environmental governance

Like many justice endeavours, the pursuit of procedural justice is aspirational. What this means is that advancing procedural justice requires always pursuing it, even once mechanisms are in place to help support it. There is no end, only a means. Key questions for designing practices for procedural justice include: (1) what principles should guide engagement and the decision-making process?; (2) who are the key communities or groups of actors that should be involved in decision-making and how can they be identified?; and (3) how should the process be structured and run? Each of these questions should be approached carefully and critically to mitigate structural conditions that may reinforce or reproduce existing injustices.

Despite the frequency with which researchers deploy concepts of **power** in environmental governance scholarship, power remains remarkably underspecified. Power has hegemonic and overt expressions like military force or market share, but it also takes on more nuanced and relational expressions like the ability of an individual to influence another because of, for example, their social ties (for a recent review see Ahlborg and Nightingale, 2018). While power can emerge from a variety of sources and take on diverse forms and expressions, at its core power is "mutual, simultaneous constitutions of agency and structures" (Marion Suiseeya and Zanotti, 2019). This means that how various constellations of individuals, communities, organizations, and institutions (e.g. norms, rules, and regulations) relate can shape how power emerges, changes, and with what effects. Thus, in the context of procedural justice, attention to how the rules of a decision-making process dictate who, what, where, when, and how representatives engage can help understand its justice possibilities.

Critically, although many decision-making processes identify representatives and recognize their authority to represent particular groups of constituencies, representation itself is a form of power that is independent of a decision-making process. Similar to power, it can take on many forms. Fundamentally, however, representation is the act of making something visible (Pitkin, 1967). It is a critical form of power for procedural justice, which is concerned with the ability to participate in and influence decision-making processes. Conceptually, power and representation are distinct from authority. Authority refers to a form of power that works through consent: i.e. those wielding authority were granted such authority by those they are governing and/or representing (Green, 2013). As such, in the context of procedural justice, attention to how **representation** emerges, by whom and with what authority is critical for assessing the abilities of different groups to influence decision-making processes. It also demands attention to who is not represented—both by those groups who are not present, but also by representatives who have no authority to speak on behalf of particular groups. An example of this is when a delegate from Canada spoke on behalf of First Nations at the Tenth Conference of Parties to the Convention on Biological Diversity, and a First Nations member quickly called the delegate out for speaking without consent or authority from First Nations Peoples (Witter et al., 2015). Only by unpacking representation and its relationships to power and authority can we begin to understand how and why procedural justice matters for environmental governance. As you progress through this chapter, keep these questions in mind, as they will help focus your attention on the politics of decision-making that are often masked by procedural tools.

To ground abstract notions of how **access**, representation, and power impact procedural justice in environmental governance, I turn to the case of shifting cultivation in Southeast Asia. Shifting cultivation, also known as swidden agriculture, is a common farming practice in upland communities. It is a technique used in mountainous areas of Southeast Asia where paddy rice farming is not possible. Traditionally, shifting cultivation involves clearing selected plots of fertile land, planting and harvesting rice and other crops, and later rotating to new plots as existing plots become less fertile. Fallow periods, when established plots are not cultivated, provide time for the land to naturally recover before it is brought back into production.

In the 1980s and early 1990s, scientists and policy-makers across Southeast Asia became increasingly concerned with transboundary air pollution, rising rates of deforestation, and land use change. The 'problem'—from their perspective—was shifting cultivation itself: by clearing agricultural plots, sometimes through burning, upland communities produced poor air quality and reduced forest cover (Forsyth and Walker, 2008). Scientists and policy-makers began to focus on shifting cultivation as a driver of deforestation. As policy-makers adopted the pejorative term 'slash and burn,' they developed policies designed to eliminate shifting cultivation as a farming practice. In 1994 in Laos, for example, the government adopted regulations to eradicate shifting cultivation by 2000. Similar policies emerged across the developing world with significant livelihood effects, generally with no input from the farmers who used this particular practice (Blaikie and Muldavin, 2014). Most scholars would point to the lack of farmer input into the policy process as the primary procedural injustice: farmers were neither consulted nor invited to participate in decision-making processes, and their perspectives on shifting cultivation were not represented. However, the possibilities for procedural justice became heavily constrained the moment that the problem of deforestation was defined as shifting cultivation. Defining it as such limited deliberations to finding ways to reduce, change, and/or eradicate shifting cultivation, therewith concentrating power to influence decisions in the hands of those defining the problem of deforestation. This reduced the potential for farmers to leverage traditional ecological knowledge as a form of power; it also diminished opportunities for representing different understandings of the problem of deforestation, such as land concessions, poverty, or infrastructure development. Thus, even in the event that a decision-making process adhered to best practices and had provided access for farmers to the process, the inability to leverage an important form of power for farmers, alongside limitations to representation, challenged the possibilities for procedural justice. Asking questions about how power is wielded from the point at which a problem is identified—the decision impetus—is thus necessary for designing approaches to procedural justice.

Participation and procedural justice

There are many perspectives from which to assess the need for procedural justice. From the perspective of practitioners, procedural justice should help improve overall project success by directly engaging frontline communities (Marion Suiseeya and Caplow, 2013). For government agencies and donors, procedural justice could enhance the legitimacy of environmental policies and regulations, thereby increasing chances of success (e.g. Chambers, 1994). From the perspective of frontline communities, e.g. those whose well-being will be most impacted by a governance initiative, procedural justice is often where environmental governance initiatives should start (Whyte et al., 2016; Wyatt et al., 2010). In fact, procedural considerations have long been essential to the environmental justice movement. Principle #7 of the *Principles of Environmental Justice* adopted October 7, 1991 at the **First National People of Color Environmental Leadership Summit** held in Washington, DC states that "environmental justice demands the right to participate as equal partners at every level of decision-making, including needs assessment, planning, implementation, enforcement and evaluation" (1991).[7]

At the global level, non-state actors, including representatives from Indigenous and local communities, have become regular and in some cases institutionalized—participants at sites of global governance. In 1992, for example, 178 governments at the United Nations Convention on Environment and Development in Rio de Janeiro endorsed Agenda 21, a sweeping plan of action for advancing progress towards sustainable development. Recognizing the "need to take a balanced and integrated approach to environment and development questions," Agenda 21 laid out principles and objectives for guiding actions in 39 distinct areas of environment and development to ensure that human and environmental well-being were advanced together. Woven throughout Agenda 21 are calls for democratic participation by all stakeholders, but especially the communities where development and environment projects would be implemented. Today, international environmental treaties from the UNFCCC to the Convention on Illegal Trade in Endangered Species include formal (although non-voting) delegate status to non-governmental organizations (NGOs), research organizations, and Indigenous Peoples and local communities, among many other non-state actors. The Convention on Biological Diversity uniquely includes at least one Indigenous delegate as co-chair of the treaty's ongoing Article 8(j) negotiations. The growth of Indigenous participation in IUCN's governance processes like the World Parks Congress reflects the growing global consensus that community involvement and participation is necessary for advancing sustainable development goals across all levels of environmental governance.

Although concerns about procedural justice have been especially visible in biodiversity conservation[8] because of the existential relationships between Indigenous Peoples, Traditional Peoples, and forest peoples and their lands, procedural justice features across environmental issue areas. The National Environmental Policy Act (NEPA) of 1970 in the United States, for example, mandates public involvement in all federal decisions that may have significant environmental impacts. This includes decisions on projects like grazing, mining, and roadbuilding on public lands; approving and siting noxious and public service facilities; and any development or activities on lands with known populations of or critical habitats for endangered species. The Aarhus Convention, with a similarly broad mandate to NEPA, is an international treaty in Europe that binds its parties to provide access to environmental information, respect the public's right to participation in decision-making, and establish and enforce the right of the public to challenge public decisions. To date, more than 100 countries have legislated rights to public involvement in environmental decision-making at national and subnational levels. Such policies are often invoked around large-scale industrial or development initiatives, such as mining and hydropower projects, which may have both negative environmental and social impacts. Central to these commitments to procedural justice is the belief that decisions about activities that have significant human welfare effects should be made with impacted communities, which also tend to be communities with the least access, representation, and power in decision-making processes.

Opportunities for individuals, communities, and other key stakeholders to engage in various governance forums range in intensity from submitting written or oral comments during a public comment period to co-leading international negotiations on maintaining and respecting traditional ecological knowledge. Ubiquitous in environmental governance initiatives across the globe, participation is the most common tool for ensuring that decision-making processes are fair. At events like the World Parks Congress, Indigenous Peoples participate alongside nation-states and NGOs to develop a ten-year biodiversity policy agenda for the world. At most sites of global environmental governance, however, like multilateral treaty negotiations, how different groups of actors participate map on to the power hierarchies that are embedded in the negotiation process. In negotiations under the United Nations UNFCCC, for example, the 197 nation-states parties to the convention are able to wield certain forms of power, such as access and authority, with greater effect than non-state actors. Secretariats of treaties can wield control over agendas and possess procedural rule enforcement powers that other actors do not have access to. Scientists hold particular forms of expertise; NGOs often wield moral authority; and Indigenous Peoples leverage a variety of other forms of power that may be less hegemonic and overt than other actors (see Marion Suiseeya and Zanotti, 2019). Participation in these contexts can only advance procedural justice when measures are taken that account for the various capabilities of different groups to leverage access and representation.

Where **participation** at sites of global environmental governance like the World Parks Congress emerged through attending, listening, presenting, sharing, and networking at different sessions throughout the event, participation at local levels was more diverse. Practices for facilitating participation have included tools to expand the diversity of voices involved in projects, livelihood initiatives, mapping activities, public outreach, consent protocols, and leadership capacity-building, among many others (see Table 4.1). In Laos, participation has always been central to environmental governance. To uphold its commitments under the Convention on Biological Diversity, and in line with the scientific guidance advanced by the IUCN and prior World Parks Congresses, for example, the government of Laos established a model national protected area system in 1993 (Prime Minister Decree 164, 1993). The system currently covers 16% of the country's land area and includes representation of 5–20% of each type of terrestrial biogeographic diversity in the country. As one of the world's poorest countries, with nearly 86% of the population dependent upon forest resources for their livelihoods, separating people from parks was never part of government policy. In fact, villages are located inside all but one of Laos' national protected areas, and participatory management is mandated by law in forest governance and widely practiced in other environmental arenas such as fisheries management, hydropower development, and mining. The government of Laos adopted this participatory approach for a number of reasons. First, agency officials point to the experiences of exclusionary conservation in neighbouring countries. Thailand, for example, violently removed ethnic minorities from forests, resulting in unrest and political instability in remote, mountainous communities. The Laos government, on the other hand, wanted to avoid rural unrest, but also wanted to demonstrate greater moral

Table 4.1 Examples of tools for participation

Tool/Practice	Description
Gender mainstreaming	Efforts to ensure that gender perspectives and impacts of an intervention are considered and that gender equality is an explicit project goal
Ethnic mainstreaming	Efforts to ensure that ethnic and minority perspectives and impacts are considered, and that equality is an explicit project goal
Free prior informed consent	An established right of Indigenous Peoples to reject or consent to activities that affect their communities and territories
Language/communication tools	Efforts to provide oral and written language supports and materials in appropriate languages to support engagement
Public outreach/information and knowledge sharing	Activities that provide information, knowledge, and opportunities for engagement to communities or stakeholders; public disclosure mechanisms
Social safeguards	Policies and mechanisms for addressing social issues in project design, implementation, and operation; includes framework for consultation and public disclosure
Public comment/input processes	Mechanisms for members of the public/stakeholders to offer comments, introduce knowledge, ask questions
Grievance mechanisms	Mechanisms for receiving and addressing complaints related to project/activity design, implementation, and operation
Participatory mapping	Creation of maps by communities to represent their spatial and experiential knowledge related to community, land, and other natural and environmental resources
Land use planning	Processes to identify and determine how to manage the land and other resources in a particular locality
Joint/participatory/community-based management	Approach to project or activity management that distributes responsibility and authority between different actor groups (e.g. government, NGO, community, etc.)
Development supports	Activities and resources that help build capacity and address structural poverty issues, e.g. water and sanitation, healthcare and education access, capacity/leadership/employment training programmes, access to capital and/or inputs for enhancing productivity and market access, small community grants, microfinance, and direct cash transfers

authority compared to its more democratic neighbour. Second, participatory conservation in many ways reflects the ideological foundations of the Lao People's Democratic Republic, a country governed by one central authority: the People's Party. Third, international donors increasingly required that integrated environment and development projects should include participation, not just in Laos but in all aid-recipient countries. Fourth, the small, dispersed, and remote population in Laos makes governing forests, land, biodiversity, and other natural resources difficult. In practical terms, decentralizing conservation through participatory approaches allowed the government to delegate responsibility for meeting its global conservation obligations to local governments and communities.[9] Thus, the design of every biodiversity conservation project in Laos, whether implemented by the government, communities, NGOs, or through public–private partnerships, includes tools for participation.

> **Box 4.2   Tyranny of participation**
>
> In 1990, James Ferguson published *The Anti-Politics Machine*. This book explores the ways in which development projects in late 1970s and early 1980s Lesotho erased the historical and political legacies that perpetuated poverty and, in doing, so further concentrated power in the hands of political elites (Ferguson 1990). As development practitioners sought to improve development outcomes, they turned towards participatory development—the same trend also emerging in conservation governance. Cooke and Kothari's (2001) research that examined the practices and impacts of participatory development demonstrated how participation was similarly apolitical and often stripped communities of their agency. The tyranny of participation refers to the ways in which participatory approaches further cultivated and concentrated illegitimate power among the elites, which negatively impacted the lives and well-being of the communities involved.

When participation is not justice

Following a steep rise in the number of NGOs participating in global environmental governance in the late 1980s and early 1990s, participation across multiple levels of governance has continued to grow in breadth and depth (see Princen et al., 1994). Not only has the sheer number of participants grown, but the modes and mechanisms for facilitating participation have also expanded. Yet, research in global environmental politics continues to demonstrate that participation does not always translate into meaningful engagement. Instead, attendance and/or presence is often considered equivalent to participation. As the World Parks Congress progressed in Sydney, the challenges of leveraging the promise of participation for generating a fair process became clear. Questions about accessibility, representation, and **capabilities** for meaningful engagement percolated through the discussions of Indigenous and community participants and allies. The Congress was paperless, which meant that access to the extensive schedule of events was only possible with a smartphone or other wired device, or by navigating to the large scrolling screen located in the main exhibition hall of the event. Not only was Sydney an expensive place for attending a meeting, but the registration fee for the Congress itself ranged from US$200—$990. Global public health threats at the time led to severe visa restrictions, reducing the number of potential participants from Ebola-impacted areas, thereby limiting representatives from western Africa. Outside of the main plenary events, the IUCN did not provide translation and interpretation services for any of the other sessions. The short 15-minute breaks in between sessions meant that participants were often running late, or simply did not go to sessions that were spread across the sprawling Sydney Olympic Park. The constant din of the soundscapes at the event, which reflected the diversity of forums for engaging, also often meant that sessions were in audio competition to make sure participants could hear the proceedings. Thus, although many Indigenous Peoples attended the World Parks Congress, there were also many who did not or could not. Among those who did attend, many could not engage because of language, mobility, ability, and technological constraints. While some participants may be able to bring their voice to decision-making venues under certain conditions, once there, they may not always be able to effectively engage or use their voice. As these cracks in the participatory framework of the World Parks Congress grew, they illuminated the disconnect between participation and procedural justice. And, although these dynamics of participation and procedural justice can often be difficult to discern at global events like the World Parks Congress, particularly if those observing the dynamics of participation are not impacted by some of the constraints, they can be quite visible on the ground in conservation communities.

Digging into the meaning of the legal mandate for participatory management in Laos similarly reveals conditions that may not only fail to advance meaningful participation, but under some circumstances may actually produce procedural injustice. Recalling that procedural justice is about the ability to participate in *and* influence decision-making processes, it is important to recognize that participation may or may not contribute to these abilities. Historically, project developers in Laos, who are most often NGOs or the government in partnership with an NGO or donor, have identified target communities from afar and arrived to inform the community that a project would be taking place. Participation commonly takes the form of village meetings—usually the domain of men—where project leaders, alongside local, district, and sometimes provincial or central government officials, disseminate information about the project. Nearly every village in Laos has had at least one, if not multiple, conservation or development projects that have mandated participation in some way, yet rarely does participation in these projects include engagement in decision-making processes. And, although not unique to Laos, in some cases, households which cannot or do not attend the meeting may not be able to benefit from the project. Thus, households and communities may face a paradox of agency: benefitting from the project both requires that one has ability and agency to participate but also requires that they cede this agency to those with decision-making power. To participate under such conditions means to sacrifice the possibility of procedural justice.

In recent years, projects have increasingly moved towards embedding **free prior informed consent** or free prior informed consultation (FPIC; see Box 4.3) protocols as a main mechanism for participation. FPIC has emerged globally as one of the most popular tools for working towards environmental justice at the community level. It is a tool used to secure the consent of communities for the implementation of a project or intervention that will impact the community. This consent must be *free*, *prior*, and *informed*, meaning that consent should not be coerced or secured after the project has begun, or without communities receiving basic information about the costs and the potential benefits, risks, and harms of the project. FPIC is based on the normative principle that communities and groups should have the right to be involved in decision-making processes—to give consent to or reject a project that may negatively impact all or part of a community. In Laos, FPIC has generally taken the form of consultations and has been tested through forest carbon projects that aim to reduce emissions from deforestation and degradation (REDD+) through enhanced protected area management, monitoring, and enforcement. The distinction between consent and consultation is conceptually significant, although in Laos it is practically insignificant. Consent grants some power or authority to communities in the decision-making process of the project, whereas consultation only requires sharing information. Because of various political and constitutional dynamics in Laos that centralize most forms of power in decision-making processes, seeking consent may seem impractical or inappropriate for the context and thus may not move participation closer to procedural justice.

Critically, the participatory turn in global environmental governance was largely in response to distributive injustices and poor environmental outcomes; it was not firmly rooted in community concerns about procedural justice. Its origins, thus, can help account for the technocratic and apolitical nature of participation, where from global to local levels being visible is sufficiently participatory from the perspective of many project proponents and policy-makers. At the World Parks Congress, such attitudes were apparent when organizers did not prioritize translation services, thus restricting the accessibility, representation, and capabilities of non-English-speaking participants in the predominantly English-language events. At the local level, these attitudes surface when sign-in sheets for village meetings are used as evidence of consent, therewith undermining household and community representation (Marion Suiseeya, 2014). In these cases when participation becomes technocratic and apolitical, rather than democratic, it can create and exacerbate environmental injustices.

Box 4.3   Free prior informed consent

FPIC is an international human rights legal principle that emerged in response to Indigenous Peoples' demands for self-determination. Although the obligation of nation-states to consult Indigenous Peoples on any decision impacting them was codified in the 1989 Indigenous and Tribal Peoples Convention of the International Labour Organization (no. 169), the principle of FPIC to guide such consultations was not broadly embraced until the 2007 United Nations Declaration on the Rights of Indigenous Peoples. In practice, FPIC is a mechanism through which Indigenous Peoples can realize their human rights. Because environmental change disproportionately impacts Indigenous Peoples (Whyte, 2014; Whyte et al., 2016; West et al., 2006), FPIC is a particularly important principle for advancing environmental justice. *Free* means consent without coercion; *Prior* means that consent is sought before a project or development is planned; and *Consent* means affirmation that a project can proceed. In procedural justice terms, FPIC addresses some of the power imbalances that are often built into large-scale development projects by requiring that the voices of those typically excluded in decision-making processes become central to decisions. In recent years, governments and other project developers have applied FPIC more broadly to other frontline communities. In some cases, project developers and funders have controversially transformed *consent* into *consultation*, which suggests that communities may still be impacted by projects that they would otherwise not consent to.

Moving towards procedural justice

Contemporary environmental governance integrates a variety of mechanisms that could contribute to advancing procedural justice. As Gordon Walker notes, we largely know environmental justice through experiences of injustice (Walker, 2012). How, then, can practitioners and policy-makers move towards procedural justice? The examples in the previous section suggest that, while participation has potential as one mechanism through which procedural justice can be advanced, it can also be harmful. Ultimately, procedural justice requires consideration of how power and influence *work* in environmental governance contexts. What is power? Where is it located? How is it cultivated? How does it emerge? What forms of power constitute and shape decision-making processes? The transformative potential for participation to be a mechanism for procedural justice requires that practitioners and policy-makers first consider the extent to which participatory approaches can confront power imbalances. While participatory management has the potential for shifting power by, for example, providing opportunities for individuals and communities to improve their well-being and access to decision-makers, this is not inevitable. In fact, participatory approaches have often resulted in further entrenchment of power imbalances (see Box 4.3). Various dimensions of participation, including accessibility, representation, and capabilities, are individually and collectively important for engaging different forms of power in decision-making processes. Accessibility refers to the structural elements of participation: what mechanisms are in place to ensure that individuals and communities are able to attend and engage? Representation refers to the constellation of identities and actors engaged in the event: to what extent do those engaged reflect and/or have authority to represent the multiple identities and needs of project constituents? Capabilities refer to the abilities of those participating to effectively engage and exercise **agency**: do participants have the political, cultural, economic, and social means to be able to participate in and influence decision-making processes? The tools described in Table 4.1 address these dimensions of participation to varying degrees and have the potential, if designed and implemented, to cultivate power and to expand the breadth and depth of participation towards procedural justice.

Although participation is perhaps often treated interchangeably with procedural justice, the examples and discussion in this chapter have shown that they are distinct approaches. Participation may or may not confront and redress the power imbalances, exclusion, and misrecognition that produce environmental injustice. By focusing on ability to participate in and influence decision-making processes, procedural justice demands attention to power in environmental governance. Different principles could guide the distribution of power in decision-making processes and impact the degree of participation and influence that different groups of actors have. For example, the principle of political equality could guide decision-making processes to remove distortions from unequal distribution of power among those in the decision-making space. Political egalitarianism takes this a bit further to address distortions in power due to social standing in order to create space for all sectors of society to be involved in decision-making spaces. **Proportionality** is a principle that distributes power corresponding in ratio to the size or population of groups in a decision-making space. Majority-based principles (e.g. simple majority, supermajority) allocate power based on the largest group or interest. Plurality attends to the politics of difference to ensure that different identities are allocated power.

One final example demonstrates how these principles could materialize in different decision-making structures. Developed by Natural Justice, an organization that supports Indigenous Peoples and local communities in securing and exercising their rights, Biocultural Community Protocols aim to situate decision-making power within community institutions to determine if, how, and when development, conservation, or infrastructure projects and programmes take hold (Bavikatte and Jonas, 2009). In practice, Natural Justice and other organizations support communities to develop the processes and rules by which the community could make such collective decisions, help them to codify a protocol that details how communities would make decisions about projects, and build capacity to implement, monitor, and revise the protocol. Once the protocol is in place, any project or programme developer would be directed to the protocol to determine how to approach the community to seek their engagement. If communities approach Biocultural Community Protocols democratically, in their development and implementation, they may have significant potential to advance procedural justice. This is because of the ways in which these protocols could ensure access, representation, and capabilities for community members to participate in and influence decision-making processes. In an ideal sense, Biocultural Community Protocols could be considered an example of environmental democracy because of the ongoing and broad-based deliberations that are required to produce its outputs (see Box 4.1). Yet, similar to participatory tools, if this approach does not continuously and reflexively consider the power dynamics within and across communities and other governance actors, there is still potential for this tool to become technocratic and apolitical—and thus not advance procedural justice.

How and why procedural justice matters in environmental governance

Reflecting that participation is the primary mechanism for advancing procedural justice, this chapter set out to examine how and why participation matters. At the core of procedural justice is the notion that, while not always the most efficient route to effective environmental governance, all peoples should have the ability to engage in and influence decision-making processes that impact their territories, communities, and well-being, regardless of the scale at which decisions are made. Because environmental injustice results from power asymmetries, leveraging decision-making processes to confront and acknowledge these imbalances can help lead to better decisions and, ultimately, outcomes. Critical, however, to this logic is that procedural justice may not always lead to better environmental outcomes. Yet, as Robyn Eckersley argues, "people must be 'free' to make ecologically bad decisions; the alternative is ecological paternalism" (Eckersley, 1996, p. 207).

The work in this chapter suggests that approaching procedural justice requires asking questions about power at each juncture in decision-making processes. *Whose* problems are identified, *how* problems are defined, and the *salience*, or importance, of particular problems are dependent on *who* constitutes the body of decision-makers and the *relative abilities* of decision-makers to influence the decisions. These factors have significant distributive justice implications because they shape *what* problems are tackled and *how* they are addressed, as well as impacting the effectiveness of governance initiatives for solving environmental problems. Ultimately, a procedural justice approach to environmental governance aims to interrogate how power and representation interact to address the potential for political displacement of frontline communities, which could have subsequent social, economic, and environmental impacts.

Follow-up questions

- How does procedural justice relate to environmental justice? From whose perspective should procedural justice be approached?
- What are some of the synergies and tensions between participation and procedural justice? What about law and justice?
- What conditions, practices, and principles should be considered when employing participation as a tool for advancing procedural justice?
- How do the different tools for participation in Table 4.1 effect the accessibility, representation, and capability dimensions of participation?

- How does consent differ from consultation? What, if any, are the procedural justice implications of this distinction?
- What scalar, temporal, and contextual dimensions should practitioners and policymakers consider when designing interventions that advance procedural justice? For example, how does procedural injustice at the global level connect with the local level? How can practitioners put justice first, rather than approaching justice in reaction to injustice? What, if any, constraints on future decision-making do current decision-making processes create, and what are the procedural justice implications?
- Some research has suggested that procedural justice is as important, if not more important, than distributive justice. Why might this be?

Notes

1 Author Field Notes, WPC 2014
2 Australia, Canada, and the United States did not sign the UNDRIP in 2007. Canada has since signed it.
3 IUCN, or the International Union for the Conservation of Nature, is a hybrid scientific organization with governmental and non-governmental members.
4 Interview with Author, WPC 2014
5 The biodiversity regime includes multiple international treaties, including the Convention on Biological Diversity, the Convention on Illegal Trade in Endangered Species, the Ramsar Convention on Wetlands, and many others.
6 On Indigenous environmental justice, see also Chapter 20 of this volume.
7 On the history of environmental justice, see also Chapter 2 of this volume.
8 On biodiversity conservation, see also Chapter 11 of this volume.
9 For an expanded discussion of this history, see Marion Suiseeya (2014).

References for further reading

- Dryzek, J. S. and Stevenson, H. 2011. Global Democracy and Earth System Governance. *Ecological Economics*, 70(11), 1865–1874.
- Taylor, D. E. 2016. *The Rise of the American Conservation Movement: Power, Privilege, and Environmental Protection*. Durham, NC, Duke University Press.
- Young, I. M. 2011. *Justice and the Politics of Difference*. Princeton, NJ, Princeton University Press.

Additional Resources

The Conservation Initiative on Human Rights, www.thecihr.org, provides downloadable reports for different practices, themes, and issues in conservation.

Cornered by Protected Areas, www.corneredbypas.com, includes case studies on protected areas and indigenous communities.

The Center for People and Forests, www.recoftc.org, provides training curriculum and reports related to FPIC, conflict transformation, and community forest and natural resource governance.

Rights and Resources Initiative, http://rightandresources.org, advocates for land tenure and policy reform.

Asia Indigenous Peoples Pact, http://aippnet.org, supports Indigenous rights, land rights, and cultural rights.

Natural Justice, www.naturaljustice.org, works with communities to develop protocols for managing relationships with donor, development, and private sector initiatives.

References used in this chapter

The First National People of Color Environmental Leadership Summit. 1991. Principles of Environmental Justice. In *Adopted at the First National People of Color Environmental Leadership Summit*, Washington, DC, October 1991, New York, NY: United Church of Christ Commission for Racial Justice.

Ahlborg, H. & Nightingale, A. J. 2018. Theorizing Power in Political Ecology the Where of Power in Resource Governance Projects. *Journal of Political Ecology*, 25, 382–401.

Bavikatte, K. & Jonas, H. 2009. *Bio-Cultural Community Protocols: A Community Approach to Ensuring the Integrity of Environmental Law and Policy*, Natural Justice, United Nations Environment Programme. www.unep.org/communityprotocols/PDF/commu-nityprotocols.pdf.

Bell, D. & Carrick, J. 2017. Procedural Environmental Justice. In *The Routledge Handbook of Environmental Justice*, London, UK, Routledge.

Blaikie, P. & Muldavin, J. 2014. Environmental Justice? The Story of Two Projects. *Geoforum*, 54, 226–229.

Brosius, J. P. 2004. Indigenous Peoples and Protected Areas at the World Parks Congress. *Conservation Biology*, 18, 609–612.

Carrick, J. & Bell, D. 2017. Procedural Environmental Justice. In *The Routledge Handbook of Environmental Justice*, London, UK, Routledge.

Chambers, R. 1994. The Origins and Practice of Participatory Rural Appraisal. *World Development*, 22, 953–969.

Clayton, S. 1998. Preference for Macrojustice Versus Microjustice in Environmental Decisions. *Environment and Behavior*, 30, 162–183.

Cooke, B. & Kothari, U. 2001. *Participation: The New Tyranny?* London; New York, Zed Books.

Dobson, A. 2007. Environmental Citizenship: Towards Sustainable Development. *Sustainable Development*, 15, 276–285.

Dryzek, J. S. & Stevenson, H. 2011. Global Democracy and Earth System Governance. *Ecological Economics*, 70, 1865–1874.

Eckersley, R. 1996. Greening Liberal Democracy. In B Doherty & M de Geus (Eds.), *Democracy and Green Political Thought*, London, Routledge, pp, 212–236.

Ferguson, J. 1990. *The Anti-Politics Machine: "Development", Depoliticization, and Bureaucratic Power in Lesotho*, Cambridge; New York, Cambridge University Press.

Forsyth, T. & Walker, A. 2008. *Forest Guardians, Forest Destroyers: The Politics of Environmental Knowledge in Northern Thailand*, Seattle, Washington, University of Washington Press.

Green, J. F. 2013. *Rethinking Private Authority: Agents and Entrepreneurs in Global Environmental Governance*, Princeton, New Jersey, Princeton University Press.

IUCN. 2014. *What Is the IUCN World Parks Congress?* [Online]. Gland, Switzerland, IUCN. Available: http://worldparkscongress.org/about/what_is_the_iucn_world_parks_congress.html [Accessed].

Marion Suiseeya, K. R. 2014. *The Justice Gap in Global Forest Governance*. Durham, North Carolina, Duke University.

Marion Suiseeya, K. R. & Caplow, S. 2013. In Pursuit of Procedural Justice: Lessons from an Analysis of 56 Forest Carbon Project Designs. *Global Environmental Change*, 23, 968–979.

Marion Suiseeya, K. R. & Zanotti, L. 2019. Making Influence Visible: Innovating Ethnography at the Paris Climate Summit. *Global Environmental Politics*, 19, 38–60.

Ministry of Agriculture and Forestry, Lao PDR. 2005. *Forest Strategy to the Year 2020 of the Lao PDR*. Vientiane, Lao PDR.

Pitkin, H. F. 1967. *The Concept of Representation*, Los Angeles, University of California Press.

Princen, T., Finger, M., Clark, M. L. & Manno, J. P. 1994. *Environmental NGOs in World Politics: Linking the Local and the Global*, New York, NY, Routledge.

United Nations. 1992. Agenda 21, United Nations Conference on Environment & Development Rio de Janerio, Brazil, 3 to 14 June 1992. http://www.un.org/esa/sustdev/documents/agenda21/english/agenda21toc.htm.

Walker, G. 2012. *Environmental Justice: Concepts, Evidence and Politics*, London, Routledge.
West, P., Igoe, J. & Brockington, D. 2006. Parks and Peoples: The Social Impact of Protected Areas. *Annual Review of Anthropology*, 35, 251–277.
Whyte, K. P. 2014. Indigenous Women, Climate Change Impacts, and Collective Action. *Hypatia-a Journal of Feminist Philosophy*, 29, 599–616.
Whyte, K. P., Brewer, J. P. & Johnson, J. T. 2016. Weaving Indigenous Science, Protocols and Sustainability Science. *Sustainability Science*, 11, 25–32.
Witter, R., Marion Suiseeya, K. R., Gruby, R. L., Hitchner, S., Maclin, E. M., Bourque, M. & Brosius, J. P. 2015. Moments of Influence in Global Environmental Governance. *Environmental Politics*, 24, 894–912.
Wyatt, S., Fortier, J.-F. & Martineau-Delisle, C. 2010. First Nations' Involvement in Forest Governance in Québec: The Place for Distinct Consultation Processes. *The Forestry Chronicle*, 86, 730–741.

5 Recognition and environmental justice

Brendan Coolsaet and Pierre-Yves Néron

Learning outcomes

- Gain understanding of the concept of recognition;
- Distinguish several traditions of thinking about recognition and the differences between them;
- Consider the reasons why recognition matters in the context of the environment.

Introduction

Of all the different ways of thinking and conceptualizing environmental justice, recognition is arguably the most neglected and under-theorized. The idea of recognition is complex and (like other approaches discussed in this volume) has a long philosophical and political history, with roots in Hegelian ethics, critical theory and post- and decolonial studies. Described by the German philosopher Axel Honneth as the "moral grammar of social conflicts" (Honneth, 1996), **recognition** not only deals with the way in which we accommodate and respect different people, their cultural practices, their identities and their knowledge systems, but is also relevant to issues of self-respect and self-worth. The right to be different is protected today in, for example, the 1948 Universal Declaration of Human Rights and the 2007 UN Declaration on the Rights of Indigenous Peoples. The latter requires parties to recognize "the right of all peoples to be different, to consider themselves different, and to be respected as such" (United Nations, 2007, p. 2).

This is important in the context of the environment because the meanings and **values** we assign to nature and the environment are always culturally defined, and people express their differences through a multitude of relations to nature and the world around them. Environmentally harmful practices may thus also be detrimental to the meanings and expression of differences attached to these environments. Conversely, protecting the environment and designing environmental policy are always influenced by culturally specific ideas about what is worth protecting in the first place. Hence, when environmental conservation is driven by dominant worldviews and disregards the meaning and value assigned to the environment by locally affected populations, it may also result in misrecognition.

This chapter looks at how scholars and movements have been dealing with such issues. As is the case with other concepts addressed in this volume, you will not find a singular definition of justice-as-recognition in this chapter. Instead, we will take the reader through different traditions of thinking about recognition and illustrate these in the context of environmental issues. By the end of the chapter, you should have a good understanding of the concept of recognition, including the key ideas of its main thinkers and the differences between them. We will also look at how the concept has been used by both movements and scholars in the context of the environment.

Theories of recognition

Hegelian inter-subjectivity

The German philosopher Georg Wilhelm Friedrich Hegel is undoubtedly the main historical source of contemporary thinking on recognition. It all started with Hegel's idea of freedom, which, according to Hegel, can only be achieved if one is recognized and respected by others (Hegel, 1991). Failing to respect a person's cultural identity, for example with regard to their relationship with nature, is a denial of their freedom to live according to their chosen belief system. Hegel saw the lack of recognition, or misrecognition, as a form of enslavement that occurs through unequal encounters in which the more powerful actor fails to recognize the concerns of the other. It represents what could be referred to as a form of asymmetrical recognition. This was famously illustrated by his master–slave dialectic (see Box 5.1).

Unsurprisingly, Marxists interpreted the master–slave relation as a metaphor for the need for class struggle and the reversal of the association. However, in line with more contemporary interpretations, Hegel was able to capture the logic of recognition. In particular, his analysis helps us grasp the crucial role of **otherness** in relations and patterns of recognition. From Hegel's point of view, the master *needs* recognition from the slave, as much as the slave needs it from the master. The critical point here is that a person can only really value their own life if others value such a life. There is a need for reciprocity. For example, it is not possible to have high self-esteem if you perceive that others treat you with contempt. This could be relevant in the context of environmental governance because a person might only be able to enjoy cultural freedom if their ways of knowing and living with nature are respected by others. The required response in such a case is therefore to move relations towards more reciprocal recognition.

Box 5.1   Hegel's master–slave dialectic

Hegel's analysis of the master–slave dialectic is probably the most famous passage of his important (and difficult) book *The Phenomenology of Spirit* (Hegel, 1977). It tells the story of two conflicting self-conscious individuals engaged in a "struggle to the death" that leads to the subsequent creation of a relation of subordination, expressed through the status of the master and the slave. By fighting each other, both subjects attempt to affirm their own freedom by proving the superiority of their status.

What matters in this story, according to Hegel, is that one can only gain self-consciousness as an autonomous subject by recognizing the other as an equally autonomous subject. Thus, while you would think that the master is entirely free from the

> slave, both are in fact deeply interdependent. In a hierarchical system based on honour, the deference of the lower orders is crucial. Hence, from Hegel's perspective, the master, too, is dependent on the slave for recognition, as he needs his (higher) status be acknowledged as such by the slave.
>
> Hegel's analysis is important because of its radical philosophical implications. In particular, it invites us to question some of the dominant tenets of the Anglo-Saxon tradition inspired by the English philosopher Thomas Hobbes. According to the Hobbesian tradition of "possessive individualism" (Macpherson, 1964), society is driven by rationally calculating actors who act out of self-interest or self-preservation. On the contrary, Hegel suggests that these actors are in fact engaged in **struggles for recognition**, which, as we will see ahead, can help us make sense of some contemporary environmental justice movements.

Moving beyond Hegel, the contemporary use of recognition gained importance through what was termed the "**cultural turn**" in critical and political theory. In a nutshell, the idea is that today's political struggles are no longer limited to calls for economic redistribution, voiced for much of the 20th century by an exploited working class. They also encompass demands for "difference-friendly" societies based on equal recognition of alternative identities, genders, races, and religions, and championed by so-called "**new social movements**" (Laclau, 1985). In other words, status and identity have supplanted class interests as the main political mobilizer. Important in the context of environmental issues, this cultural turn has also been described as the emergence of "indivisible conflicts" (Honneth, 2004, p. 352); namely that identity, culture and relationships cannot be chopped into small pieces and distributed among those asking for just treatment in the same way material resources can be distributed. We will see later that some would also argue that the same applies to the environment.

Three authors represent this cultural turn more than others: Canadian philosopher Charles Taylor, German philosopher Axel Honneth and North American philosopher Nancy Fraser. These three authors, discussed in the following subsections, have greatly influenced the inclusion of recognition in current-day environmental justice scholarship.

Charles Taylor's multiculturalism

In a series of works published in the 1980s and 1990s, Charles Taylor seeks to identify the philosophical and historical sources of the rise of political claims made in the name of "recognition" and its connections with **identity** (Taylor, 1994). Taylor, often associated with the "communitarian" school of thought (a label he rejects), might best be understood as advocating a form of "liberal multiculturalism" in which cultural minorities would enjoy strong forms of protection.

Relying heavily on Hegel's insights, Taylor argues that desire for recognition is not a frivolous demand but is fundamental to make life worth living. As he puts it, "due recognition is not just a courtesy we owe people. It is a vital human need." (Taylor, 1994, p. 26) To put it another way, following Hegel, lack or denial of recognition can be associated with a genuine psychological harm inflicted on persons and groups.

Two things are crucial to understand Taylor's theory of recognition. First, Taylor relies on a "dialogical" conception of identity, according to which our identities are shaped through a complex dialogue with **significant others** and an already existing cultural background that provides "horizons of meanings". In other words, we do not define ourselves in a societal vacuum. Therefore, to use the Hegelian logic again, identity is deeply linked to otherness.

Second, claims for recognition of one's identity emerge from what he refers to as an "ethic of authenticity". For Taylor, modernity is characterized by the rise of, on the one hand, an ethic of equal dignity, which leads to a movement of universalization based on the shared, universal dignity of all human beings; and on the other hand, an ethic of authenticity, which is linked to a movement of distinctiveness in which the unique identity of individuals and groups is being recognized.

In modern societies, the destruction of formal social hierarchies that assigned specific identities to social groups leads individuals to a quest for authenticity (Taylor, 1989, 1994). Without preassigned identities, individuals are searching for their "own true selves"; for their own, unique way to inhabit this world. In this context, misrecognition can undermine a person's ability to develop a successful relationship with themself. According to Taylor, the powerful appeal of this ideal of authenticity explains the emergence of both extreme forms of individualism and genuine claims for recognition. It contributes to the creation of "recognition" as a "problem" that can ground, articulate and structure the political struggles that surround it (Taylor, 1994).

Finally, Taylor draws a distinction between three forms of recognition, with different political implications. First, a politics of universalism aims at the equal recognition of all human beings in their common dignity. Second, a politics of "difference" aims at the recognition of the uniqueness of special features (often cultural) of one's identity. Third, Taylor highlights the importance of the recognition of concrete individualities in relationships of love, friendship and care. The challenge for a theory of liberal multiculturalism is to articulate the proper balance between these three forms, acknowledging that the latter ones have clearly been neglected.

Axel Honneth's spheres of interaction

Axel Honneth aims to present theories of recognition as the new dominant paradigm for critical theory. He argues that recognition is the "moral grammar of social conflicts" (Honneth, 1996) through which critical theorists can diagnose all "social pathologies" in what he calls a society of **disrespect** (Honneth, 2007, p. 32).

Honneth also proposed a well-known and widely discussed typology of various forms of recognition. According to him, we can distinguish three forms of recognition, connected to three different "spheres" of human interaction. First, love is a form a mutual recognition that is central to the sphere of intimacy. Second, recognition as respect is associated with the legal sphere of legitimately institutionalized interactions of universal respect for the **dignity** of people. Third, social esteem is linked to complex networks of solidarity and shared values within which the worth of members of a community can be evaluated and acknowledged. Whereas respect is for individuals a matter of equal treatment and being entitled to the same status as others, social esteem concerns what makes individuals feel different, unique or special. From this point of view, issues of recognition are not only "cultural" matters. For instance, individuals can suffer from a lack of recognition (in the social esteem sense) of their personal contribution in the workplace.

With this in mind, it is worth noting that, in Honneth's approach, the connection between recognition and justice becomes clearer. Whereas for Taylor recognition is mainly a matter of self-realization, Honneth explicitly places it (as does Nancy Fraser ahead) at the centre of a multifaceted theory of justice. Recognition matters because justice matters—and misrecognition, from this perspective, becomes the main expression of injustice in contemporary societies in which contempt, disrespect and insult are so deeply entrenched in our practices and institutions.

In Honneth's account, recognition becomes the overarching concept that one should use in practising social critique. According to him, even issues of distribution of material resources, such as those discussed in Chapter 3 of this volume, can be approached from the point of view of recognition. For instance, the highly unequal distribution of pollution across communities flows from the fact that some groups are not considered worthy of dignity and respect; in other words, the maldistribution of pollution should be understood as a form of misrecognition. From Honneth's perspective, if redistributive policies are indeed the remedy to this problem, it is because they are ultimately based on claims for better recognition. Thus, Honneth claims that we should think of "redistribution as recognition" (Fraser and Honneth, 2003).

Nancy Fraser's parity of participation

Nancy Fraser's contribution to critical theory is her ambition to combine the emerging demands for recognition with long-standing calls for economic redistribution. For Fraser, the "most general meaning of justice is **parity of participation**" (Fraser, 2005, p. 5; emphasis added). Participation-parity occurs when all adult members of society are allowed and able to interact with each other as peers (Fraser, 2001). In this context, participating in society can be impeded not only by social subordination (cultural injustice), but also through material exploitation (economic injustice) and political disenfranchisement, which are not to be subsumed to one another (Fraser, 1995; Fraser and Honneth, 2003). Gender inequality, for example, has intertwining roots in both economic arrangements (such as rules of resource access) and forms of cultural hierarchization (such as the idea that women, rather than men, should take care of the children; see Chapter 18 of this volume).

Economic injustices condition social interaction by denying the necessary resources to people to engage with others. In other words, parity of participation is inhibited when certain actors do not possess the necessary material resources to fully engage in society: "subordinated social groups usually lack equal access to the material means of equal participation" (Fraser, 1990, p. 64). Examples of maldistribution include exploitation ("having the fruits of one's labor appropriated for the benefit of others"), marginalization ("being confined to undesirable or poorly paid work or being denied access to income-generating labor altogether") and deprivation ("being denied an adequate material standard of living") (Fraser and Honneth, 2003, p. 13).

Cultural injustices, on the other hand, are rooted in the status order of society (Fraser and Honneth, 2003). Misrecognition, Fraser claims, occurs through a hierarchization of cultural values, whereby some people are seen as "inferior, excluded, wholly other, or simply invisible" and therefore cannot equally participate in social interaction (Fraser, 2000, p. 113). In opposition to Taylor and Honneth, Fraser refuses to "psychologize" issues of recognition and identity. To think about cultural injustices is to analyze not psychological reactions, but certain kinds of social relations (see Box 5.2).

These social relations can take on the form of domination ("being subjected to patterns of interpretation and communication that are associated with another culture and are alien and/or hostile to one's own"), non-recognition ("being rendered invisible via the authoritative representational, communicative, and interpretative practices of one's culture") and disrespect ("being routinely maligned or disparaged in stereotypic public cultural representations and/or in everyday life interactions") (Fraser, 1995, p. 71).

To recognition and redistribution, Fraser adds political representation as a defining factor of justice (Fraser, 2005). It is the acknowledgement that, like economic exploitation and cultural subordination, political disenfranchisement can impede people from participating fully in society. This political dimension tells us "who is included, and who excluded, from the circle of those entitled to a just distribution and reciprocal recognition" (Fraser, 2005, p. 6). Together, these three elements compose Fraser's three-dimensional theory of post-Westphalian democratic justice, according to which justice is the ideal of participation-parity.

Box 5.2 Recognition versus distribution: the Fraser–Honneth debate

Both Honneth's and Fraser's theories of recognition were further delineated through an influential debate between both authors. As noted earlier, Fraser stresses that "justice today requires *both* redistribution *and* recognition" (Fraser, 1995, p. 69; original emphasis), rejecting the monistic theories of Taylor and Honneth, in which all claims for justice, including those for economic distribution, can ultimately be understood as struggles for recognition. Fraser argues that, while important, recognition in itself (or distribution, in the same sense) is insufficient to explain the complexity of moral claims in post-modern societies.

The "status model" of recognition, as Fraser (2000) calls it, is built in opposition to Taylor's and Honneth's narrower, identity-based forms of recognition influenced by Hegelian inter-subjectivity. In their "identity model", recognition is a necessary condition to achieve self-consciousness, in that individuals only exist when recognized by others as individuals. Recognition, then, is a matter of self-realization: your psychological sense of worth is formed by how others act towards you. This, according to Fraser, is akin to blaming the victim (Fraser, 2001)

This debate resonates with a longstanding discussion in the environmental justice literature: the **race versus class** debate. In his seminal book *Dumping in Dixie*, Robert Bullard (1990) argued that race is the most salient factor in predicting the distribution of pollution across communities in the US. While confirming the importance of race, later studies have nuanced this, showing that both class and race interact in fostering environmental injustice; "the poor and especially the non-white poor bear a disproportionate burden of exposure to suboptimal, unhealthy environmental conditions" (Evans and Kantrowitz, 2002, p. 323; Figueroa, 2004). Addressing such types of injustices, thus requires that we take into account both the material aspects (class) and the status aspects (race).

Critiques of recognition

Despite the seemingly good intentions, a politics of recognition may also produce a false sense of justice. This is particularly true in post-colonial settings. As Whyte (2017, p. 120) notes, "acknowledgement and respect for difference is often a smokescreen that obscures the continuance of oppression against nondominant groups such as Indigenous people" (see also Chapter 20 of this volume).

This was famously theorized by Martinican psychiatrist Franz Fanon. In his seminal book *Black Skin, White Masks*, Fanon (1967) criticizes Hegel's understanding of reciprocal recognition for failing to account for situations of racial domination. In describing his desire to be recognized by a white woman, Fanon argues that "this is the form of recognition that Hegel never described" (1967/2008, p. 45). Moving beyond the identity versus status dichotomy discussed previously, Fanon shows how psychological processes cannot be detached from more structural, material conditions. On the contrary, psychological forms of misrecognition can themselves result from a process of **internalization** of social forces (Fanon, 1967). In other words, patterns of injustice may be continuously reproduced through the desires of those who are the victims of misrecognition, as they rest "on the ability to entice Indigenous peoples to identify [with] profoundly asymmetrical and nonreciprocal forms of recognition" (Coulthard, 2014, p. 25). As a result, Fanon argues that, in a colonial context, the slave would never gain recognition even if it were granted by the master, as the slave would have incorporated the master's image of him or herself.

At the same time, the psychological dimension also actively informs social structures through the actions of individuals. There is a dialectics between these spheres. This means that to overcome misrecognition, the sole transformation of structural conditions, as proposed by Fraser, is as insufficient as the sole transformation of the subjective sphere, as proposed by Taylor and Honneth. The identity-based approaches of Taylor and Honneth fall short of addressing some of the structural effects of colonial–capitalist exploitation, while Fraserian status-based recognition downplays the importance of the subjective dimension in overcoming injustices.

Drawing on Fanon's work, Glen Coulthard, a Canadian professor in First Nations Studies and a member of the Dene First Nation,[1] builds a compelling critique of the liberal approaches to recognition by the likes of Taylor, Honneth and Fraser. Analyzing the struggle of the Dene and the Kluane First Nations against a pipeline project, Coulthard shows how the government of Canada, through processes of deliberation guaranteeing the **participation** of minority groups, managed to transform "how Indigenous peoples now think and act in relation to the land" (Coulthard, 2014, p. 78). Over 25 years, these processes not only made Indigenous representatives accept extractive projects they had always been opposed to, but also gradually made them express their traditional relationship to their land in terms of ownership, property and monetization, concepts which are central to Western culture and capitalism but foreign to the First Nations' vision of the world. Coulthard argues that this change resulted from a smooth process of domestication through the creation of spaces of deliberation established by the Canadian state and through which the First Nations of Canada were meant to be recognized.

Recognition and environmental justice

Notwithstanding a few exceptions, most of the authors discussed above do not address environmental issues. However, their work has helped others make sense of the demands for recognition emanating from environmental justice movements. Despite the relative importance that observers have assigned to the dimensions of distribution and participation,[2] several environmental justice scholars have shown how demands for recognition have always been part of the claims of different environmental justice movements.

Misrecognition in the context of the environment has broadly played out in two opposing ways. In the first case, people who should have been treated equally have systematically been treated differently because of who they are, resulting in the injustice. This is typically the case, for example, when communities of colour face greater environmental risk than do other communities. In the second case, the source of the injustice is reversed: the differences of some people are rendered invisible when supposedly universal solutions are applied in the name of the environment. This can be observed when environmental policy initiatives involve asking people who are not responsible for the problem in the first place to halt long-standing cultural practices and renounce their ways of life (as illustrated by Box 5.3 in the context of biodiversity conservation).

By focusing on **environmental racism** (see Chapters 2 and 17 of this volume), early environmental justice movements in the US stressed that it is racism, cultural hierarchization and disrespect which lead to unfair distribution of environmental problems and exclusionary decision-making processes in the first place. In *Environmentalism and Economic Justice*, Laura Pulido (1996) suggests using a definition of environmental racism that fully incorporates not only economic inequality but also forms of cultural domination. Drawing on the study of two Chicano environmental justice struggles in the southwestern US, Pulido shows how these movements are not strictly environmental, but instead can best be understood as fighting for "ecologically and culturally appropriate economic change, confronting a racist and exclusionary political and cultural system, and establishing an affirmative cultural and ethnic identity" (Pulido, 1996, p. 193). This, she argues, is what differentiates these movements from mainstream environmentalism.

This intersection of material relations and culture resonates with Nancy Fraser's justice theory discussed previously. Drawing on Fraser's work, Robert Figueroa (2004) discusses the issue of environmental racism using a bivalent approach combining both distribution and recognition. For Figueroa, environmental injustices are always simultaneously economic and cultural. Focusing on these two aspects, he argues, allows us to fully grasp that both race-based and class-based injustices find their roots in similar institutionalized forms of subordination. Favouring one injustice over the other may be counterproductive to solving the injustices at hand. Figueroa comments: "Where the environmental racism debate evolves into a wrestle over zip codes or census tracts, race versus class, distribution versus recognition, it misunderstands the injustices, the collectivities, and the remedies." (2004, p. 34)

In *Defining Environmental Justice*, David Schlosberg (2007) too draws heavily on Nancy Fraser's work to develop what has since become an increasingly popular understanding of environmental justice in which recognition plays a key role. For many in the environmental justice movement, the struggle is "nothing less than a matter of cultural survival" (Schlosberg, 2007, p. 63). Importantly, unlike most of the theoretical approaches developed above, Schlosberg argues that what the environmental justice movement has to teach us is that different conceptions of justice are being used simultaneously, irrespective of the analytical differentiation that scholars tend to uphold.

However, this multivalence in the context of environmental justice struggles has also been found to be problematic. Drawing on Latin American **decolonial theory** (see also Chapter 7 of this volume), Álvarez and Coolsaet (2018) show that calling for a combination of both a distributive and a recognition approach can fail to account for cases that are not amenable to pluralist solutions; namely, cases in which the very idea of environmental distribution would be inconceivable. A contemporary illustration of this is the #NoDAPL movement in Standing Rock, USA. In 2016, the youth-led Indigenous movement emerged in protest at the construction of the Dakota Access Pipeline, an 1886 km long underground oil pipeline threatening the drinking water and sacred sites of the Standing Rock Indian Reservation. The protests sparked a nationwide solidarity movement, as it was seen and experienced as the latest iteration of the settler-colonial history of the US.

Relating the experiences of #NoDAPL protesters, Estes and Dhillon (2019) highlight one of the peculiarities of the movement. Under the "Water is life" banner, protesters wished to convey that the threatened Missouri River, and water in general, "is not a thing that is quantifiable according to possessive logics" (Estes and Dhillon, 2019, p. 3). The sacredness of water and land is something that can be traced back to numerous Indigenous beliefs, values and practices (McGregor, 2009). What the protesters were demanding was not a fairer distribution of the environmental impacts generated by the pipeline, as this would have implied that the water could be objectified, exploited and turned into a distributable good. Instead, they were demanding recognition of the kinship they shared with the water and the land (on kinship ethics, see also Chapter 20 of this volume). In other words, agreeing to a distributive 'solution' in Standing Rock, in which the pollution would be shared equitably across different communities, would have inevitably led to a situation of misrecognition.

This example also helps us understand how important the idea of recognition is if we wish to include non-humans as well as future and past generations as subjects of justice. It is easy to see how Charles Taylor's "significant others", through whom we shape our identities, may well be non-human others, ancestors or descendants. Interestingly, the very first of the 17 historic environmental justice principles adopted in 1991 (see Chapter 2 of this volume) affirmed "the sacredness of Mother Earth, ecological unity and the interdependence of all species".[3] While this principle has tended to be forgotten over the years, Indigenous scholars such as Winona LaDuke, Glen Coulthard, Deborah McGregor and Kyle Whyte have long stressed the importance of extending the community of justice to non-humans. As Whyte (2017, p. 122) notes, recognition is better suited than other approaches of justice to do just that, "because the very expression of cultural, economic and political differences is often rooted in particular ecosystems". The issues of Indigenous and non-human justice are further discussed in Chapters 20–21 of this volume.

Box 5.3   Recognition and biodiversity conservation

Western nature conservation approaches in the Global South have long been driven by a conception of nature as a pristine wilderness, which needed to be protected from human activity (see Chapter 11 of this volume). Given the powerful assumption that conservation is "the right thing to do", local and Indigenous practices were (and still are) often seen as harmful to nature, leading not only to the forceful physical exclusion of people from areas to protect, but also often to more subtle cultural, symbolic or psychological types of exclusion. This may amount to various forms of misrecognition through disregarding local notions of authority, ignoring local histories and symbolism, or appropriating traditional forms of knowledge.

This is not only a normative issue; recognizing and respecting difference is not just a matter of "doing the right thing". Research has shown how the recognition of cultural differences in the context of conservation can also radically improve the evidence base for useful solutions. Ignoring or excluding local and environment-specific forms of knowledge and experience undermines the possibilities for innovation.

The conservation of agricultural biodiversity is a case in point. Conserving or reintroducing new plant varieties and animal species requires agronomic knowledge, which farmers have gradually lost with modernization and the advent of "**green revolution**" type of agriculture. The dominance of industrial agriculture has standardized and centralized agricultural knowledge and practices across the globe and led to a massive decline in agricultural biodiversity.

"Obsolete" traditional varieties and local landraces are replaced by "high-performance" breeds, which are suitable for mass production. Knowledge and practices associated with the older varieties tend to be abandoned, sacrificed in the name of progress and modernity. Not only does this generate injustices by misrecognizing local ways of farming, it also creates dependence upon an industrial farming model by weakening the emergence of alternatives (Coolsaet, 2016).

In short, recognition in the context of conservation is about who gets to define "good" conservation, whose voices are heard and listened to, and whose knowledge is deemed relevant when planning conservation action. Focusing on recognition allows us to uncover the structural, institutional and psychological forms of domination which often define who benefits from conservation. It also provides a basis for looking beyond a distributive model of justice to incorporate social and cultural differences, including paying careful attention to ways of pursuing equality of status for local conservation stakeholders. This will require reflection on working practices and looking at forms of intercultural engagement that, for example, respect alternative ways of relating to nature and biodiversity.

Conclusion

This chapter has explored different theoretical approaches to recognition. It has examined concepts such as otherness, identity, cultural difference, disrespect and participation-parity to introduce the reader to the importance of recognition in understanding how people express their differences through a multitude of relations to the world around them. It has briefly reflected on some of the political limits of liberal approaches to recognition and provided examples of ways in which scholars and activists have engaged with the meanings and expression of differences attached to the environment.

As with other theoretical approaches discussed in this volume, the different approaches to recognition examined in this chapter may not be mutually exclusive. Environmental justice movements illustrate that these ideas may, in certain cases, be combined to fully capture the plurality of the injustices they are faced with. The strength of the idea of recognition is that it can apply both to individuals and to communities as a whole, focusing not only on individual and psychological harm, but also on more structurally generated issues of status subordination. In addition, recognition may be consequential for other dimensions of justice. A decision-making process is unlikely to produce a fair outcome if participants are not already committed to mutual recognition, and the conditions for distributional equity to materialize cannot be achieved without involving recognition. Above all, we have seen that alleged solutions to environmental injustices may not be solutions at all if the issue of recognition is not addressed.

Follow-up questions

- Can you think of other examples of environmentally harmful practices that generate problems of misrecognition?
- If you are involved in an initiative for the protection of the environment, how do you consider issues of recognition in your work?
- What does your relationship with nature or the environment express about who you are?
- Where would you stand on the identity versus status debate about recognition introduced in this chapter? Why?

Notes

1 The "Dene" refers to Indigenous people living in the western Canadian Subarctic, including First Nations groups such as the Chipewyan, Tlicho, Slavey, Sathu and Yellowknives.
2 On distribution and participation, see Chapters 3–4 of this volume.
3 The *Principles of Environmental Justice* (EJ) are available on http://www.ejnet.org/ej/principles.pdf

References for further reading

- Coulthard, Glen S. 2014. *Red Skin White Masks. Rejecting the Colonial Politics of Recognition*. Minneapolis/London: University of Minnesota Press.
- Fraser, N. and Honneth, A. 2003. *Redistribution or Recognition? A Political-Philosophical Exchange*. London/New York: Verso.
- Martin, A., Coolsaet, B., Corbera, E., Dawson, N.M., Fraser, J.A., Lehmann, I. and Rodriguez, I. 2016. Justice and conservation: The need to incorporate recognition. *Biological Conservation* 197, 254–261.
- Whyte, Kyle. 2017. The recognition paradigm of environmental injustice, in Ryan Holifield, Jayajit Chakraborty and Gordon Walker (eds.) *The Routledge Handbook of Environmental Justice*. Abingdon: Routledge, pp. 113–123.

References used in this chapter

Álvarez, L. and Coolsaet, B. 2018. Decolonizing environmental justice studies: A Latin American perspective. *Capitalism Nature Socialism* 31(2), 50–69 DOI: 10.1080/10455752.2018.1558272
Bullard, R.D. 1990. *Dumping in Dixie: Race, Class, and Environmental Quality*. Boulder: Westview Press.
Coolsaet, B. 2016. Towards an agroecology of knowledges: Recognition, cognitive justice and farmers' autonomy in France. *Journal of Rural Studies* 47, Part A, 165–171.
Coulthard, Glen S. 2014. *Red Skin White Masks: Rejecting the Colonial Politics of Recognition*. Minneapolis/London: University of Minnesota Press.

EJnet.org. (1991). *The principles of environmental justice*. Washington, DC, October 24–27. www.ejnet.org/ej/principles.html.
Estes, N. and Dhillon, J. 2019. *Standing with Standing Rock: Voices from the #NoDAPL Movement*. Minneapolis, MN: University of Minnesota Press.
Evans, G.W. and Kantrowitz, E. 2002. Socioeconomic status and health: The potential role of environmental risk exposure. *Annual Review of Public Health* 23, 303–331.
Fanon, Frantz. 1967/2008. *Black Skin, White Masks*. New York: Grove Press.
Figueroa, R.M. 2004. Bivalent environmental justice and the culture of poverty. *Rutgers Journal of Law and Urban Policy* 1, 1.
Fraser, Nancy. 2005. Reframing justice in a globalizing world. *New Left Review* 36.
Fraser, Nancy. 2001. Recognition without Ethics? *Theory, Culture & Society* 18 (2–3), 21–42.
Fraser, Nancy. 2000. Rethinking recognition. *New Left Review* 3, 107–120.
Fraser, Nancy. 1995. From redistribution to recognition? Dilemmas of justice in a 'post-socialist' age. *New Left Review* 1 (212).
Fraser, Nancy. 1990. Rethinking the public sphere: A contribution to the critique of actually existing democracy. *Social Text* 25/26: 56–80.
Fraser, N. and Honneth, A. 2003. *Redistribution or Recognition? A Political-Philosophical Exchange*. London/New York: Verso.
Hegel, G.W.F. 1977. *Phenomenology of Spirit*. Translated by A.V. Miller. Oxford: Oxford University Press.
Hegel, G.W.F. 1991. *Elements of the Philosophy of Right. Cambridge Texts in the History of Political Thought*. Cambridge, MA: Cambridge University Press.
Honneth, Axel. 2004. Recognition and justice: Outline of a plural theory of justice. *Acta Sociologica* 47 (4), 351–364.
Honneth, Axel. 2007. *Disrespect: The Normative Foundations of Critical Theory*. Cambridge, MA: Polity Press.
Honneth, Axel. 1996. *The Struggle for Recognition: The Moral Grammar of Social Conflicts*. Cambridge, MA: MIT Press.
Laclau, E. 1985. New social movements and the plurality of the social. *New Social Movements and the State in Latin America* 23 (3), 121–137.
Macpherson, C.B. 1964. *The Political Theory of Possessive Individualism: Hobbes to Locke*. Oxford: Oxford University Press
McGregor, D. 2009. Honouring our relations: An anishnaabe perspective on environmental justice, in J. Agyeman, P. Cole and R. Haluza-Delay (eds.) *Speaking for Ourselves: Environmental Justice in Canada*. Vancouver, BC: University of British Columbia Press.
Pulido, Laura. 1996. *Environmentalism and Economic Justice: Two Chicano Struggles in the Southwest*. Tucson: University of Arizona Press.
Schlosberg, David. 2007. *Defining Environmental Justice: Theories, Movements, and Nature*. Oxford: Oxford University Press.
Taylor, Charles. 1989. *Sources of the Self: The Making of the Modern Identity*. Cambridge, MA: Harvard University Press.
Taylor, Charles. 1994. The politics of recognition, in A. Gutmann (ed.) *Multiculturalism: Examining the Politics of Recognition*. Princeton: Princeton University Press.
United Nations. 2007. *United Nations Declaration on the Rights of Indigenous Peoples*. Resolution adopted by the General Assembly on 13 September 2007.
Whyte, Kyle. 2017. The recognition paradigm of environmental injustice, in Ryan Holifield, Jayajit Chakraborty and Gordon Walker (eds.) *The Routledge Handbook of Environmental Justice*. Abingdon: Routledge, pp. 113–123.

6 Capabilities, well-being, and environmental justice[1]

Breena Holland

Learning outcomes

- Distinguish human capabilities as a conception of human well-being.
- Understand the direct and indirect ways that the natural environment supports capabilities.
- Identify threats to capabilities as different kinds of environmental injustice.

Introduction

Environmental justice—whether conceived in terms of distributional equity, fair procedures, or recognition—[2] raises important questions about human well-being. For example, what is the best way to think about well-being in order to identify and remedy environmental injustice? Which view of well-being can reveal what is at stake in claims that unequal exposure to toxic waste is an injustice? How might we think about the effects of decisions about the environment on human well-being when the decision processes are unfair? And, finally, how can we best account for the harms to well-being resulting from decisions that disregard or fail to recognize the environment's significance and value to the cultures, traditions, and identities of different people, especially members of non-dominant groups in society?

Addressing environmental injustices requires a way of understanding how people are being harmed and why that harm is unjust. Conceptualizing well-being in terms of people's **capabilities** to do and achieve things that improve their lives provides just that. To be capable is to be effective in doing or achieving such things within the conditions and circumstances of one's life. Capabilities can vary widely, from something basic to well-being, such as being able to achieve adequate nourishment, to higher order capacities, such as being able to take part in the life of one's community or to critically reflect on one's life goals. Moreover, conceptualizing well-being in terms of a person's capabilities provides a window into what constitutes a good human life (Nussbaum, 1990, 1998), which in turn enables us to see the importance of the environment to elements of that life.

If capabilities define **well-being**, then it follows that environmental injustices impact human well-being by undermining capabilities that are regarded as valuable and worthy of protection. Conversely, environmental injustices can arise from and compound existing capability deprivation. This can occur through, for example, the unfair distribution of

> **Box 6.1 What are capabilities and functionings?**
>
> For an individual, capabilities are real opportunities to do things or achieve states of being. They are the objectives that a person *can* achieve. In the language of capabilities, the actual beings and doings are termed "functionings". Amartya Sen (1992, p. 40) refers to a person's **functionings** as "achieved well-being" and their capabilities as "well-being freedom". Thus, if hiking in nature is a functioning, then the ability to do so is the corresponding capability. The concept of a capability captures whether a person *could* hike in nature if she wanted to. Being able to hike in nature requires the personal (or internal) conditions that make hiking possible, such as being able walk over uneven terrain and being free from debilitating health problems. It also involves certain external conditions, such as the availability of a natural area for a hike and a way of getting there. Thus, the concept of a capability also captures whether a person has the internal conditions required and access to the external circumstances that enable her to take up this opportunity (see Nussbaum, 2000, pp. 84–86).

environmental harm, inequality of access and voice in environmental decision-making processes, or lack of recognition and respect for people as equals in decisions that determine whether the local environment can sustain its distinct cultures, traditions, and identities. Thus, identifying how environmental changes support or undermine well-being—defined in terms of a person's human capabilities—is a powerful way to understand the nature and extent of environmental injustice.

This chapter identifies features of human capabilities that are important for understanding and characterizing the relationship between the environment and human well-being. It then explains how we can appreciate the environment's crucial role in securing these capabilities. Finally, it shows how capabilities scholarship can inform the theory and practice of environmental justice, understood in terms of distributive fairness, just procedures, and mutual recognition.

Capabilities as well-being

Human capabilities provide a superior way of understanding well-being for the purpose of assessing the success of international development and the justness of any society (Sen, 1999; Nussbaum, 2000). This understanding of well-being has advantages for recognizing the natural environment's overarching role in human lives and for demonstrating how human impacts that alter the natural environment can produce environmental injustice.

First, capabilities define well-being in terms of what people can do or be with the resources they have and in the context of the circumstances they face (see Box 6.1). As such, capabilities can account for *inter-individual differences* that have significant implications for understanding people's relative levels of well-being. Inter-individual differences include the variations across personal conditions, and differences in the social, economic, and political circumstances that shape people's real opportunities (Sen, 1992, pp. 81–87). For example, a person who is permanently disabled or one who is pregnant has especially demanding bodily needs for all or part of her life. To achieve the same level of well-being, such a person will need more material or social resources than will someone not physically disabled or pregnant (Sen, 1982, pp. 357–358; Nussbaum, 2001). Likewise, a person living in a society that punishes violators of women's rights or that provides significant resources for social and economic support would be more likely to experience external circumstances conducive to personal and professional success. Thus, those with access to these forms of support will be more able to achieve their potential. Capabilities account for actual differences between individuals because what a person can do or achieve is in part determined by their personal conditions and external circumstances. These differences have a meaningful impact on the kind of lives people can live.

Capabilities are commonly contrasted with economic metrics of well-being assessment, such as personal incomes or holdings of goods. These are things that people possess. They provide important information about how well-off people are in material terms, but do not tell us whether they can convert these material things into opportunities or genuine improvements to their well-being (Sen, 1992, pp. 26–28). To use an environmental example, a given level of air quality is more important to the well-being of someone with a respiratory disorder than to someone without. This is because dirty air has worse health consequences for a person with a compromised respiratory system. The same level of air quality would also make those with compromised immune systems less capable of being bodily healthy. Thus, recognizing differences in susceptibility to environmental harms helps us understand why the same level of air quality does not lead to the same level of well-being for everyone. If we want to accurately understand how the benefits of environmental policies are distributed among different people, then we must assess the impact of those policies on people's capabilities to do and achieve things that improve their lives, such as having good health and being able to take outdoor recreation.

The second reason that capabilities are useful for understanding and demonstrating the environment's role in securing well-being is that capabilities define well-being as *multidimensional*: encompassing the multiple worthwhile goals and achievements a person is capable of. Capabilities can vary widely, from being able to achieve basic material conditions (e.g. nutritional health), to more sophisticated social experiences (e.g. having meaningful relationships with other individuals and communities), to political power, (e.g. effectively participating in political decisions). In a capabilities-based conception of well-being, these different dimensions of what makes a person thrive are distinct and may be irreducible. For example, Martha Nussbaum (2000, 2006) develops a "capabilities approach to justice" that defines *ten central human capabilities* that a just society must secure for its citizens (see Box 6.2) because each capability is essential to a life worthy of human dignity (see Nussbaum, 1995, 2006, pp. 70, 347). Nussbaum's theory of justice permits these central human capabilities to be specified differently, in accordance with what matters to particular people in different places, and it allows for the whole list to be revised through ongoing democratic debate (Nussbaum, 2000, pp. 77, 105). However, Nussbaum insists that each capability must be secured for each person and cannot be traded off against the others, such that less of one valuable capability could be made up for by more of another.

The multidimensionality of capabilities (see Box 6.2) is especially important for understanding how the natural environmental relates to well-being. Even though the natural environment contributes to these different dimensions of people's lives (see Holland, 2008a, pp. 323–325), its importance is often overlooked or inaccurately valued. For instance, while the provision and protection of freshwater is clearly important to a person's nutritional health, this does not capture additional ways that it is also important to individual

> **Box 6.2 Nussbaum's list of central human capabilities**
>
> 1. **Life** — Being able to live a life of normal length.
> 2. **Bodily health** — Being able to have good health, including reproductive health; to be adequately nourished; to have adequate shelter.
> 3. **Bodily integrity** — Being able to move freely from place to place; being free from physical assault and sexual violence; having opportunities for sexual satisfaction; having reproductive choice.
> 4. **Senses, imagination, and thought** — Being able to use the senses to imagine, think, and reason in a way informed and cultivated by an adequate education; having freedom of religion and expression; being able to have pleasurable experiences, and to avoid non-necessary pain.
> 5. **Emotions** — Being able to have attachments to things and people outside ourselves; being able to experience and express emotions; avoiding emotional trauma, abuse or neglect.
> 6. **Practical reason** — Being able to form a conception of the good and to engage in critical reflection about the planning of one's life; having liberty of conscience.
> 7. **Affiliation** —
>   (A) Being able to live with and toward others, to engage in various forms of social interaction; to be able to have empathy and compassion.
>   (B) Having the social bases of self-respect and non-humiliation; being able to be treated as human of equal worth; being free from various forms of discrimination (based on race, sex, ethnicity, sexual orientation, etc.).
> 8. **Other species** — Being able to live with concern for and in relation to animals, plants, and the world of nature.
> 9. **Play** — Being able to laugh, to play, to enjoy recreational activities.
> 10. **Control over one's environment** —
>   (A) *Political*. Being able to participate effectively in political choices that govern one's life; having protections of free speech and association.
>   (B) *Material*. Being able to hold property and seek employment on an equal basis with others; freedom from unwarranted search and seizure.
>
> Note: the above text is adapted and abridged from Nussbaum (2006, pp. 76–78)

well-being. A particular freshwater resource may also be the locus of meaningful relationships forged among members of a community. It might be a place of spiritual or religious practices that give people a broader sense of meaning and purpose, and an experience of cultural continuity. Because a person's capabilities include these different dimensions of well-being, they can reveal the multiple and distinct ways that the environment contributes to well-being without undervaluing the breadth and significance of its contribution. In contrast, efforts to quantify the environment's value in monetary terms (as commonly practiced in economic approaches to environmental valuation) may not accurately represent the importance of a freshwater resource as an intrinsically valuable, non-fungible good.

The multidimensional view of well-being that capabilities express also illuminates how the environment produces various forms of injustice (Schlosberg, 2012a, pp. 167–168). For instance, disproportionate exposure to environmental harms threatens the material dimensions of human well-being, such as good health, which is a component of the "bodily health" capability, as defined by Nussbaum (see Box 6.2). But that exposure may also threaten the political dimensions of well-being; for instance, by preventing effective participation in political decisions that affect the material conditions of one's life (a component of the capability for "political control over one's environment"). And, if the natural environment plays an important role in maintaining people's distinct cultures, traditions, and identities, then harmful exposure can also undermine a person's capability for "affiliation" by eliminating the possibilities for relationships that form and effectuate cultural benefits. Thus, in addition to diminished bodily health and political voice, environmental harms might also entail the loss of respect experienced when one is not treated as a dignified being whose worth is equal to that of others. It is this "internal plurality" (see Alkire, 2002, p. 9) of a capabilities-based conception of well-being that makes it possible to see a wide range of threats to human well-being as distinct harms that are relevant to justice. Before further considering the relationship between capabilities and environmental justice, let us turn to the complex ways that the natural environment supports and enables human capabilities and, therefore, the multitude of ways in which changes to the natural environment can undermine human well-being.

Capabilities and the environment

The growing literature on capabilities and the environment illustrates the strength of a capabilities approach in highlighting the comprehensive dependence of human well-being on the natural world. Anantha Duraiappah's (2004) foundational work in this area established the relationship between poverty as a form of capability deprivation (in developing countries) and ecosystems. Duraiappah identifies the capabilities people need to escape poverty and then considers how ecosystems shape those capabilities. Some of the capabilities he identifies as relevant to escaping poverty, such as "being able to be adequately nourished", are commonly referenced components of well-being. Others, such as "being able to continue using natural elements found in ecosystems for traditional cultural and spiritual practices", are more specific to particular groups of people. Duraiappah explains the critical role of ecosystems in influencing these and other dimensions of well-being and thus their significance for whether people are able to escape poverty. Several crucial points emerge from his analysis. First, ecosystems are essential preconditions of human capabilities. Second, ecosystems play a crucial and extensive role in providing the resources and ecosystem services necessary for distinct but important dimensions of people's lives. Finally, damage to ecosystems effectively dispossesses people of fundamental components of their well-being (Duraiappah, 2004, p. 11).

Following Duraiappah, many ongoing efforts to conceptualize and develop an understanding of the relationship between the natural environment and human capabilities use the concept of ecosystem services to characterize the natural environment's instrumental importance to human well-being (e.g. Holland, 2014; Pelenc and Ballet, 2015; Schultz et al., 2013).[3] Broadly speaking, **ecosystem services** refer to the many benefits people get from ecosystems (see Assessment, 2003). These benefits include both particular environmental goods (fuelwood, water, and so on), which are *directly* instrumental in the generation of capabilities and broader functions, such as climate regulation and water purification, that are *indirectly* relevant to people's capabilities. Less tangible ecosystem services, such as the provision and maintenance of features of the environment relevant to cultural heritage, are also indirectly relevant to people's capabilities. Polishchuk and Rauschmayer (2012, p. 107) argue that such services operate as "conversion factors" in determining the way people use (or convert) their available resources into valuable outcomes. A conversion factor is an external condition that determines how much a person can get out of a given resource for the purpose of achieving something valuable. For instance, a regulatory service, such as climate regulation which influences temperature variability, will play an important role in determining how much fuelwood or freshwater is needed for people to be capable of life and bodily health. In this context, the regulatory service functions as a conversion factor because it determines how much an individual can benefit from an existing amount of fuel or water in order to achieve the components of life and health capabilities, such as staying warm in winter and hydrated in summer. Thus, in addition to providing particular environmental resources that directly support human capabilities, ecosystem services are also indirectly relevant to capabilities when they determine whether and to what extent a person can translate the resources at her disposal into valuable outcomes or achievements.

Like regulatory services, cultural ecosystem services also determine whether people can convert environmental resources into the capabilities that are central to their well-being (Ballet et al., 2018). Cultural ecosystem services are generally defined as non-material benefits that people obtain from ecosystems, such as experiences that give people a sense of place, of spiritual and aesthetic inspiration, or of cultural identity. Because these experiences are often central to human well-being, it is important to relate these benefits of ecosystems to the capabilities that allow people to live in relation to nature in ways that enable experiences of cultural meaning and value (see Watene, 2016). Protecting these culturally meaningful relationships with nature often means protecting the ecosystems in which these experiences occur. Furthermore, protection or degradation of an ecosystem and the environmental media it supports (such as air or water quality) will impact people differently depending on the extent to which their cultural identity is bound up with the aspect of the environment that is being protected or degraded. For instance, water that is not clean enough to support a human diet high in fish will make Native American populations—whose cultural identities are often closely intertwined with fish consumption—less capable of maintaining those identities than groups whose cultural identity does not rely on the same kind of diet. This is why Ballet et al. (2018, p. 24) endeavour to develop an understanding of the link between ecosystems (as well as the resources within them, and the landscapes that encompass them) and a distinct "personal identity capability" that makes it possible to maintain and develop an understanding of oneself in dynamic and reciprocal social relationships.

The environment is not only important before capabilities emerge (as conditions of and inputs to capabilities) and during the transformation of a capability into an activity or achieved outcome (as a conversion factor). It is also significant after people realize their capabilities through the various activities and achievements they choose to pursue. In the literature on capabilities, a person's realized capabilities are referred to as their "functioning"—the various beings and doings that make up an individual's life (see Box 6.1). The environment is important with regard to functionings because the various things people do and achieve, in aggregate, can have a significant impact on the ecosystems that provide conditions and inputs to capabilities. For example, the aggregate impact of how people are realizing their capabilities as functionings is currently overtaxing the biogeochemical cycles through which the planet maintains a relatively stable climate. In this respect, humans and the environment are in a relationship of constant feedback in which the realization of capabilities that depend on the environment also impact the environment (see Schultz et al., 2013, p. 126). As the aggregate impact of human functionings on the environment increasingly produces conditions of environmental fragility, scarcity, and emergency, the way in which we understand a person's capabilities is also changing—specifically, by drawing attention to the need to restrict capabilities when they enable people to do things that—in aggregate—threaten the planetary life-support processes on which everyone's capabilities depend. Holland (2008b) refers to such restrictions as "capability ceilings", and Peeters et al. (2013, p. 67) refer to them as "functioning constraints".

These efforts to identify and reveal the connections between planetary systems, ecosystems, and the multiple dimensions of human well-being that capabilities can define are important for understanding who environmental changes impact, and how to address those impacts. Because people value the environment differently, depending on its direct and indirect importance to their capabilities, they will have conflicting views about what environmental protection should involve. In some cases, environmental protection may be "critical" to capabilities, in the sense of enabling people to continue to do things that they refuse to give up, even when they are offered an alternative or compensation (Pelenc and Ballet, 2015, p. 41). In such circumstances, the environment can be understood as making a *non-substitutable* contribution to well-being. For instance, in the previous example of Native American fish consumption, a level of water quality that enables a diet high in fish consumption may be making a non-substitutable contribution to well-being, at least for some Native American communities. In contrast, vegetarians might view the same level of water quality as far less important or central to the capabilities that make their lives worth living. Pelenc and Ballet (2015) argue that decisions about what ecosystems services will be prioritized and sustained should only be made through wide-ranging debate about the values, goals, and objectives of various stakeholders belonging to whole socio-ecological systems. Likewise, Holland (2014, pp. 188–195) argues that decisions about the environmental protections needed to secure people's capabilities must be undertaken through a process in which citizens identify and characterize the capabilities to which their environment contributes, while experts and stakeholders capable of exerting disproportionate political power need to be constrained in ways that limit their influence on decisions that could undermine the capabilities of the population in question.

Thus, even when environmental justice is not the overt frame of inquiry, a comprehensive and nuanced understanding of how the environment relates to capabilities is significant for achieving environmental justice. It can reveal the multiple channels through which various forms of environmental injustice occur and, therefore, the reasons why environmental damage and deprivation are sources of injustice. In providing a comprehensive picture of how the environment relates to well-being, assessing how environmental changes impact capabilities can render the structural and institutional dimensions of injustice more visible. This then raises questions such as:

- Whose capabilities to impact the environment should be limited?
- Whose views about the environment's importance to well-being should be prioritized in decisions about environmental protection?
- How can decision-making processes be structured to enable effective participation of those whose capabilities are most threatened by environmental harm?

Let us now focus more particularly on how the relationship between people's capabilities and the environment have become explicitly relevant to questions of environmental justice, conceived in terms of distribution, fair procedures, and recognition.

Capabilities and environmental justice

An understanding of the breadth and depth of the ways in which human well-being is intertwined with the natural environment is necessary for grasping why changes to it can dispossess people of components of their well-being that are central to justice. For example, not only does a capabilities-based view of well-being expose the various ways that disproportionate exposure to environmental harms threatens the material dimensions of human life, but it can also reveal impacts on a person's capability to influence political decisions and how environmental harms undermine the social bases of self-respect. While these problems of distributive, procedural, and recognition-based environmental injustice are often interwoven (see Schlosberg, 2007, pp. 73–75), scholars have drawn on capabilities to inform and transform how we understand and make claims about each.

Distributive environmental justice and capabilities

Distributive environmental justice in its most basic sense is concerned with fairness in the distribution of environmental benefits and burdens. For example, when people of certain racial or ethnic groups are disproportionately burdened by exposure to environmental risks, such as those associated with close proximity to hazardous waste facilities,[4] we can think of this as a failure of distributive justice: namely, that there is an unequal distribution of exposure to environmental harms that threaten human well-being. Likewise, when people of certain racial groups or income levels lack access to environmental benefits such as lead-free water, we can understand this as a failure of distributive justice in that there is an unequal distribution of the benefit of clean water and, consequently, the health benefits associated with clean water only accrue to some people. In the context of climate change, distributive injustice arises because those suffering the greatest harms from a warming global climate have not shared equally in the benefits of the emission-producing activities that have contributed most to this problem.[5]

In previous work (Holland, 2014), I have built on Martha Nussbaum's capabilities approach to justice to incorporate the environment into an evaluative framework that assesses the distribution of capabilities. I use this environmentally informed account of distributive justice as the basis for evaluating environmental policy decisions in terms of how they impact human capabilities. I have argued that environmental protection should be treated as a basic condition of a just society because damage to the environment can threaten the full range of capabilities that Nussbaum deems central to a life worthy of human dignity that governments are obligated to secure. In addition to securing Nussbaum's eighth capability (to have meaningful relationships with plants, animals, and the world of nature), capabilities justice requires the protection of broader ecological conditions that support and enable all the capabilities that Nussbaum's theory establishes as worthy of safeguarding in a just society. At a minimum, these "ecological conditions of justice" (Holland, 2008a, p. 328) must be protected to the extent necessary to secure a threshold level of each person's central human capabilities. Because these ecological conditions of justice are a precondition of Nussbaum's central human capabilities, I characterize them as a **"meta-capability"** of humans, which entails "being able to live one's life in the context of ecological conditions that can provide the environmental resources and services that enable the current generation's range of capabilities, [and] to have these conditions now and in the future" (Holland, 2008a, p. 324). This extension of the capabilities approach enables us to perceive damage to the environment in one place or time period that threatens the ecological preconditions of people's capabilities in another place or time period as a violation of justice.

Because the capabilities secured as a matter of justice may allow some individuals to behave in ways that—individually or in aggregate—undermine the ecological conditions needed to secure a threshold level of other's capabilities, it is essential that a capabilities approach to distributive justice is able to resolve capability conflicts. For this reason, I propose establishing "capability ceilings" that constrain people's capabilities to damage the environment in ways that cause harm to others (Holland, 2008b). Specifically, capability ceilings would be applied only to capabilities already secured at a threshold level—that is, beyond the minimum requirements of justice. The implication is that environmental justice may involve rendering some people incapable of a number of functionings in order to secure a threshold level of the capabilities central to justice for all people. For example, environmental justice may require limiting polluting activities that disproportionately harm the capabilities of asthmatic children for "bodily health, affiliation, play" and "senses, imagination, and thought" (Holland, 2014, pp. 153–159). In the context of global climate change, it may require expanding the community of consideration to include future generations, by constraining the choices of people in the present generation that may impact the ecological conditions necessary to secure the capabilities of those in future generations (see Holland forthcoming; Scholtes, 2010). In this way, I draw on human capabilities to expose and resolve distributive conflicts arising from people's connectedness through the environment, spanning both geographical space and generational time periods. Thus, capability ceilings make it possible to appreciate how the finite nature of the environment impacts the distribution of capabilities.

Procedural environmental justice and capabilities

Procedural environmental justice is concerned with the justness of the processes through which environmental decisions are made. A procedurally just process is often understood in terms of people's opportunity to participate in environmental decisions, particularly in terms of whether the processes are inclusive and fair. Inclusivity concerns who participates in decisions and can be achieved through practices that ensure the people impacted by environmental harms have a say in decisions determining their exposure to those harms (see Hunold and Young, 1998). Fairness in decision-making processes is linked to the power of the participants within these processes. It can be achieved through practices that remove barriers to expression; for example, by establishing norms that welcome a variety of forms of public expression and recognize all voices within those forums (e.g. see Shrader-Frechette, 2002, p. 28; Young, 2002).

Decision-making processes have been an important focus in the work of Amartya Sen, who, with Martha Nussbaum, first theorized justice in terms of human capabilities. Sen's distinction between the process and opportunity aspects of freedom is often referenced with regard to people's capabilities in matters of procedural justice. The *process aspect* of freedom concerns a person's agency, understood as their capability to act, while the *opportunity aspect* of freedom concerns a person's well-being (Alkire and Deneulin, 2009). Agency and well-being are connected in decision-making processes because agency involves being an active participant in change that improves well-being, according to values that are self- or collectively defined (see Crocker and Robeyns, 2010). Sen (2005, p. 158) argues that respecting human agency requires determining the capabilities that are treated as central to justice in the context of democratic decision-making processes, through which people can engage in social discussion and public reasoning about which capabilities should be protected in a just society and why. Nussbaum also recognizes the importance of just decision-making processes by including a capability for "control over one's [political] environment" as a central human capability that government is obligated to secure for each person; this entails being capable of effective participation in political decisions (see Box 6.1). While neither Sen nor Nussbaum has gone into what such commitments to public participation would mean in the context of specific environmental decisions, others have built on their ideas to address what such basic procedural commitments do or should imply.

For example, Pelenc et al. (2015) build on Sen's conception of individual agency to understand how people's capabilities can be translated into collective action to preserve and shape the management of the Campana-Peñuelas Biosphere Reserve in central Chile. In this case, the creation of a non-governmental organization (NGO) was crucial to effectively representing and pursuing the interests of people otherwise excluded from the process for making decisions about the reserve. Pelenc et al. use this environmental management context to theorize the relationship between individual capabilities and the collective capability of a group to achieve what group members define as valuable. They argue that, by bringing together individual actors with an interest in sustaining the reserve and by connecting them to regional level actors, the NGO transformed individual agency into collective agency. In this way, Pelenc et al. (2015) show how the procedural injustice of exclusionary environmental management decision-making processes can be overcome by institutional innovations that enable a collective capability for political agency, which can in turn help to ensure the environmental preconditions for individual capabilities.

In my own work (Holland, 2017), I also use capabilities to address problems of procedural injustice that result from a lack of citizen voice in and control over decisions about the environment. Building on Nussbaum's commitment to securing the capability for "political control over one's environment" (see Box 6.2), I consider what it would mean for citizens to have this capability in the context of decisions about how their communities will adapt to climate change. I argue that possessing the capability for "political control over one's environment" means that communities have the power to influence decisions about how they will adapt to climate change. This might be achieved by establishing formal control over decision-making rules and procedures according to which adaptation decisions need to be made (Holland, 2017, p. 392). In situations when vulnerable communities lack such control, securing this political capability could result in the successful application of political pressure to push decisions in a particular direction, as "Vulnerable citizens must be able to exert some control over either the content of what adaptation protects, or over the decision rules through which such content is determined" (Holland, 2017, p. 398). Based on the review of a number of case studies, I find that two variables are crucial to securing these political capabilities in the context of climate adaptation decisions: the potential for alliance with powerful stakeholders and the political relevance of conflicting expertise. Thus, political capabilities define the conditions of procedural justice that can amplify the voices of vulnerable citizens within climate adaptation decision-making processes.

Recognition-based environmental justice and capabilities

Recognition is concerned with the harms that result from failing to acknowledge or respect differences across groups of people. Recognition-based environmental injustice refers to harms suffered by (often marginalized) groups whose histories, cultures, and identities are intimately bound up with their distinct relationships to the environment (Figueroa, 2015). For instance, recognition-based environmental injustices may result from environmental relationships valued by Indigenous peoples not being treated as worthy of equal respect and formal legal protection (see Whyte, 2018). Likewise, activities destroying ecosystems that are bound up with a group's cultural integrity and spiritual practices fail to recognize and equally respect the meanings and forms of expression that distinguish group difference and make it feasible. In addition to threatening individual autonomy, identity formation, and the potential for self-realization for members of culturally or racially distinct groups, a lack of recognition can erode individual membership in one's community and participation in political decision-making contexts that determine community choices (see Schlosberg, 2007, p. 26; Young, 1990, p. 23).

Focusing on the diversity of concerns expressed in environmental justice movements, David Schlosberg (2007, 2013) has posited a capabilities approach to justice as uniquely able to integrate concerns of recognition justice with procedural and substantive inequities. This is because it conceptualizes justice as an issue of the very ability to function, be, and do. In Schlosberg's interpretation (2007, pp. 33–34), the concepts and practices of recognition are closely linked to distributional issues as part of a more holistic concern with individual functioning and flourishing. He argues that Nussbaum's inclusion of the social bases of self-respect and non-humiliation in the capability for "affiliation" puts recognition on a par with concerns of participation and distribution. Furthermore, because communities often articulate their demands for environmental justice in the language of individual and community capabilities to function, characterizing justice in terms of capabilities is one way of aligning theoretical views of justice with what claims about justice actually mean to people disproportionately suffering from environmental harm.[6] In developing this view of justice within communities, Schlosberg and Carruthers (2010) argue that it is not just the capabilities of individuals that are at stake in demands for environmental justice, but also the capabilities of whole communities to act with collective agency and reproduce socially over time. Thus, recognition justice requires securing collective capabilities because the existence and functioning of one's community is a precondition for having a political voice to express one's **identity** in relation to that community.

The protection of environmental relationships that contribute to distinct community identities is an increasingly prominent issue in the literature on capabilities justice (e.g. Heyward, 2011; Brincat, 2015). The capabilities of specific groups are threatened by environmental changes that degrade or alter land and landscapes that are closely intertwined with their cultural identities. But, in some instances, capabilities as currently conceived may be insufficient for reflecting the meaning of environment to particular groups. For example, Watene (2016) claims that Sen and Nussbaum understand capabilities in a way that is unable to fully value and protect Maori relationships with nature that are grounded in the spiritual and physical dimensions of "whakapapa" (genealogy). Because Sen does not take into account the intrinsic value of nature that is central to Maori relationships with the environment, his approach risks the loss of cultural identity that would result from damage to aspects of the environment that the Maori view as non-substitutable. In contrast, as the basis for securing human capabilities, Nussbaum adopts a conception of dignity that can be extended to include the non-human world that Maori see as intrinsically valuable, but does not ground this conception of dignity in the spirituality that is central to Maori ways of life. Consequently, Nussbaum excludes the possibility of a non-instrumental form of environmental value that deems the environment valuable for its own sake.[7] Watene argues that, without further development of the core concepts and commitments of capabilities justice, efforts to translate these accounts of justice into policy risk obscuring the meaning and value of nature itself to cultural groups, and hence, failing to protect the environments that recognition justice demands.

Conclusion

Capabilities provide a conception of well-being that is significant for environmental justice because they can take account of inter-individual differences that make some people more likely to suffer from environmental injustice. Capabilities also offer a multidimensional view of well-being that can reveal the multiple channels through which environmental harms occur and, therefore, explain why the material, political, and socio-cultural aspects of human well-being entail separate and mutually constitutive obligations of justice. Scholarship explicitly linking capabilities to environmental justice addresses what fulfilling these obligations requires, whether the relevant environmental injustice is distributive, procedural, or recognition-based. Critics of capabilities justice (such as Watene, 2016) highlight the importance of ongoing development and revision of how we conceive capabilities and capabilities justice, especially when it comes to making practical decisions about the environment's substitutability as a basis of identity for particular groups. However, the success of capabilities in expanding and deepening the focus of the lens of justice on relationships with the environment that are vital to human well-being is a significant conceptual innovation. When used to theorize and advance distributive, procedural, and recognition-based views of justice, capabilities can help us reimagine what environmental justice requires, as well as our strategies and success in achieving it.

Follow-up questions

- Which of Nussbaum's ten central human capabilities are most dependent on the natural environment?
- To what extent can technology provide substitutes for the environmental goods and services that support human capabilities?
- What human functionings that negatively impact the environment should society restrict?
- In what ways do the environmental preconditions of human capabilities differ from other preconditions of human capabilities, such as social and economic systems and relationships?

Notes

1. Acknowledgments: I would like to thank Amy Linch and Teena Gabrielson for helpful comments on this chapter. A number of productive exchanges also took place with Gareth A.S. Edwards and other participants at the Workshop on Environmental Justice and Capabilities at the University of East Anglia in February 2018.
2. On distribution, procedure and recognition, see Chapters 3, 4–5 of this volume, respectively.
3. See Crabtree (2012) as an exception in focusing on "ecological footprints".
4. On toxic legacies and environmental justice, see Chapter 10 of this volume.
5. On climate justice, see Chapter 12 of this volume.
6. Schlosberg (2012b, p. 458) discusses why framing what is at stake in terms of people's capabilities is helpful for engaging local communities in discussions about their vulnerability in the context of climate adaptation decisions and for designing adaptation politics that will protect them.
7. Winter (2019, p. 23) argues that a capabilities approach to justice can account for Maori ways of being in relation to nature by reconceiving its understanding of human dignity as a form of "Immersive Functioning Dignity."

References for further reading

- Alkire, S. and Deneulin, S., 2009. The human development and capability approach. In *An introduction to the human development and capability approach* (pp. 22–48). London, UK: Earthscan.
- Holland, B. and Linch, A., 2016. Cultivating human and non-human capabilities for mutual flourishing. In *The Oxford handbook of environmental political theory* (pp. 413–428). Oxford, UK: Oxford University Press.
- Nussbaum, M.C., 2011. *Creating capabilities*. Harvard University Press.
- Robeyns, I., 2005. The capability approach: A theoretical survey. *Journal of Human Development*, 6(1), pp. 93–117.
- Robeyns, I., 2017. *Wellbeing, freedom and social justice: The capability approach re-examined.* Cambridge, UK: Open Book Publishers.
- Schlosberg, D., 2013. Theorising environmental justice: The expanding sphere of a discourse. *Environmental Politics*, 22(1), pp. 37–55.

References used in this chapter

Alkire, S. 2002. *Valuing freedoms: Sen's capability approach and poverty reduction.* Oxford, UK: Oxford University Press.

Alkire, S. and Deneulin, S., 2009. Introducing the human development and capability approach. In *An introduction to the human development and capability approach* (pp. 22–48). London, UK: Earthscan.

Assessment, M.E., 2003. *Ecosystems and human well-being* (Vol. 5). Washington, DC: Island Press.

Ballet, J., Marchand, L., Pelenc, J. and Vos, R., 2018. Capabilities, identity, aspirations and ecosystem services: An integrated framework. *Ecological Economics*, 147, pp. 21–28.

Brincat, S., 2015. Global climate change justice: From Rawls' law of peoples to Honneth's conditions of freedom. *Environmental Ethics*, 37(3), pp. 277–305.

Crabtree, A., 2012. A legitimate freedom approach to sustainability: Sen, Scanlon and the inadequacy of the human development index. *The International Journal of Social Quality*, 2(1), pp. 24–40.

Crocker, D. and Robeyns, I., 2010. Capability and agency. In *Amartya Sen: A biography* (pp. 60–90). Cambridge, UK: Cambridge University Press.

Duraiappah, A., 2004. *Exploring the links: Human well-being, poverty and ecosystem services (United Nations Environment Program 2004).* Winnipeg, Manitoba: Published by International Institute for Sustainable Development (IISD) for the UNEP.

Figueroa, R.M., 2015. Evaluating environmental justice claims. In *Forging environmentalism: Justice, livelihood, and contested environments* (pp. 360–376). New York: Routledge.

Heyward, C., 2011. Climate justice and the capabilities approach. *Maitreyee: The E-Bulletin of the Human Development and Capability Association*, 11(18), pp. 9–11.

Holland, B., 2008a. Justice and the environment in Nussbaum's "capabilities approach" why sustainable ecological capacity is a meta-capability. *Political Research Quarterly*, 61(2), pp. 319–332.

Holland, B., 2008b. Ecology and the limits of justice: Establishing capability ceilings in Nussbaum's capabilities approach. *Journal of Human Development*, 9(3), pp. 401–425.

Holland, B., 2014. *Allocating the earth: A distributional approach to protecting capabilities in environmental law and policy.* Oxford, UK: Oxford University Press.

Holland, B., 2017. Procedural justice in local climate adaptation: Political capabilities and transformational change. *Environmental Politics*, 26(3), pp. 391–412.

Holland, B., Forthcoming. Capabilities, future generations, and climate justice. In *Oxford handbook of intergenerational ethics*. Oxford, UK: Oxford University Press.

Hunold, C. and Young, I.M., 1998. Justice, democracy, and hazardous siting. *Political Studies*, 46(1), pp. 82–95.

Nussbaum, M. C. 1990. Aristotelian social democracy. In *Liberalism and the good* (pp. 203–252). New York: Routledge.

Nussbaum, M., 1995. Aristotle on human nature and the foundations of ethics. In *World, mind, and ethics: Essays on the ethical philosophy of Bernard Williams* (pp. 86–131). Cambridge, UK: Cambridge University Press.

Nussbaum, M., 1998. The good as discipline, the good as freedom. In *Ethics of consumption: The good life, justice, and global stewardship* (pp. 312–341). Lanham, MD: Rowman & Littlefield Publishers.

Nussbaum, M., 2000. *Women and human development: The capabilities approach*. Cambridge, UK: Cambridge University Press.

Nussbaum, M., 2001. Disabled lives: Who cares? *New York Review of Books*, 48, pp. 34–37.

Nussbaum, M., 2006. *Frontiers of justice: Disability, nationality, species membership*. Cambridge, MA: Harvard University Press.

Peeters, W., Dirix, J., and Sterckx, S., 2013. Putting sustainability into sustainable human development. *Journal of Human Development and Capabilities*, 14(1), pp. 58–76.

Pelenc, J. and Ballet, J., 2015. Strong sustainability, critical natural capital and the capability approach. *Ecological Economics*, 112, pp. 36–44.

Pelenc, J., Bazile, D. and Ceruti, C., 2015. Collective capability and collective agency for sustainability: A case study. *Ecological Economics*, 118, pp. 226–239.

Polishchuk, Y. and Rauschmayer, F., 2012. Beyond "benefits"? Looking at ecosystem services through the capability approach. *Ecological Economics*, 81, pp. 103–111.

Schlosberg, D., 2007. *Defining environmental justice: Theories, movements, and nature*. Oxford, UK: Oxford University Press.

Schlosberg, D., 2012a. Justice, ecological integrity, and climate change. In *Ethical adaptation to climate change: Human virtues of the future* (pp. 165–183).

Schlosberg, D., 2012b. Climate justice and capabilities: A framework for adaptation policy. *Ethics & International Affairs*, 26(4), pp. 445–461.

Schlosberg, D., 2013. Theorising environmental justice: The expanding sphere of a discourse. *Environmental Politics*, 22(1), pp. 37–55.

Schlosberg, D. and Carruthers, D., 2010. Indigenous struggles, environmental justice, and community capabilities. *Global Environmental Politics*, 10(4), pp. 12–35.

Scholtes, F., 2010. Whose sustainability? Environmental domination and Sen's capability approach. *Oxford Development Studies*, 38(3), pp. 289–307.

Schultz, E., Christen, M., Voget-Kleschin, L. and Burger, P., 2013. A sustainability-fitting interpretation of the capability approach: Integrating the natural dimension by employing feedback loops. *Journal of Human Development and Capabilities*, 14(1), pp. 115–133.

Sen, A., 1982. Equality of what? In *Choice, welfare and measurement* (pp. 353–369). Cambridge, MA: Harvard University Press.

Sen, A.K., 1992. *Inequality reexamined*. Cambridge, MA: Harvard University Press.

Sen, A.K., 1999. *Development as freedom*. New York: Alfred A. Knopf.

Sen, A.K., 2005. Human rights and capabilities. *Journal of Human Development*, 6(2), pp. 151–166.

Shrader-Frechette, K., 2002. *Environmental justice: Creating equality, reclaiming democracy*. New York: Oxford University Press.

Watene, K., 2016. Valuing nature: Māori philosophy and the capability approach. *Oxford Development Studies*, 44(3), pp. 287–296.

Whyte, K., 2018. The recognition paradigm of environmental injustice. In *The Routledge handbook of environmental justice* (pp. 113–123). New York: Routledge.

Winter, C.J., 2019. Decolonising dignity for inclusive democracy. *Environmental Values*, 28(1), pp. 9–30.

Young, I.M., 1990. *Justice and the politics of difference*. Princeton, NJ: Princeton University Press.

Young, I.M., 2002. *Inclusion and democracy*. Oxford, UK: Oxford University Press.

7 Latin American decolonial environmental justice

Iokiñe Rodriguez

Learning outcomes

- Learn about decolonial environmental justice
- Understand the roots of decolonial environmental justice thinking
- Identify the main propositions of a decolonial environmental justice approach

---

Introduction

In Latin America, in contrast to other parts of the world, environmental justice thinking has largely developed alongside decolonial thought, which explains social and environmental injustices as arising from modernity and the ongoing expansion of European cultural values and world views. The **decolonization** of knowledge and social relations is highlighted as one of the key challenges to overcoming the history of violent oppression and marginalization in development and conservation practice in the region. Arturo Escobar (1998, 2010a, 2010b, 2010c) and Enrique Leff (2001, 2003, 2004) have been pioneers in positioning an environmental justice theory in the region with a "**decolonial turn**"[1] (Castro-Gómez and Grosfoguel, 2007). Some more recent additions to this body of knowledge from an environmental justice perspective include Alberto Acosta (Acosta et al., 2011; Acosta, 2013, 2015), Eduardo Gudynas (2010a, 2010b, 2011, 2012) and Boaventura de Sousa Santos (2008), who, although not strictly Latin American, works in close collaboration with Latin American decolonial scholars.

This strong focus on decolonization marks a difference from environmental justice thinking in the Global North, which in comparison has tended not to place as much emphasis on the colonial and epistemic roots of injustices in the Global South (Álvarez and Coolsaet, 2018, Rodríguez and Inturias, 2018). This is partly due to the fact that in its early years, environmental justice theory from the Global North had a strong focus on injustices arising from the unequal distribution of natural resource hazards in the advanced capitalist political economy. The work of Robert Bullard (1983) in particular, examining the relationship between racial discrimination and industrial and waste-dumping activities in the US, marked the start of environmental justice research in the Global North with a strong distributive focus.[2] Thanks to the work of Pellow (2007) and Schlosberg (2007, 2013), among others, over the years, environmental justice research from the Global North has been moving away from its initial focus on **distributive justice**. Schlosberg in particular, drawing upon detailed studies of environmental justice movements across different classes, races, ethnicities and gender, as well as on Fraser's (1998) three-dimensional definition of justice,[3] has made a tremendous contribution to a more pluralist understanding of the meaning of environmental justice. He has consistently debated that justice is not just about equity (distributive justice), but also includes recognition (cultural justice) and participation (procedural justice). More recently, drawing upon the work of Martha Nussbaum (2011), he added a fourth dimension to this discussion, arguing that justice is also about fulfilling community **capabilities**,[4] particularly in the context of Indigenous peoples' struggles (Schlosberg and Carruthers, 2010). Another important development in environmental justice thinking from the Global North is the increasing attention paid to the regional specificities of the histories of environmental justice movements and its scholarship across the world (Lawhon, 2013). Carruthers' (2008a) edited volume, *Environmental Justice in Latin America*, is a case in point, as the contributors focus their attention upon analyzing precisely what makes Latin American environmental justice frameworks distinct from those in other parts of the world. The authors highlight the long tradition for social justice demands and movements in different fields (political participation, land distribution, human rights and community health) as a salient feature, as well as the central importance that cultural identity (linked to *campesino*, Indigenous and women's demands for justice) has historically played in environmental justice struggles in the region.[5]

Although environmental justice literature from the Global North increasingly acknowledges the historical legacy of colonialism in environmental justice struggles in Latin America, particularly in reference to land use and distribution patterns (Carruthers, 2008b), it rarely mentions the persistence of colonial values (**coloniality**) as a cause of current injustices and violence, and the need to confront it. This is precisely what Latin American environmental justice thinking offers to environmental justice deliberations from other parts of the world, through its focus on decoloniality.

This chapter offers an insight into this perspective, focusing on two themes: i) the roots of Latin American decolonial environmental justice thinking; and ii) its main propositions.

Roots of the decolonial environmental justice perspective: a bottom-up call for the transformation of modernity

> We see Ralco as the symbolic expression of the threat of modernity and progress to the indigenous people of Chile . . . we don't want your progress to rub out our culture.
>
> Domingo Namuncura, Mapuche leader from Chile, expressing his views in 1999 about a dam built in the Bio Bio River, quoted by David Schlosberg and David Carruthers (2010)

Neocolonialism: the perpetuation of a historical process of dispossession

The Latin American decolonial environmental justice perspective owes much of its roots to Indigenous peoples' movements and their struggles against **modernity**[6] in the region (Escobar, 2017; Leff, 2017). They, along with other marginalized groups such as Afro-descent communities and women, have had a protagonist role in contemporary environmental justice struggles. This is not fortuitous. In this, the new direction that capitalism has taken upon entering a new economic phase, coined by the British geographer David Harvey (2004) as **accumulation by dispossession**, has been decisive. As pointed out by Harvey, a significant percentage of the world's capital is currently being used to deprive people of their natural wealth (waters, forests, minerals, fauna) and their ancestral knowledge—associated with the use of the commons—as part of processes of globalization and the commodification and privatization of land and natural resources. We are in the presence, according to Harvey, of a new colonialism, more rapacious than the one suffered by the Indigenous peoples of Latin America between the 15th and 18th centuries. Indigenous people know this, and that is why they are trying to free themselves from it (Escobar, 2010b; Leff, 2001).

Much to their disadvantage, Indigenous peoples in Latin America live in very resource-rich areas. As José Seane (2006, p. 86) highlighted in his text *Social Movements and Natural Resources in Latin America: Resistance to Neoliberalism, Configuration of Alternatives*:

> the peoples of Latin America and the Caribbean live in a territory in which 25% of forests and 40% of the biodiversity of the globe are found. Nearly a third of the world's reserves of copper, bauxite and silver are part of their wealth, and they are the keepers of 27% coal, 24% oil, 8% gas and 5% of uranium. Their watersheds contain 35% of the world's hydro-energy power making—from the Chiapas jungle to the Amazon—one of the largest reserves of biodiversity on the planet.

Since the late 1990s, the pressure for the use and exploitation of these and other natural resources located within Indigenous territories has been steadily increasing. This trend, which began with the implementation of neoliberal policies in most of the countries in the region, has continued to date, even under progressive governments like those of Ecuador, Argentina, Bolivia and Venezuela (Escobar, 2010d).

As resistance to this, during the last three decades there has been a sustained increase in social protest and socio-environmental conflict in the region. The Environmental Justice Atlas (www.ejatlas.org) shows that 30% of the socio-environmental conflicts registered worldwide occur in Latin America, with a large proportion of them taking place in Indigenous peoples' territories.

The most widespread conflicts are those associated with mining activities (metallic and non-metallic), oil and gas. High-income and upper-middle-income countries, such as Chile and Brazil, also experience a high percentage of water conflicts, which are linked not only to mining activities and their impact on water, but also to the building of hydroelectric dams. In addition, there are conflicts of regional scope associated with the construction of megaprojects for transport infrastructure, such as interconnected road networks and railway systems or regional hydroelectric production. Currently in the Amazon, there is strong pressure on the population from this type of project, which is home to a high proportion of Indigenous people.

Resisting and mobilizing against development as we know it

Many of the socio-environmental conflicts have turned into social and regional movements for greater environmental justice and for the defence of the commons and the environment. Throughout the region, there have been outbreaks against mining in defence of biodiversity and confrontations over the expansion of agro-business in defence of alternative forms of agricultural production and for the protection of forests. Local resistance to the privatization of water services in defence of water reserves and against the construction of dams are also common, as well as confrontations against oil exploitation in defence of the land and territories.

Some emblematic cases of the protests and social mobilizations that have arisen in the region in the last two decades as a reaction against extractive activities or development projects include:

- The resistance of the Embera-Katio of Colombia since 1993 against the construction of the Urrá hydroelectric plant on the Sinú river. Although the local people could not stop the construction, on April 23, 2000, they obtained the title of new territories and the suspension of Urrá II.
- The mobilization of the Pemon in Venezuela between 1997 and 2000 against the construction of high-voltage power lines on their lands to export electricity to Brazil (including an important part of the Canaima National Park and UNESCO's Natural World Heritage Site).
- The social mobilization and resistance of *campesinos* and the wider community of Cochabamba in Bolivia in 2000, in the so-called "Water War" against the commodification and private appropriation of water. As a result, a new phase of protests and anti-neoliberal social mobilization led to the "War of Gas" in 2003, demanding the nationalization of hydrocarbons.
- The events of Bagua in Peru, on June 5, 2009, against a decree that would allow large transnational and mining companies to use a sizeable extent of Peruvian Amazon forest for the exploitation of significant deposits of oil, gas and other minerals. This led to violent confrontations between police and residents of the Awajun and Wampis ethnic groups in the towns of Bagua and Utcubamba and a death toll of 30 people.
- The mobilization and march of Indigenous peoples throughout Argentine territory in 2010, which was the most massive in the history of the nation-state, demanding recognition of their rights to land and territory, rejection of expansive extractive activities, respect for their ancestral culture and justice in the face of the abuses of the past and the present.
- The struggle and defence of their territory by the Indigenous people of the Isiboro Secure Indigenous Territory and National Park in 2011 in Bolivia against the construction of an international highway through half of their territory.

Box 7.1 explores two of these cases in more detail.

Box 7.1 Two important recent Latin American environmental justice mobilizations

The Pemon struggle against a high-voltage power line in Venezuela. This was one of the most emblematic and widely known socio-environmental conflicts in Venezuela between Indigenous peoples and the state during the last two decades. It took place between 1997 and 2000 in the eastern sector of Canaima National Park. For five consecutive years, the Pemon fought determinedly against the project because they saw it as a threat to their cultural and environmental integrity. One of the arguments against the project was that the power line was just the beginning of a long-term plan to open up the southern part of the Bolivar state to mining. The Pemon systematically demanded territorial land rights in their struggle and were successful in temporarily suspending construction through a variety of political actions and resistance strategies. In 1999, the Pemon played a key role in forcing a change in the national constitution to include a chapter on Indigenous rights and, hence, turning Venezuela into a pluricultural nation-state. For the first time in Venezuelan contemporary history, Indigenous peoples were granted ownership rights over their traditionally occupied ancestral lands (*Constitución de la República Bolivariana de Venezuela*, 1999, Art. 119). The constitutional reform was a vital condition for reaching an agreement in which the Pemon accepted the completion of the project, conditional, among other things, on the initiation of a process of demarcation and titling of Indigenous peoples' "habitats" within a week of signing. Although, 20 years later, the Pemon are still waiting for their territorial property rights, they did gain important visibility as political actors during that struggle and made a huge contribution to the recognition of Indigenous rights in Venezuela.

The Water War in Bolivia. In 2000, the Bolivian government attempted to sanction a new Law on Privatization of Water and Sewage in Cochabamba without local consultation. This law met with strong resistance and intense mobilization on the part of *campesinos* and the Indigenous peoples of Cochabamba, to the point that it could not be approved. The Water War is renowned for the intense political and social mobilization that it generated through the development of press and media campaigns, lobbying, lawsuits and public demonstrations demanding respect for traditional water uses and customs. All these strategies helped local organizations ensure negotiations in conditions of equity and, more importantly, to halt changes in the legislation that would open the way for the commodification and private appropriation of water in Cochabamba.

Transgressing modernity—embracing the pluriverse

The struggles for greater environmental justice of the Indigenous peoples of Latin America have given rise to a regional movement that has taken up a sustainable position against the economic rationality that drives the dominant model of development and the global project of modernity. This has involved, among other things, fighting for new culturally differentiated forms of decision-making in nation-state models to acknowledge their rights to their own forms of development, self-determination and political autonomy, the property of their territories and, most significantly, the preservation of their cultural integrity. In some cases, like Bolivia, Ecuador and Venezuela, this has to some extent been achieved through the recent shift towards plurinational states, which seek to represent the interests of all cultures in recognition of the multi-ethnic nature of the country. In countries like Bolivia, with a majority of Indigenous peoples, acknowledging the multicultural nature of the population is key to developing an inclusive, equitable and just society. The **plurinational state**, in contrast to a liberal one which considers all citizens equal, assumes that indigenous peoples deserve special treatment as citizens due to their cultural characteristics and their existence in the region prior to the foundation of the Republic. This has led to the drawing up of new national constitutions and legislation, which now confer differentiated rights upon Indigenous peoples in particular aspects of social and political life.

Indigenous movements in Latin America have also been successful in the last two decades in putting across alternative views of development in national discourses and legal frameworks. For example, in the case of Bolivia and Ecuador, where the Indigenous idea of "**Buen Vivir**" or "*Vivir Bien*" was incorporated in the new national constitutions and national development plans as a guiding concept for development. The literal translation of *Buen Vivir* in Quechua and Aymara languages is "to live in plenitude". The Guari version, "*Teko porâ*", literarily means, "a good way of being" or "a good way of life". But, as stated by the Aymara writer Fernando Huanacuni Mamani (2010) from Bolivia, none of the literary translations of the term do justice to the depth of its meaning from an Indigenous perspective. *Buen Vivir* is a holistic concept rooted in principles and values such as harmony, equilibrium and complementarity, which, from an Indigenous perspective, must guide the relationship of human beings with each other and with nature (or Mother Earth) and the cosmos (see Box 7.2 for more details of the concept).

Box 7.2 Key features of the concept of *Buen Vivir*

- It is the local development pathway or horizon of a community.
- It is a state of the mind and the spirit: to be venturous, cheerful, happy and satisfied, at peace with oneself.
- It relies on the recognition and respect of all forms of diversity (including biodiversity and human diversity; and gender, generational, cultural and professional diversity, and diverse forms of knowledge). On a wider scale, it refers to creating the conditions for plural subjects.
- It speaks of an alternative form of civilization, which prioritizes the expansion of life rather than capital.
- The overall aim is the emancipation and widening of liberties, capabilities and potentialities of all human beings.

As stated by Aymara Indigenous lawyer Fernando Huanacuni Mamani (2010, p. 11), the pressure from the Indigenous movements to incorporate the concept of *Buen Vivir* in the new legal frameworks of Bolivia and Ecuador has not merely been an issue of recognition of their views of development. It is an upfront confrontation with the paradigm of modernity itself and a call for humanity to embrace an emerging alternative Indigenous paradigm, which he calls the *culture of life*:

> Indigenous people are bringing something new (to the modern world) to the forefront about how humankind should live from now on. The global market, economic growth, corporatism, capitalism and consumerism, which are the product of a Western paradigm, are to varying degrees the root causes of the serious social, economic and political crisis. From the different Indigenous communities of Abya Yala, we say that, in fact, what we are facing is a life crisis. . . .

Given this reality, the *culture of life*, which is rooted in a communal rather than individualistic paradigm, emerges as a response/proposal. This paradigm calls for the reconstitution of the vision of community (common-unit) of ancestral cultures. This legacy of the first nations considers community as structure and unity of life, that is, consisting of all forms of existence and not only as a social structure (made only by humans). This does not imply a disappearance of individuality, but rather that the individual exists in complementarity with other beings.

In these times when modernity is immersed in the individualist paradigm and humanity is in crisis, it is important to listen and practice the heritage of our forefathers: the horizon of "Vivir Bien" or "Buen Vivir", is an emerging worldview that seeks to reconstruct the harmony and balance of life with which our ancestors coexisted.

Influenced by Huanacuni Mamani's idea of humanity adopting the *culture of life* paradigm, other authors have started using a more encompassing Indigenous concept as a guide to break away from modernity: the concept of the **pluriverse**. From an Indigenous perspective, the pluriverse implies the existence of many worlds somehow interconnected. In other words, the human world is connected to the natural world and also to the spiritual world (Querejazu, 2016). On a wider scale, the pluriverse has been conceptualized as a rainbow of cosmologies, knowledges, and vital worlds (Escobar, 2018). In this context, it refers to multiple ways of being and knowing that have co-evolved in relations of power and difference and continue to do so (Paulson, 2018). Thus, the pluriverse suggests that there is not one single "universal" valid way of being and knowing in the world, as suggested by modernity, but multiple ones, which are indispensable for the long-term sustainability of people and the planet (Santos, 2014).

Nevertheless, despite the success of Indigenous people positioning a discourse of cultural difference, plurality and environmental rationality in the region, the project of modernity is a pervasive one. In Latin America, modern nation-state building has historically been premised on narratives of national identity and modernity that have sought to "assimilate" Indigenous peoples into the wider society rather than acknowledging their "**difference**". This trend has continued even within emerging pluri-cultural nation-state models, such as those currently favoured in Venezuela, Bolivia and Ecuador, where, at least nominally, and to different degrees, foundational legal frameworks such as national constitutions now acknowledge and recognize Indigenous peoples' rights. Nevertheless, while in countries such as Bolivia and Ecuador, the incorporation of alternative concepts of development, such as *Buen Vivir*, into new legal frameworks has had an important role in redefining the social dimensions of the state, it has had little impact on redefining its economic spheres (Moreano et al., 2017). The economic rationality of development and the imperative for economic growth has remained as intact as it is in the rest of the region. Thus, environmental conflicts continue to rise and only a very small percentage of them (18%, according to the *Environmental Justice Atlas*, www.ejatlas.org) show some signs of achieving greater environmental justice as a result of the social and political mobilizations against development or extractive activities.

The consolidation of an environmental justice in the region requires understanding the sophisticated ways in which hegemonic power works in the project of modernity and how this can be confronted in practice. This is what the decolonial environmental justice perspective offers.

Main propositions of decolonial environmental justice theory

Indigenous peoples' contemporary struggles for social and environmental justice have laid important empirical and conceptual foundations for a relatively new Latin American body of critical thinking known as the Decolonial Project (Lander, 2000; Quijano, 2000; Escobar, 2003; Walsh, 2007; Mignolo, 2008). Decolonial thought is distinct from other post-colonial critical theory through its focus on the Global South and for identifying mechanisms of subordination and marginalization in the project of modernity and the continual reproduction of European cultural values. Proponents of this school of thought are largely from Latin America (Quijano, 2000; Lander, 2000; Leff, 2001; Castro-Gómez and Grosfoguel, 2007; Escobar, 2003; Walsh, 2007; Mignolo, 2008), but important contributions have also come from India (Visvanathan, 1997), Portugal (Santos et al., 2008; Santos, 2010) and New Zealand (Smith, 1999), among others.

Building on Indigenous peoples' own anti-modernity agenda, Latin American environmental justice thinkers have developed a series of core ideas that are central to the way environmental justice is currently approached and perceived in region.

Justice beyond recognition: the need for the construction of "otherness"

According to decolonial theory, "colonialism" ended with political independence in the Global South, but coloniality persists through dominant Eurocentric colonial/modern values and worldviews that are institutionalized and disseminated through education, the media, state-sanctioned languages and behavioural norms. Thus, coloniality is a form of power that creates structural oppression over marginalized sectors of society, such as Indigenous peoples, whose alternative worldviews become devalued, side-lined and stigmatized in development and environmental management practice. From this perspective, coloniality is a particular mechanism and form of mis-recognition that must be confronted in order to achieve emancipation and social/environmental justice.[7]

Decolonial scholars argue that modernity leads to profound psychological harm for Indigenous peoples and other subaltern sectors of society because it erodes vital conditions for their well-being, including their cultural identity, freedom of choice and self-respect. It also has tangible impacts on the status and participation of Indigenous peoples in development and environmental management by disregarding local notions of authority and territory, frequently resulting in displacement or enforced changes to livelihoods. Structural oppression is perpetuated through a matrix operating at three levels: i) power (political and economic); i) knowledge (epistemic, philosophical and scientific); and iii) the self or ways of being (subjective, individual and collective identities).

The **coloniality of power** is exercised through two primary mechanisms: first is the codification of racial difference between Europeans and non-Europeans aimed at making the latter appear naturally inferior. This finds expression in normative rules such as definitions of development/progress. The second is the use of Western/modern institutional forms of power (like the nation-state) in non-Western societies to organize and control labour, its resources and products (Quijano, 2000). Hence, although coloniality continues to be intrinsically linked to global capitalism, it cannot be reduced to economics, as it also involves other invisible cultural mechanisms of domination.

The cultural and normative dimension of the coloniality of power also finds expression in the **coloniality of knowledge** (explained in detail in Table 7.1), through the dominance of European knowledge and symbolic systems over non-European ones. Furthermore, the coloniality of power and knowledge impacts on the individual through the **coloniality of the being**, via mechanisms of subjectivation on the life, body and mind of the "colonized" or marginalized people, to the point of stripping them of their very essence and soul. Table 7.1 illustrates how different forms of coloniality, hidden behind development and conservation projects, commonly generate different forms of environmental injustice.

Thus, responses to coloniality necessarily involve decolonizing power, knowledge and the being.[8] This involves moving away from unitary models of citizenship and civilization to one that respects different local economies, politics, cultures, epistemologies and forms of knowledge, while also forging new categories of thought, constructing new subjectivities and creating new modes of being and becoming that can lead to emancipation.

This focus on the decolonization of power, knowledge and the being marks an important divergence in environmental thinking from that of the Global North. From a Latin American decolonial perspective, environmental justice entails developing a politics of difference that is not simply based on the search for recognition or inclusion in dominant structures, such as the liberal nation-state or global economic systems, but focused rather on the construction of "**otherness**":

> an "other" process of knowledge construction, an "other" political practice, an "other" social (and State) power and an "other" society; an "other" way to think and act in relation to, and against, modernity and colonialism.
>
> (Walsh, 2007, p. 57)

Bolivia and Ecuador serve as good examples of the construction of such "otherness", through their recent shift towards becoming plurinational nation-states, their acknowledgement of differentiated cultural rights for Indigenous peoples and the institutionalization of

Table 7.1 Environmental injustices generated by different forms of coloniality

	Forms of coloniality		
Example of modernizing policy or project	Coloniality of power (political/economic)	Coloniality of knowledge	Coloniality of the being
Environmental conservation policies that prohibit certain local land use and environmental practices (e.g. fire use, slash and burn agriculture, hunting, etc.)	Denial and de-legitimization of local authorities and environmental governance systems	Denial, de-legitimization, erasing and making invisible local knowledge; cosmogonies, worldviews of nature, fire use, agriculture, hunting practices, etc.	Stigmatization of the other (local communities), weakening of self-esteem in younger generations (ethnic embarrassment) and emergence of intergenerational conflicts over values of nature
Mining projects, hydroelectric dams	Invalidation of local decision-making structures and alternative local economies, displacement, marginalization, social exclusion and environmental degradation	Erasing and making invisible local categories of the land, the territory and alternative views of the future	Erasing of local histories, identities and cultural diversity

alternative concepts of development such as *Buen Vivir* in their national constitutions. In both cases, such changes represented an important decolonial turn and a moment of epistemic rupture with modernity, which was greatly inspired and influenced by Indigenous peoples' life projects. However, as mentioned previously, they also serve as good examples of the forces at play in the project of modernity that resist the decolonial turn. The decolonizing of power cannot be achieved solely by producing changes in the political or social spheres, while maintaining the dominant economic rationality. Ultimately, a shift in values systems (knowledge) and ways of being is needed.

There is no global justice without cognitive justice

> Los Pelambres Mining Company polluted [our territory] and erased a long history of the people of Caimanes. Let's unite to start a new history for our children, for us, the family and for the future of Caimanes. Let's fight Minera los Pelambres.
>
> *Protest banner in the Los Caimanes Village, Pelambres, 2010, Chile*

A significant contribution of the decolonial environmental justice perspective is its focus on the epistemological dimension of oppression and domination. It highlights the need to engage with the invisible and extremely subtle ways in which violence is meted out in environmental justice struggles: through the imposition of particular ways of knowing the world at the expense of oppressing others; in other words, through **epistemic violence**.

As Latin American environmental justice thinkers argue, the battle of Indigenous peoples and other socio-environmental movements in Latin America is not for the re-distribution of harms and benefits in the use of the environment, as stressed in environmental justice movements in other parts of the world.[9] Rather, their struggle is for the right to live well, in accordance with their own identities, cultural imaginings and ways of knowing the world (Leff, 2017). Therefore, as suggested by Walsh (2005b) and Santos (2008), the biggest challenge for emancipation from a decolonial perspective is to move towards a situation of greater **cognitive justice** in the world, learning from, and making visible, alternative forms of knowledge.

The first author to use the term cognitive justice was Indian writer Shiv Visvanathan, in his 1997 text *Carnival of Science*, to legitimize the right of different knowledge systems to exist, but more importantly to suggest a way out of the situation of epistemic supremacy in the dominant model of knowledge production. Since then, the term has been widely adopted by decolonial scholars as one of the key pillars of emancipation in the modern world (Santos, 2008; Escobar, 2017). According to Visvanathan (2009), cognitive justice:

> demands recognition of knowledges, not only as methods but as ways of life. This presupposes that knowledge is embedded in an ecology of knowledges, where each knowledge has its place, its claim to a cosmology, its sense as a form of life. In this sense knowledge is not something to be abstracted from a culture as a life form; it is connected to a livelihood, a life cycle, a lifestyle; it determines life chances.

According to this perspective, a greater recognition of alternative knowledges in development requires changing the conditions of dialogue between knowledge systems to achieve a situation in which traditionally excluded actors, such as Indigenous peoples, do not have to fit in with the structures and standards of Western knowledge or worldviews. Far from it: research and development must be able to respond to the social, cultural, political, economic and environmental imperatives of the agendas of local and Indigenous peoples (Smith, 2012). This means that cognitive justice has to be part of a wider process of decolonization of knowledge that must start at the universities, critically questioning the "how?", and "what for?" of knowledge production (Mignolo, 2008; Santos, 2008).

The academic-activist nexus

As can be inferred from the preceding, from a decolonial perspective, academia has an important role to play in the making of the intercultural dialogues needed for emancipation and environmental justice. In fact, one of the distinctive features of Latin American environmental justice thinkers has been their commitment to understanding reality in order to transform it. In contrast to many environmental justice thinkers from the Global North, who largely engage with environmental conflicts and injustices as objects of study, Latin American environmental justice scholars have been conspicuous for taking a positive stand and active role against environmental injustices. They do so by unpacking the dominant rationality of modernization, by entering into dialogue with local movements exploring their discursive techniques and strategies of struggle and, most significantly, by using research as a vehicle to transform power asymmetries in the dominant paradigms of knowledge production and development. (Escobar, 2008; Gudynas and Acosta, 2011; Alimonda, 2011). Thus, behind decolonial environmental justice thinking, there is a political intention as much as an academic one. The long tradition of participatory action research in Latin America has been an important source of inspiration and influence in this trend (Fals Borda, 1986; Fals Borda and Brandao, 1986).

Another important aspect of the positionality of Latin American decolonial environmental justice theory is the growing acceptance of the theoretic production that takes place outside academia, specifically in activist circles and as a result of the interaction between academics and activists. Concepts such as *Buen Vivir*, which have been incorporated into the decolonial agenda and discourse, are an expression of this academic–activist interface, as is the theoretical and historical commitment of environmental justice academics to the construction of sustainable futures and other possible "worlds" (Leff, 2017).

The intercultural challenge

> We want to teach government officials who we are: our history, our ways of life, our forms of government, our rules and norms for governing nature.
> Statement collected during a workshop on socio-environmental conflict with indigenous leaders in Venezuela, October 2014 (Mirabal, 2015)

Thus, decolonial thinkers propose the ecology of knowledge (Santos, 2008), also termed dialogues of knowledge/wisdoms (Leff, 2004) or the construction of interculturality (Walsh, 2005a), as the core of a decolonial praxis.

But **interculturality** here is radically different from other more widely used functional definitions. Decolonial thinkers approach interculturality from a *critical perspective* (Tubino, 2005; Walsh, 2005a, 2005b, 2007; Santos, 2010). The term "interculturality" should not be understood as a simple contact, but as an exchange that takes place in conditions of equality, mutual legitimacy, equity and symmetry. This encounter of cultures is a permanent and dynamic vehicle for communication and mutual learning. It is not just an exchange between individuals, but also a meeting of knowledge, wisdoms and practices that develop a new sense of co-existence in their difference.

As suggested by Viaña (2009), in order to achieve this, it is necessary to change the conditions of intercultural dialogue, to ensure that the conversation is not about the right of inclusion in the dominant culture, but about the historical and structural factors that limit a real exchange between cultures in each country. Only this can help create the conditions for more symmetrical conversations about the model of development needed for *Buen Vivir*, the type of solidary economy needed for life and the participatory political system needed for the consolidation of autonomies, territories and regions that seek different forms of government and self-governance.

Thus, the "inter" space becomes an arena of negotiation where social, economic and political inequalities are not kept hidden, but are made visible and confronted. As Catherine Walsh (2005b) suggests, this entails creating new knowledge in a way that confronts existing relations of domination in hegemonic paradigms and strengthening what the people themselves understand and reconstruct as "theirs", in relation to identities, differences and knowledge. This emphasis on reconstructing, recovering and revaluing local knowledge is key to achieving cognitive justice in development practice. Local knowledge has not only been made invisible by dominant values and institutional arrangements that determine what is and is not valid knowledge in development practice, but also through structural forces linked to modernization that have had a role in erasing local identities, culture, histories and worldviews (Pilgrim and Pretty, 2013).

Therefore, a starting point for such intercultural practice is to develop a politics of knowledge that helps strengthen Indigenous peoples' own initiatives and agendas of cultural revitalization and knowledge production. Examples of such initiatives are Indigenous Universities and Life Plans, which are well underway with different degrees of success in several Latin American countries (Ancianos del Pueblo Fééneminaa, 2017; Cabildo de Guambia, 1994; Walsh, 2005a; Jansasoy and Perez-Vera, 2006; COINPA, 2008; Espinosa, 2014).

Follow-up questions

- How would decolonial environmental justice be applicable to other regions of the world beyond Latin America and Indigenous peoples?
- Can you think of ways in which you can contribute to putting into practice a decolonial environmental justice approach in your research and daily life?

Notes

1 A "decolonial turn" refers to the task of decolonising or freeing oneself and society from the legacy of colonialism and *coloniality* in its different forms and manifestations. This will be explained in detail in the second section of this chapter.
2 On distributive justice, see Chapter 3 of this volume.
3 On Nancy Fraser's justice theory, see Chapter 5 of this volume.
4 On the capabilities approach, see Chapter 6 of this volume.

5 On farmers, women and Indigenous peoples' role in environmental justice movements, see also Chapters 13, 18 and 20, respectively.
6 Modernity is a particular way of seeing the world that in the last centuries determined the division between nature and society, a colonial distinction between modern and non-modern Indigenous peoples, the myth of progress as a unidirectional linear path, and a strong confidence on Cartesian science as the valid form of knowledge (Gudynas, 2011).
7 On (mis)recognition, see also Chapter 5 of this volume.
8 Broadly speaking, "the being" is defined as the soul and essence of a person.
9 On distributive justice, see Chapter 3 of this volume.

References for further reading

Gudynas, E. 2011. Buen Vivir: Today's tomorrow. *Development* 54(4): 441–447, Society for International Development 1011–6370/11 www.sidint.net/development.
Quijano, A. 2000. Coloniality of power and eurocentrism in Latin America. *International Sociology* 15: 215–232.
Rodríguez, I. & Inturias, M. 2018. Conflict transformation in indigenous peoples' territories: doing environmental justice with a 'decolonial turn'. *Development Studies Research* 5(1): 90–105, DOI: 10.1080/21665095.2018.1486220
Santos, B. de Sousa. 2008. *Another knowledge is possible: Beyond northern epistemologies*. London: Verso.
Smith, L. 1999. *Decolonizing methodologies: Research and indigenous peoples*. London: Zed Books.

References used in this chapter

Acosta, A. 2013. *El Buen Vivir. Sumak Kawsay, una oportunidad para imaginar otros mundo*. Barcelona: Icaria editorial.
Acosta, A. 2015. Extractivismos y subdesarrollo. La maldición de la abundancia. *Rebelión. 04–11–2015. Disponible en línea.* Available at: www.rebelion.org/noticia.php?id=205247
Acosta, A., Gudynas, E., Houtart, F., Ramírez Soler, H., Martínez Alier, J. & Macas, L. 2011. *Colonialismos del siglo XXI. Negocios extractivos y defensa del territorio en América*. Barcelona: Icaria Editorial.
Alimonda, H. 2011. *La colonialidad de la naturaleza. Una aproximación desde la Ecología Política Latinoamericana*. En: Alimonda Hector (Ed) *La naturaleza colonizada. Minería y Ecología Política en América Latina*. Buenos Aires: CLASO.
Álvarez, L. & Coolsaet, B. 2018. Decolonizing environmental justice studies: Perspective. *Capitalism Nature Socialism*, DOI: 10.1080/10455752.2018.1558272
Ancianos del Pueblo Fééneminaa. 2017. *Ancianos que caminan y cuentan historias*. Colombia: Consejo Regional Indígena del Medio Amazonas (CRIMA) y Forest Peoples Programme (FPP).
Bullard, R.D. (Ed). 1983. *Confronting environmental racism: Voices from the grassroots*. Boston: South End Press.
Cabildo de Guambia. 1994. *Plan de Vida del Pueblo Guambiano*. Colombia: Popayan.
Carruthers, D. (Ed) 2008a. *Environmental justice in Latin America: Problems, promise and practice*. Cambridge, MA: The MIT Press, 329 p.
Carruthers, D. 2008b. Introduction: Popular environmentalism and social justice in Latin America. In Carruthers, D. (Ed). *Environmental justice in Latin America: Problems, promise and practice*. Cambridge, MA: The MIT Press, 329 p.
Castro-Gómez, S. & Grosfoguel, R. (Eds). 2007. *El giro decolonial Reflexiones para una diversidad epistémica más allá del capitalismo global*. Bogotá: Siglo del Hombre Editores; Universidad Central, Instituto de Estudios Sociales Contemporáneos y Pontificia Universidad Javeriana, Instituto Pensar, 308 p.
COINPA. 2008. *Plan de Vida Pueblos Huitoto e Inga*. Documento de avance. Colombia, Abril: Consejo Indígena de Puerto Alegría (COINPA).

Escobar, A. 1998. Whose knowledge, whose nature? Biodiversity, conservation, and the political ecology of social movements. *Journal of Political Ecology* 5: 53 82.

Escobar, A. 2003. "Mundos y conocimientos de otro modo" El programa de investigación de modernidad/colonialidad Latinoamericano. *Tabula Rasa. Bogotá—Colombia* 1: 51–86, enero-diciembre de 2003.

Escobar, A. 2008. *Territories of difference, Place, movements, life, networks*. Durham and London: Duke University Press.

Escobar, A. 2010a. *América Latina en una encrucijada ¿Modernizaciones alternativas, postliberalismo o postdesarrollo?* En Breton Víctor (Ed) *Saturno Devora a sus hijos. Mirada crítica sobre el desarrollo y sus promesas*. Barcelona: Icaria Editorial.

Escobar, A. 2010b. Una ecología de la diferencia: igualdad y conflicto en un mundo glocalizado. En: Escobar Arturo (Ed) *Más allá del Tercer Mundo. Globalización y Diferencia*. Bogotá: Instituto Colombiano de Antropología e Historia.

Escobar, A. 2010c. Epistemologías de la naturaleza y colonialidad de la naturaleza. Variedades de realismo y constructivismo. En *Revista Cultura y Naturaleza* Bogotá: Jardín Botánico de Bogotá José Celestino Mutis, pp. 49–71.

Escobar, A. 2010d. América Latina en una encrucijada ¿modernizaciones alternativas, postliberalismo o postdesarrollo? En Breton, Victor (Ed) *Saturno Devora a sus hijos. Mirada crítica sobre el desarrollo y sus promesas*. Barcelona: Icaria Editorial.

Escobar, A. 2017. Desde abajo, por la izquierda y con la tierra: la diferencia de Abya Yala/Afro/Latino/America. En Alimonda, H., Toro-Perez, C. & Martin, F. (Cords). *Ecologia Política Latinoamericana. Pensamiento crítico, diferencia latinoamericana y rearticulación epistémica*. 1ra edición, Ciudad Autónoma de Buenos Aires: CLACSO; México: Universidad Autónoma de Buenos Aires.

Escobar, A. 2018. *Designs for the pluriverse: Radical interdependence, autonomy, and the making of worlds.* Durham, NC: Duke University Press.

Espinosa, O. 2014. Los planes de vida y la política indígena en la Amazonía peruana. *Anthropologica* [online] 32(32): 87–114.

Fals Borda, O. 1986. *La investigación-acción participativa: Política y epistemología*. En Camacho G. Álvaro (Ed) *La Colombia de hoy*. Bogotá: Cerec, pp. 21–38.

Fals Borda, O. & Brandao, C.1986. *Investigación participativa*. Montevideo: Instituto del Hombre.

Fraser, N. (1998). *Social justice in the age of identity politics: Redistribution, recognition, participation*. Discussion paper, Wissenschaftszentrum Berlin für Sozialforschung, Forschungsschwerpunkt Arbeitsmarkt und Beschäftigung, Abteilung Organisation und Beschäftigung, No. FS I 98-108, http://hdl.handle.net/10419/44061.

Gudynas, E. 2010a. La senda biocéntrica: valores intrínsecos, derechos de la naturaleza y justicia ecológica. *Tabula Rasa. Bogotá—Colombia* 13: 45–71, julio-diciembre 2010

Gudynas, E. 2010b. Tensiones, contradicciones y oportunidades de la dimensión ambiental del Buen Vivir. En Farah, I.H. y Vasapollo, L. (cords). *Vivir bien: Paradigma no capitalista?* La Paz: CIDES-UMSA y Plural.

Gudynas, E. 2011a. Buen Vivir: Today's tomorrow. *Development* 54(4): 441–447. Society for International Development 1011–6370/11 www.sidint.net/development

Gudynas, E. 2011b. Desarrollo y sustentabilidad ambiental: diversidad de posturas, tensiones persistentes. En Matarán, A. y López, F. (Eds). *La Tierra no es muda: diálogos entre el desarrollo sostenible y el postdesarrollo*. Granada: Universidad de Granada, pp. 69–96.

Gudynas, E. 2012. Estado compensador y nuevos extractivismos. Las ambivalencias del progresismo sudamericano. *Nueva Sociedad* 237.

Gudynas, E. & Acosta, A. 2011. La renovacion de la crítica al desarrollo y el buen vivir como alternative. *Utopia y Praxis Latonoamericana* 16(53): 71–83.

Harvey, D. 2004. The 'new' imperialism: accumulation by dispossession. *Socialist Register* 40: 63–87.

Huanacuni Mamani, F. 2010. *Vivir Bien/Buen Vivir*. La Paz: Convenio Andrés Bello, Instituto Internacional de Investigación y CAOI.

Jansasoy, J. & Perez-Vera, A. 2006. *Plan de Vida: Propuesta para la supervivencia Cultural, Territorial y Ambiental de los Pueblos Indigenas*. The World Bank Environment Department, 30 p.

Lander, E. (Ed). 2000. *La colonialidad del saber: eurocentrismo y ciencias sociales*. Buenos Aires: CLACSO.

Lawhon, M. 2013. Situated, networked environmentalism: A case for environmental theory from the South. *Geography Compass* 7(2): 128–138.

Leff, E. (Ed). 2001. *Justicia ambiental: Construcción y defensa de los nuevos derechos ambientales culturales y colectivos en América Latina*. México: UNEP.

Leff, E. 2003. *Latin American environmental thought: A heritage of knowledge for sustainability*. ISEE Publicación Ocasional, South American Environmental Philosophy Section, No. 9.

Leff, E. 2004. Racionalidad Ambiental y Diálogo de saberes: significancia y sentido en la construcción de un futuro sustentable. *Polis*, Revista de la Universidad Bolivariana, Santiago de Chile, año/vol 2, numero 007.

Leff, E. 2017. Las relaciones de poder del conocimiento en el campo de la Ecología Política: una mirada desde el Suro. En Alimonda, H., Toro-Perez, C. & Martin, F. (Cords). *Ecologia Política Latinoamericana. Pensamiento crítico, diferencia latinoamericana y rearticulación epistémica*. 1ra edición, Ciudad Autónoma de Buenos Aires: CLACSO; México: Universidad Autónoma de Buenos Aires.

Mignolo, W. 2008. Epistemic disobedience and the decolonial option: A manifesto. *Subaltern studies: An interdisciplinary study of media and communication*, 2 February. Available at: http://subalternstudies.com/?p=193.

Mirabal, G. 2015. Reflexiones de los Pueblos Indígenas de Venezuela sobre la conflictividad socioambiental y la construcción de interculturalidad en nuestros territorios. En Rodriguez, I., Sarti, C. & Aguilar, A. (Eds). *Transformación de conflictos socio ambientales e interculturalidad. Explorando las interconexiones*. Norwich: Grupo Confluencias, Grupo de Trabajo de Asuntos Indígenas de la Universidad de los Andes y Organización de los Pueblos Indígenas de Amazonas (ORPIA). Swallowtail Print.

Moreano, M., Molina, F. & Bryant, R. 2017. Hacia una Ecología política Global: aportes desde el Sur. En Alimonda, H., Toro-Perez, C. & Martin, F (Cords). *Ecologia Política Latinoamericana. Pensamiento crítico, diferencia latinoamericana y rearticulación epistémica*, 1ra edición, Ciudad Autónoma de Buenos Aires: CLACSO; México: Universidad Autónoma de Buenos Aires.

Nussbaum, M. 2011. *Creating capabilities: The human development approach*. Cambridge, MA: Harvard University Press.

Paulson, S. 2018. Pluriversal learning: Pathways toward a world of many Worlds. *Nordia Geographical Publications* 47(5): 85–109.

Pellow, D. 2007. *Resisting global toxics: Transnational movements for environmental justice*. Cambridge, MA: The MIT Press.

Pilgrim, S. and J. Pretty, editors. 2013. *Nature and culture. Rebuilding lost connections*. Abingdon: Earthscan.

Querejazu, A. 2016. Encountering the pluriverse: Looking for alternatives in other worlds. *Revista Brasileira de Política Internacional* 59(2): e007.

Quijano, A. 2000. Coloniality of power and eurocentrism in Latin America. *International Sociology* 15: 215–232.

Rodríguez, I. & Inturias, M. 2018. Conflict transformation in indigenous peoples' territories: doing environmental justice with a 'decolonial turn'. *Development Studies Research* 5(1): 90–105, DOI: 10.1080/21665095.2018.1486220

Santos, B. de Sousa. 2008. *Another knowledge is possible: Beyond northern epistemologies*. London: Verso.

Santos, B. de Sousa. 2010. *Descolonizar el saber reinventar el poder*. Montevideo: Ediciones Trilce.

Santos, B. de Sousa. 2014. *Epistemologies of the South: Justice against Epistemicide*, Boulder, CO: Paradigm Publishers, pp. 200–240.

Santos, B. de Sousa, Arriscado, J., & Meneses, M.P. 2008. Introduction: Opening up the canon of knowledge and recognition of difference. En Boaventura De Sousa (Ed) *Another knowledge is possible: Beyond northern epistemologies*. London: Verso, pp. vx–ixii.

Seane, J. 2006. Movimientos sociales y recursos naturales en América Latina: resistencias al neoliberalismo, configuración de alternativas. en *Sociedade e Estado*, Brasília, Vol. 21, No. 1, pp. 85–107, January/April.

Smith, L. 2012. *Decolonizing methodologies: Research and indigenous peoples*. London: Zed Books.

Schlosberg, D. 2007. *Defining environmental justice: Theories, movements and nature*. Oxford: Oxford University Press.

Schlosberg, D. 2013. Theorising environmental justice: The expanding sphere of a discourse. *Environmental Politics* 22(1): 37–55.

Schlosberg, D. & Carruthers, D. 2010. Indigenous struggles, environmental, justice, and community capabilities. *Global Environmental Politics* 10(4), November.

Tubino, F. 2005. La interculturalidad crítica como proyecto ético-político. En *Encuentro continental de educadores agustinos*. Lima, pp. 24–28, de enero de 2005. Available at: http://oala.villanova.edu/congresos/educación/lima-ponen-02.html.

Viaña, J. 2009. *La Interculturalidad como Herramienta de Emancipación*. Bolivia: Instituto Internacional de Integración/Convenio Andrés Bello.

Visvanathan, S. 1997. *A carnival for science: Essays on science, technology and development*. London: Oxford University Press.

Visvanathan, S. 2009. *The search for cognitive justice*. Available at: www.india-seminar.com/2009/597/597_shiv_visvanathan.htm, January 2, 2020.

Walsh, C. 2005a. *La interculturalidad en la educación*. Lima: Ministerio de Educación DINEBI.

Walsh, C. 2005b. Interculturalidad, conocimientos y decolonialidad. *Signo y pensamiento. Perspectivas y Convergencia* 46(24): 31–50.

Walsh, C. 2007. Interculturalidad y colonialidad del poder. Un pensamiento y posicionamiento "otro" desde la diferencia colonial. En Castro-Gómez, Santiago y Grosfoguel, Ramón (Eds) *El giro decolonial Reflexiones para una diversidad epistémica más allá del capitalismo global*. Bogotá: Siglo del Hombre Editores; Universidad Central, Instituto de Estudios Sociales Contemporáneos y Pontificia Universidad Javeriana, Instituto Pensar, 308 p.

8 Degrowth and environmental justice
An alliance between two movements?

Julien-François Gerber, Bengi Akbulut, Federico Demaria, Joan Martínez-Alier

---

Learning outcomes

- Understand what post-growth and degrowth are, and how they relate to environmental conflicts and justice.
- Grasp why degrowth and environmental justice can be seen as 'natural' allies.
- Gain familiarity with the similarities and complementarities between the two movements.

Introduction

One does not need to search too long before reaching a rather sombre view, to say the least, of the current state of the global environment: essential resources, such as oil, are peaking; absorption capacities of the atmosphere and the oceans are overdrawn; and growth rates in industrialized countries are declining or stagnating, while some large countries, such as China and India, have been doubling their GDPs every ten years (Steffen et al., 2011). Within this context, overconsumption and overdevelopment are being acknowledged as problems by an increasing portion of the world's population and new voices for 'prosperity without growth' or even for 'degrowth' have strengthened (Drews et al., 2019). In addition, various forms of conflict over access to natural resources, the burdens of pollution and the use of ecosystem services are on the rise worldwide. The key questions that we would like to discuss in this chapter are the following:

- Is there a collective alternative vision emerging from the millions of people involved in environmental conflicts worldwide?
- Are the people involved in these conflicts the promoters and practitioners of more sustainable economies?
- If they oppose forms of economic growth, are they the 'natural' allies of the post-growth movement?

While the answers to these questions remain, of course, ambivalent and open to debate, it is important and increasingly necessary to explore them. Calls for a radical rethinking of economic growth have a long tradition within **ecological economics** (Daly, 1973, 1996; Georgescu-Roegen, 1971; Victor, 2008). Not only have the pioneering contributions on limits to growth and alternative conceptions of well-being come from ecological economists, but the very questioning of economic growth—paralleled with the study of the uneven distribution of its costs and benefits—has, in fact, been a constitutive element of the field. This has given rise to the development by ecological economists of key concepts, such as 'social metabolism', 'valuation languages', 'cost-shifting' and 'ecologically unequal exchange', some of which have in turn been imported and used by environmental justice activists (Healy et al., 2013). The intellectual roots and foundations of ecological economics thus make it eminently well suited for exploring the links between post-growth and environmental conflicts.

Martínez-Alier (2012) was possibly the first author to identify a link between such conflicts and a radical alternative to the existing economic regime—the so-called post-growth project. Our objective in this chapter is to elaborate on this relationship, namely the possible connection between the (still largely intellectual) post-growth movement and the mounting (grassroots) environmental conflicts that are allegedly giving rise to a global movement for environmental justice (Martínez-Alier et al., 2016). We will argue that the environmental justice movements and the degrowth movement are not merely materialist but wider in scope than that, and also that both degrowth and environmental justice seek a political reconfiguration of the ways our economies use resources. Perhaps more fundamentally, we will try to show that degrowth and environmental justice are deeply complementary in that environmental justice has not yet developed a unified and broad theoretical roadmap, while degrowth has largely failed so far to connect with a wider social movement. After some definitions of the main concepts used throughout the text, we will discuss the similarities and complementarities between degrowth and environmental justice, before concluding with some remarks.

What is post-growth? What are ecological distribution conflicts?

The '**post-growth**'—or 'beyond growth'—research agenda has become one of the major contributions of ecological economics over the past few decades. It has generated substantial research and has broadly differentiated into three main currents: 'degrowth', 'a-growth' and 'steady-state economics'.

The first of these, **degrowth**, not only challenges the hegemony of growth, but also calls for a democratically led redistributive downscaling of production and consumption—especially in industrialized countries—as a means to achieve environmental sustainability, social justice and well-being (Martínez-Alier, 2009; D'Alisa et al., 2014; Kallis et al., 2018). It is usually associated with the idea that smaller can be beautiful. However, the emphasis should not only be on less: degrowth promotes a society with a smaller metabolism (see Box 8.1) but, more importantly, strives for a society with a metabolism that is differently structured and serves new functions. Degrowth was launched into the political arena as a provocative slogan by environmental activists at the beginning of the 2000s, and it soon became a social movement and a concept debated in academic circles.

The second, **a-growth**, is agnostic about growth, meaning that welfare and sustainable targets should be carefully defined, and whether these targets require growth in market-based exchanges is irrelevant (van den Bergh, 2011).

Finally, **steady-state economics** promotes non-growing societies, based on a consistent material and energy throughput and stable populations (Daly, 1991).

Within these three different currents, we will focus on degrowth because we believe it has the greatest potential to be transformative and broaden into a social movement (Demaria et al., 2013). For instance, by December 2018, the network of groups claiming a link with degrowth included over 100 organizations with 3,000 active members, mostly located in Europe but also in North and South America, the Philippines, Tunisia, Turkey, etc.[1]

The term '**ecological distribution conflict**' (hereafter EDC), coined by Martínez-Alier and O'Connor (1996), has subsequently become a central concept both in ecological economics and political ecology. It denotes social conflicts arising not only over the unequal distribution of environmental benefits, such as access to natural resources and ecosystem services, but also those involving unequal and unsustainable allocations of environmental burdens, such as pollution or waste.[2] EDCs typically contest activities and projects like new roads, airports, dams, nuclear power stations, mines, plantations, fossil-fuel extraction, landfills or incinerators for waste disposal, and urban pollution (Temper et al., 2015; Martínez-Alier et al., 2016; Scheidel et al., 2018).[3] EDCs can—and do—overlap with agrarian conflicts over land resources and with labour movements over the environmental conditions of work.[4] They may thus also overlap with conflicts based on gender, race, class or caste differences.[5]

EDCs can be differentiated into three broad branches: 'environmental justice movements', 'environmental conflicts' and 'NIMBY (not in my backyard) mobilizations'. Environmental justice movements include an ethical or moral dimension that goes beyond environmental conflicts merely involving distributional aspects. While environmental justice movements typically problematize issues of participation, power and recognition (Schlosberg, 2013), what we call here '**environmental conflicts**' usually focus on a single issue and do not explicitly include questions of social justice, nor a quest for broader societal alternatives. They do not (yet) form organized networks and use common slogans. Finally, the **NIMBY** label, first used in the United States, implies that people have narrow, selfish, misinformed, emotional and/or irrational views of the situation (Burningham, 2000). While the term is often used as a way to discredit activists, NIMBY attitudes can also be seen as an essential starting point and an on-going component of full-fledged social movements. It may, for example, turn out that NIMBY protesters actually do have a good grasp of hazards ignored by authorities, thereby serving a broader public interest. Hence, NIMBY protests may and do (in the parlance of the environmental justice movement of the United States) turn into NIABY movements: not in anyone's backyard.

In this chapter, we will focus on environmental justice movements because this current is, of the three, the one that has the greatest potential for social transformation and for connecting with degrowth.

What are the connecting points between degrowth and environmental justice? What brings them together or pulls them apart? Why would an alliance between the two be desirable from a social justice and sustainability perspective? And how could such an alliance be articulated? It is these analytical, but also strongly normative, questions that we will endeavour to answer.

Both environmental justice and degrowth start with material concerns...

What is the main motive behind environmental justice and degrowth? The quality, quantity and distribution of environmental burdens and benefits are obviously among the prime concerns of these movements, albeit to different extents depending on particular cases. In our understanding, this characteristic would correspond to what some sociologists have labelled 'old' social movements, as they mainly refer to ownership, distribution and material issues (e.g. Touraine, 1981; Della Porta and Diani, 2006).

In the industrialized world of the 1960s, the environmental movement was largely born from deeply 'material' concerns, such as the risk posed by nuclear energy and other health issues related to the 'green revolution' (e.g. Carson, 1962). Contrary to Ronald Inglehart's thesis, there is nothing 'post-material' about such matters (Inglehart, 1990). Inglehart considered 'materialist' concerns those related to money, incomes and employment (i.e. working-class 'old' social movements) and labelled as 'post-materialist' those based on other more salient issues after 1968—such as human rights, women's rights and environmental issues, typically manifested in the growth of organizations such as Amnesty International and Greenpeace (i.e. middle-class, 'new' social movements). But **post-materialism** was a misnomer, as the likes of DDT and nuclear radiation continue to entail grave material risks.

The distinction between old and new social movements is focused mainly on the Global North, but different political geographies must also be considered. In the Global South, far from being the typical concerns of the rich, some environmental issues also mobilize—and perhaps more so—the poor and the Indigenous (and their supporters) because of their need to maintain direct and often customary channels of access to natural resources and services for livelihood purposes (Martínez-Alier, 2002). For instance, opposing eucalyptus plantations (called 'green deserts' in Brazil) because of their high consumption of water, as evidenced in many places around the world (Gerber, 2011), is a 'materialist' position; as is opposing open-cast mining of coal, copper or gold, or oil exploration and drilling, such as that taking place in the Niger Delta or the Amazon territory in Ecuador (under the slogan 'leave oil in the soil' promoted by Oilwatch since 1997).

Material concerns have similarly occupied a central place in the texts of the founding figures of degrowth, such as André Gorz, Cornelius Castoriadis and Nicholas Georgescu-Roegen. In response to the worsening of the environmental crisis in general, and the publication of the seminal *Limits to Growth* report (Meadows et al., 1972) in particular, degrowth authors began developing what appeared to be the most realistic response to the new situation: namely, the selective downscaling of production and consumption in order to reach a smaller social metabolism that could be organized differently.

Box 8.1   The politics of the social metabolism

Ecological sustainability depends on the interactions of humans with nature. One rigorous approach to the analysis of ecological sustainability is to use a socio-metabolic lens. Akin to the study of the metabolism of living organisms by physiologists, ecological economists study the metabolism of societies (Fischer-Kowalski, 1998). The metabolic analogy is rooted in the observation that biological systems (organisms and ecosystems) and socio-economic systems (human economies, companies, households and cities) heavily depend on a continuous throughput of energy and materials in order to maintain their internal structure (Fischer-Kowalski and Haberl, 2007). Therefore, the concept of social metabolism refers to the processes of appropriation, transformation and disposal of materials and energy by society in order to maintain itself and evolve. Methods such as material and energy flow accounting (MEFA) and multi-scale integrated analysis of societal and ecosystem metabolism (MuSIASEM) aim to quantify the social metabolism (Gerber and Scheidel, 2018). Clearly, different societies, regions and sectors have distinct metabolisms, which sometimes coexist and are always changing over time. These metabolisms can be characterized both by their *biophysical dimension* (i.e. the amount and composition of materials and energy they consume) and their *social, political and economic dimensions* (i.e. their political economy and the institutions that define the sources and types of extraction, as well as the distribution and disposal of materials and energy across the members of a given society). To understand the implications for sustainability, one must look not only at the quantification of metabolic flows (e.g. are they too large?) but also into the power relations that shape the metabolism (e.g. who controls it?). Ultimately, the coevolution of materiality and political economy transforms/shapes the social metabolism, and, as a result, political opportunities are fostered or foreclosed. The social metabolism is thus a central battlefield for environmental justice and degrowth activists.

Both the degrowth and the environmental justice movements—even if they do not use the term—seek to modify the social metabolism and, hence, the politico-institutional structure that governs it, in order to reach a higher level of ecological sustainability. In this sense, the environmental justice movement and degrowth can be seen as organic allies.

. . . but both are also clearly not just materialist in scope

Having said this, it would be a mistake to characterize environmental justice and degrowth as contesting exclusively—or even primarily—the material conditions of production and reproduction (as in the social metabolism). The different languages of valuation used and developed by environmental justice movements around the world are not limited to material and economic concerns, but include cultural, ethical, aesthetic and spiritual elements, as well (Martínez-Alier, 2002). The widespread call in Latin America for the right to hold local consultations or referendums with regard to mining or fossil fuel projects, thereby appealing to local democracy and/or Indigenous territorial rights, combines concerns for avoiding damage to local land and water resources with a proud display of autonomy (Urkidi and Walter, 2011; Walter and Urkidi, 2017).

Likewise, the degrowth movement is also concerned with notions such as autonomy, democracy and conviviality that extend far beyond the mere material. Indeed, as we have seen, degrowth not only promotes "the reduction of energy and material throughput" (Demaria et al., 2013, p. 209); it is also a political project that seeks greater equality and justice. Cattaneo and Gavaldà (2010) have criticized a limited understanding of degrowth based solely on metabolism reduction in terms of material and energy flows. They argued instead that the degrowth project is fundamentally a democratizing process, namely a collective choice for a '**good life**'. These aspects, we contend, correspond to the characteristics of 'new' social movements.

The centrality of justice might appear more obvious in the case of the environmental justice movement, which builds on the disproportionate shouldering of environmental burdens (toxic pollution, degradation) by the ethnically/racially marginalized and the poor. Yet, justice is no less a fundamental basis for the degrowth movement. In a nutshell, the latter advocates both the degrowth *of* injustice and degrowth *for* justice. Indeed, injustice is one of the main drivers of growth: on the one hand, inequality in consumption and relative poverty can motivate consumerism; on the other, growth is often used as an argument against redistributive justice with its promise of 'a rising tide' that will 'lift all boats'.

We argue that environmental justice and degrowth understand justice both consequentially (i.e. focusing on the outcome of an action), as well as deontologically (i.e. focusing on judging the actions themselves). In other words, justice is not only associated with the distribution of given outcomes (such as monetary compensation), but also includes the questions of recognition, difference and participation.[6] Accordingly, Bullard and Johnson (2000, p. 558) defined environmental justice as the "fair treatment and meaningful involvement of all people regardless of race, color, national origin, or income with respect to the development, implementation, and enforcement of environmental laws, regulations, and policies". Similarly, degrowth goes beyond the **'polluter pays principle'** at the global scale and emphasizes the historically unfair relationship between the Global North and South enabled by the former's unscrupulous appropriation of resources through systems of economic and ecologically unequal trade and disproportionate use of environmental sinks for greenhouse gases (Hornborg, 1998; Warlenius et al., 2015).

Cultural resistance to hegemonic actors is an important motivation for many environmental justice movements, as is the population's inner relationship to specific places/resources identified as sacred. In some cases, such as First Nations protests in North America, these factors may even be more effective in mobilizing local people against extractive industries or commercial monocultures than the actual material impacts such activities have on people's livelihoods. In a parallel sense, degrowth also advocates nothing less than a 'cultural revolution', which—unlike its Maoist forerunner—would aim at redefining the 'good life' towards forms of voluntary simplicity, a return to the 'essential' and the potential for non-material quests (e.g. having more time for relational, political, caring, artistic or intellectual pursuits). Along the same lines, Latouche (2007) has emphasized the need to 'decolonize the imaginary'.[7]

The existential critique of growth within environmental justice and degrowth is apparent in the involvement of progressive spiritual figures in both movements. The pursuit of inner growth through outer degrowth is a recurrent theme across religious traditions worldwide. Gandhi, for example, is a well-known source of inspiration for both Indian environmental justice movements and degrowth activists. Moreover, a number of key degrowth (proto-) theorists were in fact actively engaged in their own spiritual pursuits (for example Thoreau, Tolstoy, Kumarappa, Illich, Schumacher). The language of sacredness often appears in environmental justice conflicts and in cases of the deaths of 'environmental defenders', such as Berta Caceres in Honduras on 2 March 2016, while defending a river sacred to her Lenca community against a hydroelectric project.

Thus, environmental justice movements and degrowth proposals have both a materialist and non-materialist agenda, in that they both portray a combination of 'old' and 'new' social movements, engaging in 'old' and 'new' structural conflicts (Della Porta and Diani, 2006). In Marxist terms, one could say that, while most mobilizations may originate in the (material) base of economic relations of (re)production, they rapidly level up to incorporate the (cultural) superstructure as well: it is not only people's biophysical (exterior) relationship to natural resources that shapes such mobilizations, but also their psychological–spiritual (interior) relationship with the natural world, as linked to their values, beliefs and emotional lives.

The complementarity between degrowth and environmental justice

Beyond their similarities, environmental justice and degrowth are also, in many ways, complementary. In short, environmental justice proposes a large-scale force of resistance, whereas degrowth theorizes a route towards social and ecological sustainability.

Taken as a whole, the myriads of EDCs represent one of the most powerful socio-political forces in the Global South today. Martínez-Alier (2002, p. 1) has even compared the current explosion of environmental justice struggles with the beginning of the socialist movement and the 'First International'. What is more, the general occurrence of such conflicts is on the rise, as the metabolism of industrialized regions requires ever more energy and materials and as the commodity frontier advances spatially, as well as structurally.

However, this political–practical strength has so far failed to translate into an equal strength in theoretical production, despite the fact that many creative concepts have been forged through environmental justice activism, such as **biopiracy, land-grabbing, ecological debt, climate justice** and **Indigenous territorial rights**.[8] Indeed, it is still the case that there appears to be no common theoretical basis for what has been called the global environmental justice movement (Martínez-Alier et al., 2016).

This is not to say that environmental justice movements lack conceptual frameworks within which the dynamics and relationships they emerge from are interpreted. Sarayaku's resistance against Amazon oil exploration in Ecuador is a well-known example, as this community became the cradle of the recent use of the concept of *Sumak Kawsay* or **Buen Vivir**,[9] which was then incorporated into the country's new constitution. In India, a Gandhian worldview has been mobilized, while in Indigenous lands particular cosmologies can be invoked for advocating a just order. Yet, overall, many grassroots environmental justice movements remain local or regional in their conceptual scope—which can be both a strength and a weakness. Concepts like **food sovereignty**[10] (from Via Campesina), or more recently energy democracy or energy sovereignty, have the potential to become universal.

This conceptual fragmentation can nevertheless obstruct wider synergies and the broader societal alternatives that can be imagined and constructed. In contrast, the labour movement, for instance, has given rise to rich (and at times competing) theoretical traditions, which have the potential to nourish debates and political strategies.[11] The same applies to the feminist movement, which is decentralized and at times fiercely divided, yet perhaps more effective than its labour (and the environmental justice) counterpart.[12]

This is where the contribution of degrowth can be crucial. The degrowth movement has so far largely been an intellectual endeavour, albeit with numerous local experiments; but a good theory can be a powerful weapon for fostering understanding and action. We would therefore like to briefly review some of the key degrowth ideas applicable to environmental justice movements.

One basic starting point of degrowth is the 'impossibility theorem', namely that a Western-style mass consumption economy for a world of 7.5 billion people is neither possible nor desirable (Daly, 1991). Against metabolically naïve 'Green New Dealers', degrowth reminds us that with current technologies and institutions it would be impossible to reach Western levels of consumption for everyone based on renewable energy alone (Kallis, 2018). A wind-hydro-solar economy could only support much smaller economies, and a transition to renewables would thus have to be a degrowth transition. It is therefore essential to tackle consumption levels, and the proper way to rethink them can only be world-systemic and class-based.

Capital has become so mobile that it has been able—with more or less success—to reorganize production worldwide in accordance with profit-maximizing opportunities and resource locations. World-system theorists have thus argued that a single, transnational global system has emerged, largely administered by a global ruling class whose members share a similar lifestyle and comparable consumption patterns (Robinson, 2004). Accordingly, the degrowth critique applies to the global middle and upper classes, regardless of whether they are located in the Global North or South. As for the '**global poor**', a post-growth scenario would not only leave some environmental space for them to determine their own futures, but also address the issue of the ecological debt that the 'global rich' historically owe to the rest of the world (Box 8.2).

In the Global South, the target of post-growth public policies should be wealth redistribution and the satisfaction of basic needs, rather than the pursuit of GDP growth per

---

Box 8.2   Do we need growth to fight poverty?

It is often claimed—sometimes by environmental justice activists themselves—that "we still need growth to fight poverty". This statement must be carefully examined from a variety of perspectives. Who are 'we'? And what does 'growth' really mean at the grassroots level? What kind of 'wealth' does it create, and for whom? In reality, it turns out that high growth rates are not necessarily good news for the poor. Take India, for example (Gerber and Raina, 2018). "[India's] pattern of dazzling economic growth", noted Walker (2008, p. 561), "has been propelled by a powerful 'reinforcing mechanism' through which 'inequality drives growth and growth fuels further inequality'". Such mechanisms must be explained in order to demystify the current common-sense assumption that growth continuously leads to more welfare. *First of all*, India's massive growth rates have helped the rich rather than the poor: while middle and upper classes now enjoy Western levels of overconsumption, nearly 80% of the population lives on 20 rupees a day (USD 0.30); and, in 2017, the richest 10% had almost 80% of the total wealth of the country. *Second*, growth has largely been a jobless phenomenon, especially in the countryside where growth is often synonymous with mechanization. *Third*, far from eliminating poverty, growth has relied on the poor to provide cheap arms and land, as the quote earlier in this box shows. *Finally*, growth creates new poverties, undermining meaningful local activities and fostering accumulation by dispossession, contamination and commodification. The 'growth against poverty' argument is therefore not a straightforward one, and, on top of that, the post-growth alternative is also often misunderstood. Degrowth has never been about imposing 'austerity' everywhere and shrinking 'everything' indiscriminately. Some items (like local products and services) will surely be consumed and produced much more in a post-growth society, and many economic activities will also increase in those circumstances—for example, agroecology, urban gardening, repairing, caring and so on. A post-growth society, because of its local, egalitarian and democratic nature, is therefore, unlike today's world, likely to be in a better position to tackle poverty.

se. But, of course, needs-based targets must be the object of careful collective reflection (Max-Neef, 1991). What are our needs, and who are they for? How can we distinguish 'real' needs from detrimental ways of channelling desires (i.e. 'false consciousness')? Clearly, these are fundamental yet difficult questions that any degrowth project has to tackle (Gerber and Raina, 2018). Furthermore, these questions can only be addressed within the greatest possible political, economic and cultural autonomy. Table 8.1 summarizes some of the key complementarities between the environmental justice and degrowth movements.

Referring to Martínez-Alier (2012), Kallis (2018, pp. 179–180) notes that:

> the small movement for degrowth . . . finds natural allies in movements against extraction and for environmental justice in the Global South (movements that confront in practice, rather than in theory, the growth of the insatiable metabolism that supports the imperial mode of living) as well as among Indigenous groups who profess values of sharing, sufficiency and common ownership, in their own language and with their own significations.

This alliance can not only foster socio-political activism, but could also nurture conceptual cross-fertilization. Concepts that are part of the degrowth vocabulary, like autonomy, simplicity and care, are mobilized in environmental justice struggles; and, vice versa, activist notions such as land-grabbing, ecological debt and biopiracy (see footnote 2) are now used by degrowth researchers (Martínez-Alier et al., 2012, 2014).

However, some authors have been more sceptical about the 'naturalness' of this alliance. Scheidel and Schaffartzik (2019), for example, argue that environmental justice protesters and degrowthers do not have exactly the same aims: while the former often seek to protect "traditional livelihoods and ways of living", the latter seek "new livelihoods and new ways of living, within alternative societies" (Scheidel and Schaffartzik, 2019, p. 332). The fact is that many grassroots resistance movements do indeed start with the defence of a local 'status quo' as their goal; hence, the normative quest for a radical theory able to transcend this. The key point, from a degrowth perspective, is to transform NIMBY movements into NIABY ones (not in anyone's backyard) which are actively seeking broader transformations.

Table 8.1 Complementarities between environmental justice and degrowth

	Environmental justice movement	Degrowth movement
Size	Huge	Tiny
Main location	'Global South'	'Global North'
Scope	Mainly local, but increasingly deploying global concepts (e.g. climate justice, food sovereignty)	Global, but many local experiments (e.g. degrowth communes, transition towns)
Actors	Lower (and middle) classes, racial minorities, Indigenous communities, mainly rural	Middle class, mainly urban
Combativeness	At the grassroots level	At the theoretical level
Weakness	No inclusive theoretical roadmap	No broad popular basis

Concluding remarks

This chapter starts from the observation that both degrowth and environmental justice movements share a common quest for profound socio-ecological transformations towards justice and sustainability, and that an alliance among these research and activist communities is essential. Of course, this relationship remains tenuous, and its future consolidation cannot be taken for granted. Yet, we see it as an 'organic' one, since degrowth and environmental justice contest the same fundamental processes—in short, the nature and impacts of our economies' relentless expansion—in a complementary and synergetic way. *Without a degrowth strategy, environmental justice movements will never fully succeed, and vice versa.*

It was, after all, André Gorz, an activist-scholar deeply concerned with environmental injustices, who coined the term degrowth (Demaria et al., 2013). Modern degrowth was thus directly born out of concerns for environmental (in)justices. However, whether environmental justice and degrowth will eventually converge in the future depends to a large extent on a number of practical barriers related to differences in the languages, ontologies, geographies and class backgrounds of the different activists. Rodríguez-Labajos et al. (2019) have argued that there are "significant differences" between the two movements and that more attention should be paid to the specificities of places and histories before any substantial alliance can be built.

At least one successful example of such a convergence has already taken place in Germany via *Ende Gelände* ('Here and no further'), a large civil disobedience movement seeking to limit global warming through the phasing-out of fossil fuels. Every year since 2015, up to 4,000 activists have carried out direct actions to prevent the operation of open-pit coal mines and coal-fired power stations. In parallel, remarkably, an annual degrowth summer school has also been organized, explicitly linking 'degrowth in action' and climate justice. The summer schools are run by *Konzeptwerk Neue Ökonomie*, an NGO of young scholars and activists that is sharply focused on the alliance between the two movements.

This example shows that the work of concrete articulation has already begun and that similar convergences are likely to gain importance as the 21st century unfolds, carrying with it a high risk of further multi-dimensional crises. Of course, other social movements will have to be included—such as those for the **commons**, spiritual ecology, post-development and **ecofeminism**—and more research and action will be called for to strengthen their combined impacts.

Follow-up questions

- What are the alternative visions emerging from the myriads of environmental justice movements worldwide?
- Under what conditions would NIMBY (not in my backyard) movements transform into NIABY (not in anyone's backyard) movements, and what would be the implications?
- Is degrowth doomed to remain a Northern middle-class movement?

Notes

1 See: www.degrowth.info/en/map.
2 On distributive (in)justice, see also Chapter 3 of this volume.

3 See the Environmental Justice Atlas for an inventory of almost 3,000 EDCs: www.ejatlas.org.
4 On labour movements and environmental justice, see Chapter 19 of this volume.
5 On the roles of race and gender in shaping environmental justice, see Chapters 17–18 of this volume, respectively.
6 On participation and recognition, see Chapters 4–5 of this volume, respectively.
7 On decolonial and Indigenous environmental justice, see also Chapters 7 and 20 of this volume, respectively.
8 *Biopiracy* refers to the appropriation of genetic resources (in medicinal or agricultural plants) without recognition of the original ownership of the peoples who have been using them. *Land-grabbing* refers to the new and brutal waves of land acquisitions in the Global South, typically for export plantations. The *ecological debt* refers to the compensation that rich nations or regions owe to poor nations or regions for the environmental damage embodied in their growth and for exploiting the raw materials and ecosystems of poor countries, sometimes over centuries (Martínez-Alier et al., 2014). On climate justice, see Chapter 12 of this volume.
9 On the concept of *Buen Vivir*, see Chapter 7 of this volume.
10 On food sovereignty see also Chapter 14 of this volume.
11 On labour unions and environmental justice, see Chapter 19 of this volume.
12 On the role of women in the environmental justice movement, see Chapter 18 of this volume.

References for further reading

D'Alisa, G., Demaria, F., Kallis, G. (Eds) (2014) *Degrowth: a vocabulary for a new era*. Routledge, London.

Martínez-Alier, J. (2012) Environmental justice and economic degrowth: an alliance between two movements. *Capitalism Nature and Socialism*, 23(1): 51–73.

Special issue of the journal *Ecological Economics* on the relationship between degrowth and environmental justice, introduced by: Akbulut, B., Demaria, F., Gerber, J.-F., Martínez-Alier, J. (2018) Five theses on the relationships between degrowth and the environmental justice movement. *Ecological Economics*, submitted.

References used in this chapter

Bullard, R.D., Johnson, G.S. (2000) Environmental justice: grassroots activism and its impact on public policy decision making. *Journal of Social Issues*, 56(3): 555–578.

Burningham, K. (2000) Using the language of NIMBY: A topic for research, not an activity for researchers. *Local Environment*, 5(1): 55–67.

Carson, R. (1962) *Silent spring*. Houghton Mifflin, New York.

Cattaneo, C., Gavaldà, M. (2010) The experience of rurban squats in Collserola, Barcelona: what kind of degrowth? *Journal of Cleaner Production*, 18: 581–589.

D'Alisa, G., Demaria, F., Kallis, G. (Eds) (2014) *Degrowth: a vocabulary for a new era*. Routledge, London.

Daly, H.E. (Eds) (1973) *Toward a steady-state economy*. WH Freeman, San Francisco.

Daly, H.E. (1991 [1977]) *Steady-state economics* (second edition). Island Press, Washington, DC.

Daly, H.E. (1996) *Beyond growth: the economics of sustainable development*. Beacon Press, Boston.

Della Porta, D., Diani, M. (2006) *Social movements: An introduction*. Blackwell, Malden.

Demaria, F., Schneider, F., Sekulova, F., Martínez-Alier, J. (2013) What is degrowth? From an activist slogan to a social movement. *Environmental Values*, 22: 191–215.

Drews, S., Savin, I., van den Bergh, J. (2019) Opinion clusters in academic and public debates on growth-vs-environment. *Ecological Economics*, 157: 141–155.

Fischer-Kowalski, M. (1998) Society's metabolism: The intellectual history of materials flow analysis, Part I, 1860–1970. *Journal of Industrial Ecology*, 2(1): 61–78.

Fischer-Kowalski, M., Haberl, H. (Eds) (2007) *Socioecological transitions and global change: trajectories of social metabolism and land use*. Edward Elgar, Cheltenham.

Georgescu-Roegen, N. (1971) *The entropy law and the economic process.* Harvard University Press, Cambridge.
Gerber, J.-F. (2011) Conflicts over industrial tree plantations in the South: Who, how and why? *Global Environmental Change,* 21(1): 165–176.
Gerber, J.-F., Raina, R.S. (Eds) (2018) *Post-growth thinking in India.* Orient Blackswan, New Delhi.
Gerber, J.-F., Scheidel, A. (2018) In search of substantive economics: Comparing today's two major socio-metabolic approaches to the economy—MEFA and MuSIASEM. *Ecological Economics,* 144: 186–194.
Healy, H., Martínez-Alier, J., Temper, L., Walter, M., Gerber J.-F. (Eds) (2013) *Ecological economics from the ground up.* London: Routledge.
Hornborg, A. (1998) Toward an ecological theory of unequal exchange: articulating world system theory and ecological economics. *Ecological Economics,* 25(1): 127–136.
Inglehart, R. (1990) *Culture shift in advanced industrial society.* Princeton University Press, Princeton, NJ.
Kallis, G. (2018) *Degrowth.* Agenda Publishing, Newcastle Upon Tyne.
Kallis, G., Kostakis, V., Lange, S., Muraca, B., Paulson, S., Schmelzer, M. (2018) Research on degrowth. *Annual Review of Environment and Resources,* 43(1): 291–316.
Latouche, S. (2007) *Le pari de la décroissance.* Fayard, Paris.
Martínez-Alier, J. (2002) *The environmentalism of the poor: a study of ecological conflicts and valuation.* Edward Elgar, Cheltenham.
Martínez-Alier, J. (2009) Socially sustainable economic de-growth. *Development and Change,* 40: 1099–1119.
Martínez-Alier, J. (2012) Environmental justice and economic degrowth: an alliance between two movements. *Capitalism Nature and Socialism,* 23(1): 51–73.
Martínez-Alier, J., Anguelovski, I., Bond, P., Del Bene, D., Demaria, F., Gerber, J.-F., Greyl, L., Haas, W., Healy, H., Marín-Burgos, V., Ojo, G. U., Porto, M., Rijnhout, L., Rodríguez-Labajos, B., Spangenberg, J., Temper, L., Warlenius, R., Yánez, I. (2014) Between activism and science: grassroots concepts for sustainability coined by environmental justice organizations. *Journal of Political Ecology,* 21: 19–60.
Martínez-Alier, J., Healy, H., Temper, L., Walter, M., Rodríguez-Labajos, B., Gerber, J.-F., Conde, M. (2012) Conclusion—Between science and activism: learning and teaching ecological economics and political ecology with EJOs. In: H. Healy et al. (Eds), *Ecological economics from the ground up.* Routledge, London.
Martínez-Alier, J., O'Connor, M. (1996) Ecological and economic distribution conflicts. In: R. Costanza, J. Martínez-Alier, O. Segura (Eds), *Getting down to Earth: practical applications of ecological economics.* Island Press/ISEE, Washington, DC.
Martínez-Alier, J., Temper, L., Del Bene, D., Scheidel, A. (2016) Is there a global environmental justice movement? *Journal of Peasant Studies,* 43(3): 731–755.
Max-Neef, M. (1991) *Human scale development.* Apex Press, New York.
Meadows, D., Meadows, D., Randers, J., Behrens, W. III (1972) *The limits to growth.* London: Pan.
Robinson, W.I. (2004) *A theory of global capitalism.* Johns Hopkins University Press, Baltimore.
Rodríguez-Labajos, B., Yánez, I., Bond, P., Greyl, L., Munguti, S., Uyi Ojo, G., Overbeekh, W. (2019) Not so natural an alliance? Degrowth and environmental justice movements in the global South. *Ecological Economics,* 157: 175–184.
Scheidel, A., Schaffartzik, A. (2019) A socio-metabolic perspective on environmental justice and degrowth movements. *Ecological Economics,* 161: 330–333.
Scheidel, A., Temper, L., Demaria, F., Martínez-Alier, J. (2018) Ecological distribution conflicts as forces for sustainability: an overview and conceptual framework. *Sustainability Science,* 13(3): 585–598.
Schlosberg, D. (2013) Theorising environmental justice: the expanding sphere of a discourse. *Environmental Politics,* 22(1): 37–55.
Sengupta, A. (Ed) (2007) *Report on conditions of work and promotion of livelihoods in the unorganised sector.* New Delhi: National Commission for Enterprises in Unorganised Sector.

Steffen, W., Grinevald, J., Crutzen, P., McNeill, J. (2011) The Anthropocene: conceptual and historical perspectives. *Philosophical Transactions of The Royal Society A*, 369: 842–867.

Temper, L., Del Bene, D., Martínez-Alier, J. (2015) Mapping the frontiers and frontlines of environmental justice: the EJAtlas. *Journal of Political Ecology*, 22: 255–278.

Touraine, A. (1981) *The voice and the eye: an analysis of social movements*. Cambridge University Press, Cambridge.

Urkidi, L., Walter, M. (2011) Dimensions of environmental justice in anti-gold mining movements in Latin America. *Geoforum*, 42: 683–695.

van den Bergh, J. (2011) Environment versus growth—A criticism of "degrowth" and a plea for "a-growth". *Ecological Economics*, 70(5): 881–890.

Victor, P. A. (2008) *Managing without growth: slower by design, not disaster*. Edward Elgar Publishing, Northampton.

Walker, K. (2008) Neoliberalism on the ground in rural India: predatory growth, agrarian crisis, internal colonization, and the intensification of class struggle. *Journal of Peasant Studies*, 35(4): 557–620.

Walter, M., Urkidi, L. (2017) Community mining consultations in Latin America (2002–2012): the contested emergence of a hybrid institution for participation. *Geoforum*, 84: 265–279.

Warlenius, R., Pierce, G., Ramasar, V. (2015) Reversing the arrow of arrears: the concept of "ecological debt" and its value for environmental justice. *Global Environmental Change*, 30: 21–30.

9 Sustainability and environmental justice

Parallel tracks or at the crossroads?

Julie Sze

Learning outcomes
- Consider the history of sustainable development and its relationship with environmental justice.
- Identify the barriers to and opportunities for sustainability as a frame for environmental justice policies.
- Examine cases illustrating a situated sustainabilities framework.

Introduction

Sustainability emerges out of a sense—and empirical documentation—of environmental crises. The International Geosphere–Biosphere Programme publishes what it calls the "Great Acceleration" indicators, which have shown an uptick in human activity since 1950. Its data show that "human activity, predominantly the global economic system, is now the prime driver of change in the Earth System (the sum of our planet's interacting physical, chemical, biological and human processes), according to a set of 24 global indicators." These include, for example, carbon emissions, ocean acidification, and tropical land loss (IGBP, 2015). The rapid growth of environmental problems maps onto socio-economic trends (rising populations and increasing urbanization, associated with increased energy use, and fertilizer consumption).

The occurrence of environmental crises has been linked to global economic growth over the last fifty years. The intensification of social inequalities in the last decade has exacerbated already dangerous conditions of life and land. In the latter half of the 20th century, the biological, physical, and ecological sciences and economics dominated sustainability research and action at the relative expense of the humanities and humanistic social sciences. Despite the rapid expansion of sustainability research since the turn of the 21st century, fundamental questions about core concepts and values of sustainability as it relates to social justice have, far too often, remained largely unexplored (Sze et al., 2018).

Sustainability is thus an aspiration, amidst global economic systems that extract natural resources and leave ruined landscapes in their wake. Sustainability emerges out of crisis, in much the same way that environmental justice is an aspirational goal amidst crises of environmental racism, inequality, and injustice. Sustainability *and* environmental justice remain elusive, even as it has become increasingly clear that each domain is unattainable without the other. Simply put, unsustainable practices diminish social justice. To illustrate, the effects of animal extinctions, toxic waste, and air pollution alike have fallen disproportionately on the poor. Meanwhile, efforts at achieving sustainability in the industrialized West and Global South have often aggravated social inequities—for examples, when Indigenous people have been displaced to create wildlife or nature reserves, or when governments have mandated expensive new environmental management technologies that exacerbate the burden of the poor. As a result, sustainability has historically been, and often still is, perceived as an elite, technologically driven project in an increasingly diverse world. Opposition to environmental reform can thus find a solid footing among the expanding ranks of the world's working and impoverished peoples.

The contested nature of sustainability—bottom-up versus top-down, technologically driven versus grassroots—is complex, yet unsurprising. Such political conflicts are largely predictable, given the colonial histories based on natural resource extraction and how economic development policies have often reiterated hierarchies within and between nation-states. Sustainability is so vague a concept that it invites immediate scepticism. Environmental scientist Lucas Seghezzo argues that it is the ambiguity of the term "sustainability" that has contributed to its large-scale acceptance as a framework for environmental and social action (2009). A number of scholars have explored these tensions at length.

In the introduction to a co-edited volume entitled *Sustainability: Approaches to Environmental Justice and Social Power*, I argue with colleagues from the social sciences, environmental sciences, and humanities that sustainability is enhanced by focusing squarely on social justice and by using interdisciplinary, multi-scalar lenses of space and time (Sze et al., 2018). Specifically, we asked framing questions and used them to shape the individual empirical case studies in the volume which demonstrate how a justice-oriented and interdisciplinary frame enhances the cases studied, rather than forming the basis for the opposition between sustainability and environmental justice.

Collectively, and through empirical demonstration, we argue simply that sustainability is enhanced through the foregrounding of questions of power and the use of interdisciplinary methods drawn from the humanities and humanistic social sciences. We add these perspectives to economistic, environmental science, and technocratic approaches that predominate within sustainability research. These questions raise and name what is often held as *implicit* in discussions of sustainability. Simply put, as environmental scientists Mary Cadenasso and Stewart Pickett ask, "Why is a shared vision of sustainability so elusive, especially given that it is so widely invoked?" (2018, p. 31).

A genealogy of the term sustainability illuminates the connections and divergences with environmental justice, as a parallel complex and contested field of knowledge and practice. This intellectual history illustrates how sustainability is connected to ideas of nature and sustainable development. These debates parallel the complex politics of naming environmental justice, specifically, as **environmental racism**. To better integrate sustainability with environmental justice requires linking these critical conversations more explicitly. This bridging work is evident in social justice frames applied to sustainability, such as just, critical, and situated sustainabilities.

Thus, these frameworks are similar—only differing in emphasis rather than in kind—with situated sustainability being the most significant. The first to appear, just sustainability, emerged soon after environmental justice appeared in the 1990s. Critical sustainability then took up the questions posed by just sustainability from a perspective of critical sociology and geography. Situated sustainability further expands these two earlier frameworks, placing environmental humanities and questions of interdisciplinarity at the fore. These frameworks show how sustainability and environmental justice have been on intersecting, rather than parallel or divergent, tracks of scientific knowledge, public policies and social movement activism. In other words, I am adhering to David N. Pellow's view in Chapter 22 of this volume, **critical environmental justice** studies, in which he identifies overlapping fields. His first generation of environmental justice studies focuses on problem documentation, while the second generation expands in topics and approaches. These generations, or waves, impact what he describes as critical environmental justice, not previously highlighted. Pellow focuses on liberation movements that link humans, ecosystems, and nonhuman animal species as one of his "key pillars" of critical environmental justice studies. The other three are **intersectionality**, scale, and **indispensability** (contrasted with disposability). In outlining environmental justice "generations", he is not using a teleological narrative which ends with critical environmental justice. Rather, he is tracing how moments and movements build on each other, at their overlapping edges. In the same vein, this piece traces the histories of fields that connect sustainability and environmental justice—just, critical and situated sustainabilities—and how they position themselves ideologically, vis-à-vis the hegemony of the market and narrow (and arguably racist) views of nature.

Sustainability, sustainable development, and nature

Sustainability is a flexible concept, with no standard definition. Its counterpart, the *lack* of sustainability, is probably better understood (Robinson, 2004). Rather than attempting to provide a single definitive explanation of "sustainability", placing it in the context of time, place, scale, and politics can help to enact debates surrounding it. The recent history of sustainability is linked closely to a 1972 report entitled *The Limits to Growth* (Meadows et al., 1972). The report—written for the Club of Rome, a global think tank that distributed over thirty million copies—suggested that sustainability was regarded as an "antonym" for collapse. This report was the first to suggest that there are indeed limits to growth and to a world system capable of supporting human life, and it suggested action (Grober, 2012).

The report was followed by many others, as well as by international gatherings. Seghezzo (2009) focuses on the *limitations* of the sustainability framework put forth by the World Commission on Environment and Development (WCED) for sustainable development. The WCED's 1987 report, *Our Common Future*, and the development of the UN Conferences on Environment and Development (beginning in 1992) were signal achievements in the field (WCED, 1987). "**Sustainable development**" is defined in the 1987 report as development that meets the needs of the present without compromising the ability of future generations to meet their own needs. Seghezzo (2009) identifies key limits of the WCED paradigm of sustainable development: it is anthropocentric, the role of the economy is overstated, it is incompatible with intergenerational justice, it neglects space and time, and it ignores personal aspects of development. Sustainability consists of three pillars popularly referred to as the 3 'E's: economy, environment, and equity.

Despite these recent accounts, sustainability's roots go much further back. Sustainability is implicitly connected to the term "**nature**" and emerging conceptions of natural systems that support life and land. These conceptions of nature were formalized by economists, philosophers, scientists, and social reformers in the 18th century in Western Europe. However, this story of sustainability as a product of 18th-century Western European science and philosophy is incomplete. First, indigenous societies have always practised sustainability to ensure their survival. Second, this periodization of sustainability from the 18th century ignores the earlier contexts of colonialism and capitalism which depended on control of nature. In addition to this dominance over nature, capitalism and, by extension, **colonialism** cheapened it. As world systems scholars Raj Patel and Jason Moore explain: "Nature is not a thing but a way of organizing—and cheapening—life. It is only through real abstractions—cultural, political, and economic all at once—that nature's activity becomes a set of things." (2018, p. 47). They call this "Cheap Nature", which is the foundational process that enables the cheapening of money, work, care, food, energy and lives. The classification and abstraction of nature to be controlled and cheapened can thus be traced back to the history of colonialism and modernity itself.[1]

Most historical accounts of sustainability focus on the Western European and US efforts to study the limits of "natural" systems in the contexts of emergent fields of environmental science and ecology. But what is considered a natural system is always also a political and cultural object of concern and analysis. As the socialist culture critic Raymond Williams famously argued, ideas about nature "contain an extraordinary amount of human history" (1980). From Williams onward, many scholars, principally environmental and policy

---

Box 9.1 Sustainability from an ecological science perspective

Cadenasso and Pickett (2018) discuss how sustainability is multidimensional and explain how these dimensions can be identified as metaphor, meaning, and model, using as a case study a water clarity project (in the Chesapeake Bay in Maryland) that aims to sustain aquatic life and important fisheries. This site is a complex area to research, encompassing rural, suburban, and urban locales, and faces a myriad of pollution sources. Based on their long-term engagement with a major ecosystem services research project—the Baltimore Ecosystem Study (BES), which began in 1997—they describe how sustainability can be operationalized. In other words, to address Seghezzo's critique of the vagueness of the sustainability concepts, they seek to bring precision to the term from an ecological science perspective. First, they write, meanings must be articulated. Then, specific models can be generated to "test, apply and refine" with a specific socio-spatial context. Next, models are further refined "because new relationships among components may be found or the relative strength of different relationships discovered" (p. 32). This reflexive concept is continually being refined. Cadenasso and Pickett then complicate the 3 'E's—economy, environment, and equity—into the corresponding theoretical realms of ecosystem services, specifically resilience and environmental justice, respectively. Grounded within ecological scientific research, they nonetheless explicitly tackle the lack of precision around the term. Thus, they are significantly involved, in effect, in translation work between domains long thought to be difficult to traverse, specifically science, policy, and community concepts. They write: "Sustainability is not solely a scientific concept. It draws on ecological concepts and processes but incorporates social perceptions and values. The three pillars of sustainability—human well-being, ecological integrity and equity—resonate with the specific theoretical realms of ecosystem services, resilience, and environmental justice" (p. 48). Ultimately, they write, "sustainability is not a fixed state or something achievable *per se*. Sustainability is more usefully thought of as a trajectory along which the balancing of tradeoffs and synergies among goals results in making systems relatively more sustainable" (p. 49).

historians, have attempted to outline "human history" in nature, while others have focused on the intellectual history of the connection between political economy and nature and the exchange of ideas between European and American intellectuals (Grober, 2012).

Earth and its resources have been understood as finite and subject to collapse since at least the 18th century. This understanding was originally drawn from forestry, but was far broader than just forests. The first generation of forestry research in Western Europe set the stage for successive concepts of sustainability discourses, research, and practices. Sustainability and its closely linked cousin, sustainable development, are thus complex and highly contested terms.

Both are simultaneously radical and reformist, dependent on the ways in which the concepts are perceived and operationalized. Sustainability and sustainable development are potentially radical because they posit that there may be environmental limits to economic growth. Sustainability and sustainable development contain a critique of mainstream economics, and its dogma of endless growth itself. At the same time, these concepts are reformist in that they presuppose the existing capitalist system will be able to "fix" itself. This line of reasoning is most clear in ecological economics where it attempts to quantify the "natural capital" of "**ecosystem services**". In other words, the extraction and exploitation of nature is based on the assumption that natural resources are infinite. To save them requires "pricing" the costs.

However, unlike Patel and Moore's Marxist critique of the cheapening of nature, attempts to quantify natural resources actually bring them further into the capitalist system. Philosophically and ethically, Indigenous communities "value" nature in fundamentally different ways. For example, in the battle against the Dakota Access Pipeline at the Standing Rock Sioux Reservation in the US, #NoDAPL activists spoke of *Mni Wiconi—Water Is Life*. Native peoples spoke of water as a relative, rather than a resource. In contrast, discourses and policies that use the notion of environmental resources and ecosystem services suggest that "environment" is based on utilitarian conceptions of markets and exchange value.[2]

Other sceptics take aim at reformist efforts to ameliorate, rather than eliminate, the sources of environmental destruction. For example, literary scholar Leerom Medovoi asks: "What is it about sustainability that enables the particular word . . . to express the ecological hopes and fears of so diverse and antagonistic an array of social actors?" (129–130). He suggests that the "discourse of sustainability is a new intensification in the exercise of **biopower** . . . of what is the value of what we kill when we extract value from what remains living" (142).

Naming and framing problems

Sustainability/sustainable development discourses first emerged mainly in policy circles, evolving in temporal alignment and on parallel tracks related to the demands/history of the environmental justice movement.[3] What this means in practice is that the critique of sustainability and sustainable development evolved at the same time as the critique of environmental justice movements against mainstream environmentalism. Both lines of critique focused (and continue to focus) on the capture of the concept of sustainability/environmentalism by elite actors which prioritize scientific narratives and market-based approaches to "solve" the problems of environmental crises that capitalism and colonialism has produced.

During the late 1980s and early 1990s, at the same time that environmental historians were analyzing "nature" as a racialized and historically produced category, environmental justice scholars were exploring what these constructions of nature meant for people of colour in the US. The social movement and academic field of environmental justice offers another challenge to the "nature of nature" as a category unmarked by race, class or gender. The environmental justice movement famously defined the "environment" as where we "live, work, and play". The movement did so to complicate the view of environment-*as*-nature and to put a more complex set of spaces into environmentalism (e.g., urban, housing, workplace, Native lands).

This discursive expansion also brought additional and diverse bodies and issues into environmental spaces—the lead-poisoned, the occupationally injured, the food insecure.[4] But environmental injustice was always considered as also **more-than-human**.[5] This emphasis on interconnectedness and Mother Earth was embedded in the *Principles of Environmental Justice*, a movement manifesto produced in 1991 (EJNET.org, undated). As David N. Pellow writes in his account of critical environmental justice, the movement was never purely anthropocentric but also includes animals, the watershed, and the air basin (2017).[6]

This interconnectedness as a worldview carries over to the disciplinary frameworks for understanding environmental problems and their roots. Generally, contemporary sustainability research and policy programmes rely heavily on mathematical models and economic methods, whereas humanities and social science agendas tend to draw from studies of culture, history, and the philosophy of values. To illustrate, scientific and economic modelling of future climates dominates climate change analysis. But the profound inequalities associated with rapid changes in climate—including massive population displacement, catastrophic disruptions in food systems, and political instability—require scholarship that can uncover the characteristic human responsibilities for and responses to climate change.

This separation between the environmental sciences and the humanities and humanistic social sciences is a persistent problem within both sustainability studies *and* environmental justice praxis. Geographer John Robinson clarifies the role of science in sustainability research, writing that:

> good science is necessary, but . . . in the end, sustainability is an issue of human behaviour and negotiation over preferred futures, under conditions of deep contingency and uncertainty. It is an inherently normative concept, rooted in real world problems and very different sets of values and moral judgments.
>
> (2004, p. 379)

In other words, he continues: "Science can inform but not resolve, scientific analysis embeds values, other forms of knowledge" (p. 380).

Dynamic conversations around sustainability and social justice have rarely been the subject of the inclusive inquiry that places interdisciplinarity and questions of social and environmental justice at its centre. This task has precedents in just, critical, and situated sustainabilities frameworks.

Just, critical, and situated sustainabilities

Sociologists Julian Agyeman, Robert Bullard, and Bob Evans argue in their ground-breaking *Just Sustainabilities* that a sustainable society is a just society. They write of the linked relationship between poverty and environmental problems, the disproportionality of pollution exposure among disenfranchised populations (i.e., low-income and/or racial minorities), and the need to move toward sustainable development. They defined "just sustainabilities" as "the need to ensure a better quality of life for all, now and into the future, in a just and equitable manner, whilst living within the limits of supporting ecosystems" (Agyeman et al., 2003, p. 5). Each scholar is impacted by different intellectual and activist traditions. For example, in several of his pieces, Agyeman (citing Campbell) highlights how sustainability can be a "powerful and useful organizing principle for planning" but only if the various conflicts (property, development, and resources) are balanced and understood (2013, p. 298). Agyeman further discusses the four principles of **just sustainabilities**: improving our quality of life and well-being; meeting needs of present and future generations (equity); focusing on recognition and process; and living within ecosystem limits.

In recent years, sociologist Miriam Greenberg, with a number of collaborators, has articulated what she calls **critical sustainabilities**, with a focus on Northern California. Like Agyeman, Bullard, and Evans, Greenberg and those associated with critical sustainabilities highlight the variegated and contested notions found within "sustainability" discourse and practice. In this formulation, sustainability can be explained as a variegated concept that draws upon distinct repertoires, specifically the eco-oriented, market-oriented, justice-oriented, vernacular and/or utopian (Greenberg 2013, 2018). Thus, a market-oriented sustainability can align with environmental extraction, while drawing heavily upon sustainability discourses, or what was termed, in an earlier political moment, "greenwashing". Critical sustainability draws from practice and policy, but moves more squarely into critical sociological and environmental theoretical terrain. It is also more grounded in a place (Northern California) and place-making, more generally, vis-à-vis art and public space.

In a recent volume, a number of scholars call for a **situated sustainability** framework which centralizes interdisciplinarity and social justice (Sze, 2018). This focus on interdisciplinarity is two-fold, both within the academy (specifically the humanities, humanistic social sciences, and science, technology, and engineering), and between the academic and the public spheres. Segregating academic knowledge from lived/community knowledge—especially the knowledge of politically and culturally marginalized people—is a form of epistemic exclusion, as the community development scholar Jonathan London, urban ecologist Mary Cadenasso, and I have argued elsewhere (2018). Indeed, the move for collaboration across disciplines needs to be accelerated, according to the sustainability scientist Thaddeus Miller and his colleagues. They suggest that sustainability science and knowledge as currently generated by academic institutions is inadequate for actually transitioning to a sustainable society (2011). Miller's call for sustainability scientists to take seriously cultural values and other works of knowledge—including different disciplinary approaches—is echoed in the work of environmental humanists, linguists, and historians (Adamson, 2017; DeLoughrey and George Handley, 2011; Bird Rose et al., 2012).

If, as Robinson argues, sustainability is both about human behaviour and negotiation over "preferred futures", then the question of sustainable futures is tied implicitly to storytelling and imagination. Science fiction writers have been plumbing the real world to imagine dystopian and utopian pathways along the lines of those described by authors like Octavia Butler (Streeby).

Understanding culture through storytelling, art, and history is thus crucial for developing diverse ways of adding to our knowledge about environmental issues. We can understand the future, in part, by understanding the past, as well as through epistemic claims that differ from the dominant perspectives that frame sustainability. Different methods of environmental knowing, living, and thinking in interdisciplinary and attentively political ways therefore make sustainability stronger. In turn, making sustainability more socially just benefits those who are the most environmentally harmed.

The situated sustainability framework argues that we need to take history seriously, especially from **postcolonial** and **decolonial** perspectives.[7] The historical dimension is not a large focal point in just and critical sustainabilities. Situated sustainability centralizes environmental history and justice. These perspectives have tended to be overlooked, yet they are crucial in emergent discussions on the **Anthropocene**, the era during which human impacts have begun to shape geologic time. Despite the considerable scientific discussion in the last two decades about the Anthropocene, there has been far less research on how history, justice, and inequality are related to climate change.[8]

The particular contribution of social science and humanities-based historical research is to better centre politics, power, and global flows, or what the Indigenous scholar Makere Stewart-Harawira calls the "challenge to knowledge capitalism" (2013). Although she does not specifically address the Anthropocene/climate change or environmentalism/sustainability issues, there are many important points in her work to be considered, in particular regarding the centrality of Indigenous knowledge, long overlooked and degraded by mainstream institutions and disciplines. The historian Dipesh Chakrabarty examines the limits of dominant paradigms of the Anthropocene from the standpoint of colonial history in his seminal work, "Climate and Capital: On Conjoined Histories" (2014). In his overview of the collisions of "histories" of "the earth system, history of life, history of industrial civilization (mostly, capitalism)", he identifies issues of spatial and temporal justice (p. 1).

"Situated" sustainability is based on four factors. First, it draws from sustainability and sustainable development. Second, it emerges out of just sustainability and environmental justice research. Third, it is indebted to the environmental humanities/radical interdisciplinarity and to developing cross-sector knowledge coproduction with communities and knowledge makers. Finally, it differs from earlier works in all these fields in how it balances out the first three, and it centralizes issues of gender, race, and indigeneity and, to some

---

Box 9.2 Sustainable forestry and the Sustainable Development Institute on Menominee tribal lands

Kyle Whyte, et al. (2018) discuss the sustainable forest harvesting on Menominee tribal lands in Wisconsin as an effective example of Indigenous planning. Menominee forestry practices are a powerful example of sustainable forestry that places Indigenous worldviews, experiences, and perspectives at its core. They discuss the Sustainable Development Institute (SDI), housed at the College of the Menominee Nation, which has provided one of the first Indigenous-run higher education programmes in sustainability. They articulate how Indigenous ecologies "physically manifest Indigenous governance systems through origin, religious and cultural narratives, ways of life, political structures, and economies" (p. 159). The SDI, through its Indigenous planning framework, highlights how Indigenous communities adapted under settler colonialism while keeping core values intact. SDI is also focusing squarely on climate change and its impacts on Native communities through a reflective research praxis.

degree, gestures toward anti-capitalism. It emerges from, and transforms, environmental and climate justice agendas. Situated sustainability embraces core values that link environmental justice scholarship—a focus on praxis and social/racial justice and a fundamentally respectful appreciation of history and other modes of knowing and engaging in the world—including art and the humanities.

Situated sustainability implies an awareness of the multiple ways in which sustainability is marshalled and deployed in social and political life. Sustainability and environmental justice research provide one important crossroads. At the same time, an even more expansive sustainability and environmental justice research agenda does not necessarily address the *fundamental* political conditions that set the parameters for why and how vulnerability (environmental or otherwise) is disproportionately distributed, one of the key questions in environmental justice research. Radical critiques from major environmental justice scholars like Laura Pulido (2016) and David N. Pellow (2017) join many others in insisting that ideologies inherent in sustainability be named as such, specifically racial capitalism and hierarchies more broadly.

Conclusion

Scholars broadly interested in sustainability from across the disciplinary divides have and must continue to collaborate in a critical and constructive way, and with broader environmental publics. To better move toward robust and justice-oriented sustainabilities, we need a better vocabulary and analytic, not only to diagnose problems but also to understand how existing efforts for "sustainability" merely replicate existing epistemological and political problems. These are not new efforts—earlier scholars in just and critical sustainabilities have been making similar arguments for well over a decade. Situated sustainabilities build upon these earlier projects to show how central interdisciplinarity and environmental justice continue to be to the unrealized political and cultural projects for sustainability.

Environmental justice researchers, activists and policy-makers have been expanding their domains and fighting multiple battles for almost forty years. Although sustainability as a formalized framework in the US has been around for roughly the same amount of time, these two fields have developed largely on parallel tracks. However, now more than ever, their aspirational and political agendas demand more convergence and engagement. Given the scale of the environmental problems that hit the poor and powerless particularly hard, sustainability and environmental justice advocates, scholars, and policy-makers need more, not fewer, allies.

To better dislodge inequality, sustainability researchers and advocates require better tools, rather than reusing poor ones or tinkering around the edges of systems on the verge of collapse. Unjust environments are rooted in racism, capitalism, militarism, colonialism, land theft from Native peoples, and gender violence. The status quo is too deeply invested in the institutional forces and ideological structures that exacerbate already existing conditions of environmental and social injustice. States, corporations, markets, and their constituent institutions seek to maintain their dominance through a number of methods—including particular forms of sustainability. In order to move toward a more environmentally just sustainability, we must situate sustainability explicitly, as environmental justice research has already (32) done, in both interdisciplinary and political terms.

> **Follow-up questions**
> - What are the ideological functions of sustainability discourse, practice, and politics?
> - Does sustainability address race, class, gender, and other axes of power and identity? If so, how, and if not, then why not?
> - What are the limits and advantages of using sustainability as a framework for understanding the roots of environmental problems and defining responses?
> - What do the history of the definitions of sustainability mean in the search for environmental justice?

Notes

1 On colonialism and modernity, see also Chapter 7 of this volume.
2 On Indigenous environmental justice, see Chapter 20 of this volume.
3 On the history of the environmental justice movement, see also Chapter 2 of this volume.
4 On environmental justice issues related to toxic waste, food, and workers, see Chapters 10, 14 and 19 of this volume, respectively.
5 On justice beyond humanity, see Chapter 21 of this volume.
6 On critical environmental justice, see Chapter 22 of this volume.
7 On post- and decolonial environmental justice, see Chapter 7 of this volume.
8 For an Indigenous critique of the Anthropocene concept, see the work of Indigenous philosopher Kyle Whyte (2017) and Chapter 20 of this volume.

References for further reading

Estes, N., and Dhillon, J., eds. (2019), *Standing with Standing Rock: Voices from the #NoDAPL Movement* (Minneapolis: University of Minnesota).
Gómez-Barris, M. (2017), *The Extractive Zone: Social Ecologies and Decolonial Perspectives* (Durham, NC: Duke University Press).
Hoover, E. (2017), *The River Is in Us: Fighting Toxics in a Mohawk Community* (Minneapolis: University of Minnesota Press).
Kelley, R.D. (2002), G. *Freedom Dreams: The Black Radical Imagination* (Boston: Beacon Press).
Klein, N. (2018), *The Battle for Paradise: Puerto Rico Takes on the Disaster Capitalists* (Chicago: Haymarket Books).
Pauli, B. (2019), *Flint Fights Back: Environmental Justice and Democracy in the Flint Water Crisis* (Cambridge, MA: MIT Press).
Pellow, D.N. (2017), *What Is Critical Environmental Justice?* (Cambridge: Polity).
Sze, J., ed. (2018), *Sustainability: Approaches to Environmental Justice and Social Power* (New York: New York University Press).
Voyles, T.B. (2015), *Wastelanding: Legacies of Uranium Mining in Navajo Country* (Minneapolis: University of Minnesota Press).

References used in this chapter

Adamson, J. (2017), "Introduction: Integrating Knowledge, Forging New Constellations of Practice," in Adamson, J. and Davis, M. eds. *Humanities for the Environment: Integrating Knowledge, Forging New Constellations of Practice* (New York: Routledge), pp. 3–19.

Agyeman, J. (2013), *Introducing Just Sustainabilities: Policy, Planning, and Practice* (London: Zed Books).
Agyeman, J., Bullard, R.D. and Evans, B. eds. (2003), *Just Sustainabilities: Development in an Unequal World* (Cambridge, MA: MIT Press).
Bird Rose, D. et al. (2012), "Thinking through the Environment, Unsettling the Humanities," *Environmental Humanities*, vol. 1, pp. 1–5.
Cadenasso, M. and Pickett, S.T.A. (2018), "Situating Sustainability from an Ecological Science Perspective: Ecosystem Services, Resilience, and Environmental Justice," in Sze, J. ed. *Sustainability: Approaches to Environmental Justice and Social Power* (New York: New York University Press), pp. 29–52.
Chakrabarty, D. (2014), "Climate and Capital: On Conjoined Histories," *Critical Inquiry*, vol. 41, no. 1, pp. 1–23.
Critical Sustainabilities (undated), retrievable at https://critical-sustainabilities.ucsc.edu/sustainabilities-2/.
DeLoughrey, E. and Handley, G. eds. (2011), *Postcolonial Ecologies: Literatures of the Environment* (Oxford: Oxford University Press).
Greenberg, M. (2018), "Critical Sustainabilities: Towards a Situated, Urban Environmental Analysis," in Sze, J. ed. *Sustainability: Approaches to Environmental Justice and Social Power* (New York: New York University Press), pp. 180–195.
Greenberg, M. (Winter 2013), "What on Earth Is Sustainable? Toward Critical Sustainability Studies," *Boom: A Journal of California*, vol. 3, no. 4, pp. 54–66. DOI: 10.1525/boom.2013.3.4.54
Grober, U. (2012), *Sustainability: A Cultural History*, trans. R. Cunningham (Devon: Green Books).
International Geosphere–Biosphere Programme (January 15, 2015), "Planetary Dashboard Shows 'Great Acceleration' in Human Activity since 1950," *Global Change*, retrievable at www.igbp.net.
London, J.K., Sze, J. and Cadenasso, M.L. (2018), "Facilitating Transdisciplinary Conversations in Environmental Justice Studies," in Holifield, R., Chakrborty, J. and Walker, G. eds. *The Routledge Handbook of Environmental Justice* (New York: Routledge), pp. 252–263.
Meadows, D.H., Meadows, D.L., Randers, J. and Behrens, W.W. (1972), *The Limits to Growth: A Report for the Club of Rome's Project on the Predicament of Mankind* (New York: Universe Books).
Miller, T., Munoz-Erickson, T. and Redman, C.L. (2011), "Transforming Knowledge for Sustainability: Towards Adaptive Academic Institutions," *International Journal of Sustainability in Higher Education*, vol. 12, no. 2, pp. 177–192.
Patel, R. and Moore, J. (2018), *A History of the World in Seven Cheap Things: A Guide to Capitalism, Nature, and the Future of the Planet* (Berkeley: University of California Press).
Pellow, D.N. (2017), *What Is Critical Environmental Justice?* (Cambridge: Polity).
Principles of Environmental Justice, retrievable at https://www.ejnet.org/ej/principles.html.
Pulido, L. (2016), "Flint Michigan, Environmental Racism and Racial Capitalism," *Capitalism Nature Socialism*, vol. 27, no. 3, pp. 1–16.
Robinson, J. (2004), "Squaring the Circle? Some Thoughts on the Idea of Sustainable Development," *Ecological Economics*, vol. 48, no. 4, pp. 369–384.
Seghezzo, L. (2009), "The Five Dimensions of Sustainability," *Environmental Politics*, vol. 18, no. 4, pp. 539–556.
Stewart-Harawira, M. (2013), "Challenging Knowledge Capitalism: Indigenous Research in the 21st Century," *Socialist Studies/Études Socialistes*, vol. 9, no. 1, pp. 39–52.
Sze, J., Rademacher, A., Beamish, T., Grandia, L., London, J.K., Middleton, B.R., Warren, L. and Ziser, M. (2018), "Introduction," in Sze, J. ed. *Sustainability: Approaches to Environmental Justice and Social Power* (New York: New York University Press), pp. 1–25.
Whyte, K.P. (2017), "Indigenous Climate Change Studies: Indigenizing Futures, Decolonizing the Anthropocene," *English Language Notes*, vol. 55, no. 1–2, pp. 153–162.
Whyte, K.P., Caldwell, C. and Schaefer, M. (2018), "Indigenous Lessons about Sustainability Are Not Just for 'All Humanity'," in Sze, J. ed. *Sustainability: Approaches to Environmental Justice and Social Power* (New York: New York University Press), pp. 149–179.
Williams, R. (1980), *Problems in Materialism and Culture: Selected Essays* (London: Verso).
World Commission on Environment and Development (1987), *Our Common Future* (Oxford: Oxford University Press), retrievable at www.un.org.

Part II
Issues of environmental justice

10 Toxic legacies and environmental justice

Alice Mah

Learning outcomes

- Understand the role of anti-toxic struggles in the environmental justice movement.
- Appreciate the environmental justice legacies of particularly significant toxic disasters and anti-toxic movements within the United States and globally.
- Reflect critically on the uneven public attention given to particular toxic injustices, through the themes of "Black Love Canals" and "slow Bhopals".
- Understand the conceptual relevance of "slow violence" and "expendability" for examining toxic injustices around the world.

Introduction

Toxic landfills, waste, and industrial sites permeate our landscapes. We live in a toxic world (Boudia and Jas, 2014), and late modernity is itself toxic (Pellow, 2007). According to the anthropologist Kim Fortun (2014), the 1984 Bhopal disaster marked the start of "late industrialism", whereby toxicity persists in both new and disused forms of industry: contaminated industrial ruins, heaps of electronic waste, and leaking old pipelines full of accumulated residue. Yet, the burdens of widespread toxicity are unequally felt, heavily concentrated in low-income and minority communities around the world (Bullard, 1990; Edelstein, 2004; Lerner, 2010; Pellow, 2007; Taylor, 2014).

Just as toxicity has proliferated, so too have cases of toxic injustice. In this chapter, there is not enough space for the number of illustration boxes needed to summarize all the major cases of toxic disasters, injustices, and struggles. Stacked from floor to ceiling, all the required "boxes" would fill several large industrial warehouses. Of course, some toxic disasters are particularly infamous, such as Bhopal, Chernobyl, Love Canal, and the BP Deepwater Horizon spill, while others have been less visible, or have faded from public memory. By foregrounding some cases, others are neglected. As Robert Bullard (1990) argues, there are multiple "Black Love Canals" that are missing from the historical record. Yet, to continue the collective struggles for global environmental justice, it is important to highlight the toxic legacies and unjust burdens of our world, and we have to start somewhere.

First, this chapter situates anti-toxic struggles within the context of the environmental justice movement, examining the legacies of particularly significant toxic disasters in the United States and globally. Next, it critically reflects on two themes that demonstrate the uneven public attention to cases of **toxic injustice**: "Black Love Canals" (Bullard, 1990) and "slow Bhopals" (Zavestoski, 2009), drawing on environmental justice concepts of "slow violence" (Nixon, 2011) and "expendability" (Pellow, 2018; see also Chapter 22 of this volume). The chapter concludes with a discussion of the enduring global problem of toxic inequality.

Anti-toxic struggles and environmental justice

The first major issue in the environmental justice movement in the United States was the disproportionate siting of hazardous waste in poor communities of colour.[1] In 1982, African American leaders, activists, and residents protested the siting of a toxic polychlorinated biphenyl (PCB) landfill in a majority-black community in **Warren County**, North Carolina. Even though the protests did not prevent the landfill from going ahead, Warren County is nevertheless widely cited as the beginning of the environmental justice movement. As David N. Pellow (2018, p. 9) argues, there is a "long environmental justice movement" dating back centuries, featuring environmental inequalities and the struggles against them. However, Warren County was significant because it gave momentum to the first national environmental movement led by people of colour. The injustice of toxic dumping on a poor, black community resonated with the experience of many communities of colour who were living near toxic hazards.

The Warren County protests led to the landmark United Church of Christ (1987) study *Toxic Wastes and Race in the United States*, which identified national patterns of disproportionate toxic waste facility siting in relation to race. The problem of environmental racism was examined further in Bullard's (1990) significant study, *Dumping in Dixie*, and discussed at the first **National People of Color Environmental Leadership Summit** in 1991. Since the 1990s, the environmental justice movement has expanded to address issues beyond waste facility siting, and across a broader range of geographical and temporal scales (Agyeman et al., 2016; Martinez-Alier et al., 2016; Nixon, 2011; Pellow, 2018; Schlosberg, 2013; Whyte, 2018). However, despite the widening scope of environmental justice, the problem of discriminatory toxic waste siting has remained a central concern.

The follow-up study, *Toxic Wastes and Race at Twenty* (Bullard et al., 2007), revealed that communities of colour and poor communities were still being used as dumping grounds for all kinds of toxic contaminants. The authors discovered evidence that the clustering of environmental hazards, in addition to single sources of pollution, presented significant threats to communities of colour. Furthermore, the research showed that polluting industries frequently singled out communities of colour in siting decisions, countering the "**minority move-in hypothesis**": the claim that people of colour voluntarily move into contaminated communities rather than being targeted in situ by dirty industries. The key finding in the study was that, despite the growth of the environmental justice movement, race still matters in the siting of toxic waste.

Anti-toxics movements in the United States

The history of the anti-toxics movement in the United States is related to that of the environmental justice movement, yet is also distinct from it. The Love Canal disaster (Box 10.1) was the seminal moment within the **anti-toxics movement**. In 1978, a leaking toxic dump with over 20,000 tonnes of chemical waste was discovered buried beneath an elementary school in the white middle- and low-income residential community of LaSalle in Niagara Falls, New York. The health effects for the residents were staggering, with high incidences of cancer, miscarriages, rare diseases, and birth defects. It was the first US state of emergency to be declared over a human-made disaster. Love Canal is widely cited as the beginning of the anti-toxics movement and led to the creation of the Superfund Act of 1980, national legislation to tax corporations to clean up hazardous waste sites throughout the United States.

Box 10.1   Love Canal

Between 1942 and 1953, Hooker Chemical Corporation buried approximately 20,000 tonnes of toxic industrial waste in the abandoned Love Canal, named for the entrepreneur William Love who had created the canal as a dream project. The company sealed the waste with a clay cap but did not line the canal to prevent leakage. In 1953, Hooker Chemical sold the 16-acre site to the Niagara Falls Board of Education for a nominal $1, with the clause that it would not be held responsible for the harmful effects of the waste. In the late 1950s, an elementary school and one hundred houses were built on top of the dump, and the white working-class community of LaSalle grew up around this area.

The new residents were not aware of the area's toxic history, but they became suspicious when chemical liquids began erupting in their basements and backyards following a particularly harsh winter in 1977. Concerned first about their housing values, residents only became seriously worried when a number of miscarriages, birth defects, and illnesses started to occur in their community. Under the leadership of Lois Gibbs, residents fought a long protracted battle for relocation and compensation (Gibbs, 1998). President Jimmy Carter declared it a state of emergency, and residents were evacuated from the immediate vicinity. Love Canal hit media headlines, leading to the creation in 1980 of the federal "Superfund" programme, which taxed polluting industries and set up a $1.6 billion trust fund for cleaning up hazardous waste sites. The Superfund is managed by the US Environmental Protection Agency and has since been used to clean up hundreds of toxic sites. Love Canal was officially removed from the Superfund list in 2004, but controversies continue over compensation and the enduring toxic legacies of the site (Telvock, 2016).

There are parallels between the grassroots anti-toxic struggles in Warren County and Love Canal, yet each had a different political significance. Love Canal gained national notoriety as a warning about the public dangers of toxic pollution and corporate neglect. The fact that this disaster could happen in an ordinary white community in America hit a nerve. The publication of Rachel Carson's *Silent Spring* in 1962 had already brought public attention to the disastrous health effects of chemicals, but Love Canal made the lessons of toxic pollution more palpable. Love Canal showed the consequences of corporate negligence and the power of grassroots communities to fight against toxic hazards. By contrast, Warren County was the first anti-toxic protest by African Americans to gain widespread attention in the United States. It demonstrated the power of the civil rights movement to address environmental hazards as manifestations of social and racial inequalities. The public authorities responded differently in each case. More than 400 people in Warren County were arrested, while in Love Canal, the residents were relocated and many were given compensation.

The environmental justice movement in the United States brought together a wide range of grassroots groups, including civil rights, anti-toxics, community health, and environmental activists, amongst others. For example, the Great Louisiana Toxic March in 1988 along the Mississippi Chemical Corridor, or "**Cancer Alley**" (Box 10.2) from Baton Rouge to New Orleans involved labour union, environmental, and civil rights activists, who slept on African American church floors along the route (Mah, 2014, p. 198). Despite some successful environmental justice coalitions, there have also been tensions between white communities and communities of colour in anti-toxic struggles, given their unequal political power and systemic environmental racism (see Epstein, 1997). Environmental justice scholars and activists continue to debate the relative importance of race, class, gender, and other social categories of difference, versus multiple and intersecting forms of inequality (Pellow, 2018). These tensions account for the overlapping yet separate collective histories of the anti-toxic and environmental justice movements.

Box 10.2 Cancer Alley

One of the key battlegrounds over toxic injustice and environmental racism is the Mississippi Chemical Corridor, also known as "Cancer Alley", an 85-mile stretch of former plantation land along the Mississippi River between New Orleans and Baton Rouge with a high concentration of petrochemical facilities and oil refineries. In the early to mid-20th century, numerous plantations were sold to petrochemical companies, which were attracted by cheap natural resources and weak labour and environmental regulations. Many poor, rural African Americans, descendants of slaves, continue to live in "fenceline" communities adjacent to the petrochemical sites, with high levels of cancer and other illnesses (Allen, 2003; Davies, 2019; Singer, 2011; Ottinger, 2013).

Several environmental campaigns have been launched against these injustices, particularly in relation to environmental racism. Some notable victories have even been won in African American "fenceline" communities along the petrochemical corridor in Louisiana, such as Diamond and Convent (Lerner, 2006; Ottinger, 2013). However, these communities face powerful resistance from corporations and government lobbies, and the state continues to invest heavily in the petrochemical industry. Recent controversies have revolved around the construction of the Bayou Bridge oil pipeline between Lake Charles and St James Parish, a 95% African American community, and further petrochemical developments throughout Cancer Alley related to the fracking boom (Davies, 2019). In 2018, activists renamed Cancer Alley "Death Alley" to highlight the continuing environmental injustices in the area.

Global anti-toxics movements

The global anti-toxics movement emerged in the 1980s, coinciding with the US anti-toxics and environmental justice movements, in response to particular disasters. The Bhopal gas tragedy in 1984 is still recognized as the world's worst industrial disaster, resulting in the immediate deaths of between 3,800 and 15,000 people (conservative estimates) (Zavestoski, 2009). Bhopal survivors have since suffered from illnesses, lack of compensation, and early deaths. Global anti-toxic activists mobilized in response to Bhopal, with the phrase "no more Bhopals", to draw attention to Bhopal as a worst-case scenario and to advocate for international chemical regulations. However, Steve Zavestoski (2009, p. 391) argues that the anti-toxics movement's calls for "no more Bhopals" had little meaning for Bhopal survivors who were dealing with chronic health problems and abandonment by the state. Reflecting on the historical significance and enduring toxic legacies of Bhopal, Fortun (2014, p. 312) writes:

> Over half a million people were exposed; death figures remain contested, ranging from five thousand to more than twenty thousand. And the exposure continues. The Union Carbide factory in Bhopal hasn't operated since 1984, but the waste produced by it remains on site, underground, and in open ponds. Nearby water wells, still used by local communities, have high chemical as well as bacterial contamination.

Chernobyl represents another infamous incident in the late industrial era. It was the worst nuclear disaster in history, transcending national borders and geopolitical imaginaries (Kuchinskaya, 2014; Petryna, 2002). The death toll from Chernobyl remains contested, with illnesses from radiation near the Chernobyl Exclusion Zone still occurring today. In the years following the disaster, over 350,000 people were displaced as "environmental refugees" and forcibly relocated to different parts of the Soviet Union (see Gill, 2010, p. 861). Contaminated villages were razed, and 700,000 cleanup workers known as "liquidators" were recruited to detoxify the radioactive remains of the catastrophic meltdown inside "the Zone" (Petryna, 2002, Davies and Polese, 2015). Like Bhopal, Chernobyl was held up as a cautionary example in transnational anti-nuclear and anti-toxic struggles, but these debates did little to address the everyday struggles near the Chernobyl disaster site over routine exposures to radiation, state neglect, and inadequate compensation (Petryna, 2002; Davies and Polese, 2015).

Transnational anti-toxic movements have also mobilized around issues beyond particular disasters. For example, Pellow's (2007) *Resisting Global Toxics* charts transnational networks of opposition to the export of hazardous waste to poor communities and communities of colour around the world. Pellow exposes Global North-to-Global South flows of toxic disposal, through dumping and trade, as practices of institutional racism. He cites an infamous 1991 leaked internal memo on trade liberalization from Lawrence Summers, Vice President of the World Bank, which advocated the economic benefits of dumping toxic waste in developing countries.

Recently, global resistance to the export of hazardous waste has regained attention in the aftermath of the marine plastics crisis, which was catapulted into the public and political spotlight following the final episode of David Attenborough's *Blue Planet II* in December 2017. The timing coincided with China's National Sword Policy ban on the import of post-consumer waste, which came into effect in January 2018. These events sparked a global public and political backlash against single-use plastics. The report *Plastic and Health* (Azoulay et al., 2019), co-authored by a number of environmental research organizations, is the first comprehensive study to trace the human health impacts of plastic across its lifecycle, showing how "individually, each stage of the plastic lifecycle poses significant risks to human health" (p. 6). In May 2019, governments extended the Basel Convention on the Control of Transboundary Movements of Hazardous Wastes and their Disposal to include plastic waste.

Black Love Canals and slow Bhopals

Anti-toxic environmental justice scholars have highlighted the multiple social inequalities of living in a toxic world. In this section, I critically examine the legacies of toxic disasters and anti-toxic struggles through two related themes: "Black Love Canals" (Bullard, 1990) and "slow Bhopals" (Zavestoski 2009). Through considering these themes, I raise questions of structural inequality, drawing on Rob Nixon's (2011) concept of "slow violence" and Pellow's (2018) critical environmental justice concept of "expendability" (see also Chapter 22 of this volume).

In the aftermath of the Love Canal disaster, the Environmental Protection Agency warned in 1979: "This is not really where the story ends. Quite the contrary. We suspect that there are hundreds of such chemical dumpsites across this Nation." During my research on the toxic legacies of the chemical industry around Niagara Falls, in both New York and Ontario (Mah, 2012), I interviewed residents and workers in two such places: the Highland Avenue African-American community in Niagara Falls, New York, and the working-class community of Glenview-Silvertown, Niagara Falls, Ontario. Both low-income residential communities were living next to chemical brownfields with unknown or disputed levels of contamination. But despite the legacy of Love Canal in Niagara Falls, neither community supported strong environmental justice movements campaigning for change. In fact, many local residents viewed Love Canal as a source of stigma that had brought about devastating industrial decline. The few concerned voices were scattered and often had more immediate concerns to keep them occupied than the insidious environmental deterioration around them, such as getting jobs and maintaining weakening social communities.

Bullard (1990) has highlighted the problem of **"Black Love Canals"** throughout the United States, where issues of environmental injustice are deeply connected with environmental racism. For example, Bullard highlights the case of toxic DDT water contamination in the African American community of Triana, Alabama. In 1978, in the midst of the national media attention focused on Love Canal, residents in Triana raised complaints over ill-health effects and contaminated fish and waterfowl. Lawsuits in Triana against the Olin Corporation continued throughout the 1980s. Although the case is noted within environmental justice histories (see Taylor, 2014), it is not widely recognized or commemorated. Indeed, there have been numerous cases of environmental injustice in the United States which overwhelmingly burden minority and low-income communities, but which have not received the same national media attention as Love Canal. There are multiple Black Love Canals in the United States alone, and many more around the globe.

However, there has only been one Bhopal: it still remains the worst industrial accident in history. Zavestoski (2009) first made a different but related point about the unequal attention within the global anti-toxics movement following Bhopal. Second, he pointed to the wider problem of many **"slow Bhopals"**, places marked by "chronic low-level releases of toxic chemicals" (Zavestoski, 2009, p. 402). This observation resonates with Nixon's concept of **"slow violence"**: "a violence that occurs gradually and out of sight, a violence of delayed destruction that is dispersed across time and space, an attrition violence that is typically not viewed as violence at all" (Nixon, 2011, p. 2). Nixon poses an important question about the role of media attention in rendering violence visible:

> In an age when the media venerate the spectacular, when public policy is shaped primarily around perceived immediate need, a central question is strategic and representational: how can we convert into image and narrative the disasters that are slow moving and long in the making, disasters that are anonymous and that star nobody, disasters that are attritional and of indifferent interest to the sensation-driven technologies of our image-world?
>
> (p. 3)

Many researchers have observed themes of slow violence in "sacrifice zones" (Lerner, 2010), low-income neighbours with disproportionately high levels of chemical pollution (see Wiebe, 2016; Davies, 2019). The unfolding of slow violence is connected with uncertainty about the hazards and risks of chronic exposure to toxic chemicals. In her research on Indigenous mobilization against toxic pollution in Canada's Chemical Valley in Sarnia, Ontario, Wiebe (2016) argues that the Aamjiwnaang First Nation has faced the slow violence of "everyday exposures".

Some communities living with toxic pollution have mobilized and protested over environmental injustice. However, in many cases of slow violence, the risks of living with pollution are normalized. For example, recent research on peri-urban petrochemical areas in Nanjing, China (Mah and Wang, 2019, p. 1961), highlights the "accumulated injuries of environmental injustice", the unequal, multilayered, social effects of living and working in polluted environments in that country. Similarly, a study of a polluted Argentinian shantytown, *Flammable* by Auyero and Swistun (2008), examined the lived experience of toxic hazards, using the concept of "toxic uncertainty" to describe how communities live with confusion about toxic pollution, where "daily life is dominated by ignorance, error, and doubt regarding the sources and effects of toxicity" (p. 360).

Since Bhopal, there have been several major chemical explosions, notably in China with the disasters in Tianjin in 2015, which killed 173 people and injured hundreds, and Jiangsu in 2019, which killed seventy-eight people and injured more than 600. These catastrophes were reported in the local and international media, although quickly downplayed within the Chinese press. They circulated in headlines but had nowhere near the same global impact as Bhopal or Love Canal. In fact, even some disasters have been "fast", with immediate deaths, injuries, and media attention, but most eventually fade into the everyday realities of "slow violence", evident in the lingering toxic legacies in Warren County, Love Canal, Bhopal, Chernobyl, and multiple "Black Love Canals" and "slow Bhopals" around the world. In the aftermath of the BP Deepwater Horizon disaster in 2010 in the Gulf of Mexico, Nixon warns of this eventuality, asking: "What will be the long-term cascade effect of the slow violence, the mass die-offs, of phyloplankton at the food chain base? It is far too early to tell" (Nixon, 2011, p. 21).

Underpinning the slow, structural violence (see Galtung, 1969; Davies, 2019) of unequal and unjust toxic exposures is the problem of "expendability" (Pellow, 2018; see also Chapter 22 of this volume). Pellow (2018) proposes that indispensability is a key pillar of critical environmental justice studies (alongside intersectionality, scale, and state power). This idea builds on the work of critical race and ethnic studies scholar John Marquez (2014) on **"racial expendability"** to argue that, within a white-dominated society, people of colour are typically viewed as expendable. A critical environmental studies perspective strongly opposes dominant ideas of "socio-ecological expendability", arguing that "excluded, marginalized, and othered populations, beings, and things . . . must not be viewed as expendable but rather as *indispensable* to our collective futures" (p. 26). Provocatively, Pellow (2018) challenges the environmental justice idea of the **"sacrifice zone"** (Lerner, 2010), suggesting that the "implication of a 'sacrifice zone' is that one could presumably move away to safety, but the implication of expendability is that there is no escape" (p. 17).

The water contamination crisis in Flint, Michigan, which began in 2014, is one of the most recent national US scandals of toxic environmental injustice. A federal state of emergency was declared over lead contamination in the Flint water supply in January 2016. In March 2016, the Flint Water Taskforce published a report which found that: "Flint residents, who are majority Black or African American and among the most impoverished of any metropolitan area in the United States, did not enjoy the same degree of protection from environmental and health hazards as that provided to other communities." (Davis et al., 2016, p. 54) Since the crisis began, the city has worked slowly to replace the lead pipes, with 2,500 lead pipes still operating as of April 2019 and remaining local concerns over the safety of the water supply (Ahmad, 2019). National and international media headlines followed the Flint water crisis story as it unfolded, but, after the initial shock, Flint faded from media attention. It shifted from being a spectacular disaster to a case of slow violence. This parallels the dynamics of public memory surrounding many toxic disasters, struggles, and legacies.

Conclusion

In the first two decades of the 21st century, frequent toxic disasters have become the norm. The Indian Ocean Tsunami in 2004, Hurricane Katrina in 2005, the BP Horizon Spill in 2010, the Fukushima nuclear disaster in 2011, the Tianjin explosions in 2015, the Fort McMurray Tar Sands fire in 2016, the marine plastics waste scandal that went viral in 2017: the list continues. Around the world, there are mounting public discourses of ecological crisis and catastrophe, with climate crisis, plastic waste, smog in cities, global e-waste and hazardous waste problems, floods, forest fires, ice storms, and ever-increasing frequent toxic and extreme weather disasters. The threat and extent of toxicity encroaches on ordinary populations and generates fears. There is also a collective sense of being overwhelmed and desensitized, as toxic disasters and health scandals have become the norm. Public concerns over toxic injustices and inequalities risk being eroded in a world saturated with apocalyptic images of environmental devastation. **Ecological catastrophism** obscures the "long environmental justice movement" (Pellow, 2018, p. 9) and can stymie the possibility for critical environmental justice scholarship and action.

There has been some considerable international momentum behind the issues of climate justice and hazardous plastic waste, much of it online, involving many young people. Will these new environmental justice movements have enduring significance in the same way that Warren County, Love Canal, Bhopal, and Chernobyl changed public perceptions of industrial risk? For decades, environmental justice and anti-toxic activists have fought against toxic dumping across multiple scales and forms of social and environmental inequalities. There have been several success stories against these systemic injustices. As Bullard (1990) urges grassroots activists, it is important to celebrate citizen victories. However, many people in Black Love Canals and slow Bhopals around the world continue living in conditions of slow violence and toxic exposures, without recognition.

> **Follow-up questions**
> - What are some similarities and differences between the legacies of anti-toxic and environmental justice movements?
> - Why do Love Canal and Warren County have different political significance for grassroots movements?
> - How do the themes of "Black Love Canals" (Bullard, 1990) and "slow Bhopals" (Zavestoski 2009) relate to each other? Can you think of other examples?
> - How do the concepts of "slow violence" (Nixon, 2011) and "expendability" (Pellow, 2018) address questions of scale and power?

Note

1 On the history of the environmental justice movement, see Chapter 2 of this volume.

References for further reading

- Bullard, R.D., 1990. *Dumping in Dixie: Race, Class, and Environmental Quality*. Boulder: Westview Press.
- Nixon, R., 2011. *Slow Violence and the Environmentalism of the Poor.*. Cambridge, MA: Harvard University Press.
- Pellow, D.N., 2007. *Resisting Global Toxics: Transnational Movements for Environmental Justice*. New York: MIT Press.
- Taylor, D.E., 2014. *Toxic Communities: Environmental Racism, Industrial Pollution, and Residential Mobility*. New York: New York University Press.
- Zavestoski, S., 2009. The Struggle for Justice in Bhopal: A New/Old Breed of Transnational Social Movement. *Global Social Policy*, 9(3): 383–407.

References used in this chapter

Agyeman, J., Schlosberg, D., Craven, L. and Matthews, C., 2016. Trends and Directions in Environmental Justice: From Inequity to Everyday Life, Community, and Just Sustainabilities. *Annual Review of Environment and Resources*, 41: 321–340.

Ahmad, Z., 2019. Roughly 2,500 Lead Service Lines Left to Replace in Flint. *The Flint Journal* 11 Apr. www.mlive.com/news/flint/2019/04/roughly-2500-lead-service-lines-left-to-replace-in-flint.html Accessed: 3 July 2019.

Allen, B.L., 2003. *Uneasy Alchemy: Citizens and Experts in Louisiana's Chemical Corridor Disputes*. Cambridge, MA: MIT Press.

Auyero, J. and Swistun, D.A., 2008. The Social Production of Toxic Uncertainty. *American Sociological Review*, 73(3): 357–379.

Azoulay, D., Villa, P., Arellano, Y., Gordon, M., Moon, D., Miller, K. and Thomson, K., 2019. *Plastic and Health: The Hidden Cost of a Plastic Planet*. February ed. Washington, DC, and Geneva: Center for International Environmental Law.

Boudia, S. and Jas, N., 2014. Introduction: The Greatness and Misery of Science in a Toxic World. In Boudia, S. and Jas, N. [eds] *Powerless Science? Science and Politics in a Toxic World*. New York: Berghahn Books, pp. 1–28.

Bullard, R.D., 1990. *Dumping in Dixie: Race, Class, and Environmental Quality*. Boulder, CO: Westview Press.

Bullard, R.D., Mohai, P., Saha, R. and Wright, B., 2007. *Toxic Wastes and Race at Twenty, 1987–2007*. New York: United Church of Chris.

Carson, R., 1962. *Silent Spring*. Boston, MA: Houghton Mifflin Harcourt.

Davies, T., 2019. Slow Violence and Toxic Geographies: 'Out of Sight' to Whom? *Environment and Planning C*. Early view online: https://doi.org/10.1177/2399654419841063.

Davies, T. and Polese, A., 2015. Informality and Survival in Ukraine's Nuclear Landscape: Living with the Risks of Chernobyl. *Journal of Eurasian Studies*, 6(1): 34–45.

Davis, M., Kolb, K., Reynolds, L., Rothstein, E. and Sikkema, K., 2016. *Flint Water Advisory Task Force Final Report*, 21 Mar. Commissioned by the Office of Governor Rick Snyder, State of Michigan.

Edelstein, M.R., 2004. *Contaminated Communities: Coping with Residential Toxic Exposure*, 2nd ed. Boulder, CO: Westview Press.

Epstein, B. 1997. The Environmental Justice/Toxics Movement: Politics of Race and Gender. *Capitalism Nature Socialism*, 8(3): 63–87.

Fortun, K., 2014. From Latour to Late Industrialism. *HAU: Journal of Ethnographic Theory*, 4(1): 309–329.

Galtung, J., 1969. Violence, Peace, and Peace Research. *Journal of Peace Research*, 6(3): 167–191.

Gibbs, L., 1998. *Love Canal: The Story Continues*. Twentieth anniversary revised edition. Gabriola Island: New Society Publishers.

Gill, N., 2010. 'Environmental Refugees': Key Debates and the Contributions of Geographers. *Geography Compass*, 4(7): 861–871.

Kuchinskaya, O., 2014. *The Politics of Invisibility: Public Knowledge about Radiation Health Effects After Chernobyl*. Cambridge, MA: The MIT Press.

Lerner, S., 2006. *Diamond: A Struggle for Environmental Justice in Louisiana's Chemical Corridor*. Cambridge, MA: The MIT Press.

Lerner, S., 2010. *Sacrifice Zones: The Front Lines of Toxic Chemical Exposure in the United States*. Cambridge, MA: The MIT Press.

Mah, A., 2012. *Industrial Ruination, Community, and Place: Landscapes of Urban Decline*. Toronto: University of Toronto Press.

Mah, A., 2014. *Port Cities and Global Legacies: Urban Identity, Waterfront Work, and Radicalism*. Basingstoke: Palgrave Macmillan.

Mah, A. and Wang, X., 2019. Accumulated Injuries of Environmental Injustice: Living and Working with Petrochemical Pollution in Nanjing, China. *Annals of the Association of American Geographers*, 109(6): 1961–1977. DOI: 10.1080/24694452.2019.1574551

Márquez, J.D., 2014. *Black-Brown Solidarity: Racial Politics in the New Gulf South*. Austin: University of Texas Press.

Martinez-Alier, J., Temper, L., Del Bene, D. and Scheidel, A., 2016. Is There a Global Environmental Justice Movement? *The Journal of Peasant Studies*, 43(3): 731–755.

Nixon, R., 2011. *Slow Violence and the Environmentalism of the Poor*. Cambridge, MA: Harvard University Press.

Ottinger, G., 2013. *Refining Expertise: How Responsible Engineers Subvert Environmental Justice Challenges*. New York: New York University Press.

Pellow, D.N., 2007. *Resisting Global Toxics: Transnational Movements for Environmental Justice*. New York: MIT Press.

Pellow, D.N., 2018. *What Is Critical Environmental Justice*. Cambridge, UK: Polity.

Petryna, A., 2002. *Life Exposed: Biological Citizens after Chernobyl*. Princeton, NJ: Princeton University Press.

Schlosberg, D., 2013. Theorising Environmental Justice: The Expanding Sphere of a Discourse. *Environmental Politics*, 22(1): 37–55.

Singer, M., 2011. Down Cancer Alley: The Lived Experience of Health and Environmental Suffering in Louisiana's Chemical Corridor. *Medical Anthropology Quarterly*, 25(2): 141–163.

Taylor, D.E., 2014. *Toxic Communities: Environmental Racism, Industrial Pollution, and Residential Mobility*. New York: New York University Press.

Telvock, D., 2016. Landfill with Love Canal Legacy Still Poses Danger. *Investigative Post* 10 Feb. www.investigativepost.org/2016/02/10/landfill-with-love-canal-legacy-still-poses-danger/. Accessed: 3 July 2019.

United Church of Christ. 1987. *Toxic Wastes and Race in the United States.* New York: UC Commission for Racial Justice.

Whyte, K., 2018. The recognition paradigm of environmental injustice. In Holifield, R., Chakraborty, J. and Walker, G. [eds] *The Routledge Handbook of Environmental Justice.* London: Routledge, pp. 113–123.

Wiebe, S.M., 2016. *Everyday Exposure: Indigenous Mobilization and Environmental Justice in Canada's Chemical Valley.* Vancouver: UBC Press.

Zavestoski, S., 2009. The struggle for justice in Bhopal: A new/old breed of transnational social movement. *Global Social Policy,* 9(3): 383–407.

11 Biodiversity
Crisis, conflict and justice

Adrian Martin

Learning Outcomes
- Consider some of the main direct and underlying causes of biodiversity loss.
- Understand the different ways in which biodiversity loss can lead to environmental injustices.
- Identify some of the different ways in which humans value nature and justify biodiversity conservation.
- Learn to use a justice analysis to evaluate different ways of responding to the biodiversity crisis.

Introduction: from fear of crisis to living with crisis

The term '**biodiversity**' was coined and promoted mainly by US and British scientists and conservationists in the late 1980s (Redford and Mace, 2019). In its most common definition, it refers to the variation of life on earth at different scales, including genetic diversity, species diversity and ecosystem diversity.

There has been concern about protecting wildlife for many centuries, with areas dedicated for conservation by Indigenous and local people the world over. The spread of formal, legally protected areas took off in the late 19th century with the designation of national parks in the United States. However, concern about loss of biodiversity, as opposed to species and landscapes, is more recent. It is a concern that goes beyond loss of aesthetics or recreation, and beyond duty to other species. It is the realisation that anthropogenic change is causing a mass extinction event that threatens the future viability of human life on earth. The scientific evidence for this biodiversity crisis has been carefully evaluated by scientists worldwide and was published in a major United Nations assessment in 2019. This finds that 25% of species are currently threatened, with a million species facing extinction, many within decades. Furthermore, the drivers of extinction are still growing, and the rate of species loss is accelerating (Díaz et al., 2019). This accelerating loss of biodiversity is undermining the functioning of ecosystems in ways that threaten fundamental requirements of human well-being such as food security.

This rapid diminution of the diversity of life on earth has not yet set off the alarm bells in the way that the climate emergency has; and, similarly, '**conservation justice**' has received only a fraction of the attention from environmental justice movements compared to 'climate justice', developed in the next chapter. But loss of biodiversity is a matter for environmental justice. First, as Bryan Norton (2003) argues, we will not find a more compelling environmental ethic than the obligation for each generation to pass on the ecological conditions that provide the options necessary for future freedom and well-being. This is a primary role of biodiversity in **inter-generational justice**. Second, harms to current people are unequally and unfairly distributed. These harms include cultural loss, whereby nature-based local knowledge and practices cannot be sustained, and economic loss, for example when soils become less productive or crops become more vulnerable to pests, diseases or climate change. Such unequally distributed burdens are central to biodiversity's role in **intra-generational justice**.

There is a tendency to express global environmental challenges as shared problems, for example as threats to 'our common future' that require collective action to fashion a 'safe operating space for humanity'. But, in the short to medium term at least, there is no common future; no shared level of safety for humanity. Those who are currently enjoying the ephemeral pleasures of over-consumption will not be the ones struggling to protect their children from hunger. That fate is already falling to the small farmers and fishers who depend on biodiversity's contribution to ensuring good and reliable harvests on land and at sea. It is also falling to the multitude of Indigenous peoples and local communities whose cultures are defined and reproduced through place-bound relationships with the non-human world, for whom a life worth living is not limited to living 'from' nature, but also to living 'with', 'in' and even 'as' nature (O'Connor and Kenter, 2019).

Whilst previewing the types of harms arising from biodiversity loss, we should remember that costs arise not only from reduced biodiversity but also from efforts to mitigate those losses. A good rule of thumb is that all environmental governance interventions, however benign they may appear at first sight, are likely to create winners and losers and therefore give rise to claims about what is more or less just (Sikor, 2013). For example, **protected areas** have been the flagship of area-based conservation worldwide but have not always been innocent protectors of a common future. From the creation of Yellowstone National Park in 1872 to the Chagos Marine Protected Area in 2010, parks have too often involved the denial of local and Indigenous territories, including evictions and loss of access to resources (Brockington and Igoe, 2006). For smallholders who live around the borders of parks in the tropics, wild animals are a very common threat to human life, crops and livestock. Echoing the earlier point, there is no 'common' experience of biodiversity conservation: wealthier people tend to enjoy the benefits of nature conservation—for example, through the experience of wildlife tourism or nature documentaries—whilst the poor have often had to bear the costs (Adams et al., 2004). As we move towards a greater sense of an extinction crisis, with ever louder calls for emergency responses, it is therefore vital to be attentive to winners and losers, and to conservation justice.

The starting point of this chapter is therefore that both biodiversity loss and the ways in which we respond to it are matters of social as well as ecological justice. Inter-generational injustice, especially through erosion of biodiversity option value, is a crucial aspect of this sector. But intra-generational injustice is also highly problematic, involving harms across economic and cultural dimensions of human well-being that are disproportionately felt by less powerful social groups. The environmental justice movement has paid particular attention to how the costs of environmental harm fall on certain social groups. This is discrimination—overt or insidious—in that harms to some people are tolerated more than harms to other groups because of arbitrary social characteristics such as class, gender, ethnicity or nationality. It is environmental injustice.

The extinction crisis

According to the *Living Planet Report*, 60% of wild animals have been lost since 1970 (Barrett et al., 2018). Biologists describe the current period as the sixth age of mass extinction, in which rates of species loss exceed even the most dramatic of past mass extinction events. The loss of apex species, such as the decline of African elephants during the 21st century, has gained most media attention. Losses of apex herbivores and carnivores has a cascading effect on ecosystems (Estes et al., 2011)—for example, where loss of elephants disrupts the dispersal of fruit seeds, leading to loss of habitat for frugivores such as primates—with further repercussions through the food web. The term 'empty forest syndrome' (Redford, 1992) describes intact forests where species loss has cascaded through trophic levels. More recently, reports of devastating declines in insects have been grabbing headlines. A long study of flying insects in protected areas in Germany found a 76% decline between 1989 and 2016 (Hallmann et al., 2017). A subsequent global review found that such devastating losses of insects is commonplace, occurring across many insect taxa, in both terrestrial and aquatic systems and with rates of decline many times higher than for vertebrates (Sánchez-Bayo and Wyckhuys, 2019). As with the loss of apex mammals, massive loss of insects leads to cascading effects across trophic levels of ecosystems and ultimately reduces both the stability and productivity of the benefits from nature that are relied upon by humans. For example, 23% of terrestrial areas are already categorised as degraded to the point of reduced productivity, whilst annual crop losses from pollinator decline are estimated at between $235 billion and $577 billion (Díaz et al., 2019). In the United States, large- and medium-sized beekeepers no longer make most income from selling honey—their biggest revenue stream is now transporting bee colonies around on trucks to rent out their pollination services for almonds and other plantation crops (Ferrier et al., 2018).

Causes

The direct drivers of biodiversity loss are increasingly well known (Sánchez-Bayo and Wyckhuys, 2019, Díaz et al., 2019):

1. *Land and sea use change*, the biggest direct driver of biodiversity loss in terrestrial and freshwater ecosystems, including the extension and intensification of arable and livestock farming, urbanisation, mining and infrastructure.
2. *Direct resource use*, including fishing (the biggest driver of biodiversity loss in marine ecosystems), hunting, logging and other harvesting.
3. *Climate change*, which threatens biodiversity in even the most protected of waterscapes and landscapes.
4. *Pollution*, including the massive use of agricultural and other chemicals that directly kill insects (pesticides) and plants (herbicides) and overload water courses with nitrogen and phosphorous (fertilisers).
5. *Biological factors*, including invasive and alien species, that lead to loss of native species (such as the introduction of Nile perch and tilapia in Lake Victoria) or degrade habitat diversity and prevent regeneration (such as lantana in India).

These direct drivers of biodiversity loss are themselves shaped by indirect (or underlying) drivers such as major demographic, economic and cultural trends. It may seem obvious that major global trends such as consumption growth are core underlying drivers of biodiversity loss. However, to be critical of such deeply locked-in measures of societal progress is highly political and has therefore often been avoided; for example in David Attenborough documentaries and some multilateral environmental reports. However, these spaces for more critical questioning have been gradually opened up by political ecologists and environmental justice scholars. The need for transformative societal change that addresses such root causes is now making it into globally agreed reports such as the UN's 2019 Global Sustainable Development Report.

This is not the place to rigorously explore underlying causes, but a brief consideration of consumption provides a useful glimpse into why some injustices are produced with regularity. Global increases in demand for food crops, biofuel crops, fish, water and minerals are not discouraged by states. To the contrary, virtually all current states, regardless of political ideologies, consider economic growth and rising consumption as core measures of societal progress. The appearance of environmental constraints has had no real effect on this view of progress, other than to necessitate the rhetoric that consumption growth can be decoupled from environmental impact through technological innovation and efficiency. In practice, that rhetoric has proved wrong, partly because of the sheer scale of consumption demand in growing economies such as China, India and the United States, and partly due to the 'rebound effect', a phenomenon whereby production efficiencies themselves lead to increased demand, thus reducing expected environmental savings (Vivanco et al., 2016). For example, despite the availability of more energy efficient and technologies and renewables, global carbon emissions from the energy sector rose by 2.9% during 2018, the biggest rise in seven years (BP, 2019).

Any understanding of the biodiversity crisis really must confront the reasons why it is so difficult to challenge consumption growth as a societal goal and ideology. This is important for this chapter in two crucial respects: first, and most straightforwardly, because infinite consumption growth is proving incompatible with biodiversity conservation; second, and more complicatedly, because the prioritisation of consumption growth (rather than e.g. sufficiency and distribution) has been linked to the perpetuation of inequality. In *The Enigma of Capital*, David Harvey (2010) explains that the recent history of capitalism is characterised by its dependence on both economic growth and inequality. In an unequal world, perpetual economic growth is essential because it maintains the perception that everybody is benefiting—i.e. even the poor can consider themselves materially better off when a rising tide lifts all boats, even if wealth gaps are widening. Economic growth therefore helps to maintain good social relations and reduce social conflict.

Just as sustaining **inequality** requires economic growth, so sustaining economic growth requires inequality. One of the inherent contradictions (flaws) of capitalism is that compound growth periodically runs up against the kind of environmental constraints, such as the current biodiversity crisis, that require significant reconfigurations of global production systems. Looked at historically, the responses to these crises have been dependent on inequalities of power. In the 19th century, there was crisis of falling soil fertility in Europe and the United States. The response was to use economic and military power to extract and export the guano deposits of coastal Chile and Peru, with the help of indentured Chinese labourers (Clark and Foster, 2009). The resultant flows of nitrate temporarily fixed the constraint on growth in global superpower states, at the cost of ecological burden and 'guano wars' in the colonies. In the 19th century, the United Kingdom had also depleted its forests to the point that it lacked the timber to build the ships that protected its empire. The solution was to use the forests of India, Burma and elsewhere. The fixing of the energy crisis of the 21st century similarly requires the ability to open up new ecological flows to the centres of power, through new frontiers of extraction for lithium, biofuels and other 'green' energies (Muradian et al., 2012). The case of biodiversity conservation is not exempt from such criticism. We should remind ourselves that protected areas in the tropics have often been deeply unpopular with host communities and thus only possible due to power asymmetries and the framing of morality in terms of 'common futures' rather than, for example, current rights or equality.

As stated, the essence of environmental justice analysis is to explore the intersection between social and environmental inequalities. A focus on underlying causes reveals that this intersection is at least partly structured by the prevailing political economic system and the associated distributions of power. Seen through a conservation justice perspective, the current system is not only economically structured to produce material injustices (through the growth and inequality imperatives outlined previously) but also ideologically structured in ways that only recognise and respect some people's knowledge and values whilst marginalising others. This is sometimes referred to as the use of **'discursive power'**, whereby dominant actors determine how problems are framed in public debate, setting the substantive and moral terms of debate and, crucially, placing boundaries on what knowledge and values are deemed salient. As the example of 'sustainable intensification' (Box 11.1) shows, the typical effect is to foreground ideas that align with the incumbent regime whilst rendering invisible alternative framings rooted in other knowledge and value systems. For many Indigenous peoples, for example, nature is not separate to society (some do not even have a word for nature) and cannot be conceived as something purely instrumental to human progress—the variety of life on earth cannot be understood and valued in terms of its role in supporting benefits for human well-being. The marginalisation of these alternative ways of knowing and valuing nature is a form of injustice in itself (a failure of recognition), but is also a barrier to sustainability because it ensures that few heads are at the table at a historical moment when we clearly need new ideas and massive support for resolving the crisis.[1]

Box 11.1 Food systems, biodiversity and recognition injustices

Underlying systemic barriers to addressing biodiversity decline are well illustrated by challenges of transitioning to more sustainable and biodiverse food systems (e.g. Coolsaet, 2016). Agricultural expansion and intensification is the biggest driver of terrestrial biodiversity decline, directly replacing plant diversity with monoculture, replacing wildlife with livestock, consuming huge quantities of fossil fuels and polluting ecosystems with agro-chemicals. And, yet, efforts to reveal and address the fundamental failings and insecurities of the global food regime are consistently resisted by incumbent regimes of government and corporate actors. Discursive power is exercised to legitimise heavily contested terms such as 'sustainable intensification', which has secured its status as a legitimate 'green' strategy in the Sustainable Development Goals (Pretty et al., 2018), despite evidence that most existing intensification practices are damaging to biodiversity (Rasmussen et al., 2018). Sustainable intensification is argued to be readily acceptable to the incumbent regime because it aligns with the existing growth agenda and legitimises opportunities for capital investment (Newell and Taylor, 2018). For example, Li (2014) explores how the need to intensify agriculture for sustainable development has legitimised land acquisitions (or 'land grabs', as they are often termed) in the Global South by investors who can promise to enhance yields.[2] The World Bank is also accused of pursuing this agenda in its 2019 report on *Enabling the Business of Agriculture*, a policy approach that Mousseau (2019) considers to be unequivocally promoting the spread of large-scale industrial agriculture.

The framing of the agenda through discursive power is being used to present processes of land accumulation and intensification as 'sustainable development' solutions. Newell and Taylor (2018) identify one specific strategy used by government and corporate actors which is to blur the distinction between sustainable intensification and potentially more challenging alternatives such as 'climate smart agriculture' and **'agroecology'**. Blurring the distinction serves to co-opt and blunt the power of alternatives and, perhaps, gives the impression that 'sustainable intensification' is a more radical agenda than it really is. The dominance of such discourse of intensification, supported by incumbent regimes that include producers of fertilisers and other agricultural inputs (ACBIO, 2014), has facilitated a highly successful evasion of commitment to address the impact of agriculture on biodiversity loss and climate change.

Conceptualising conservation injustices

Justice and injustice are not fixed and universal categories, but are conceived in conjunction with diverse human cultures and values that produce a variety of notions of right and wrong. Western environmentalism has for more than a century debated between two different basic framings of the injustices arising from degradation of biodiversity and ecosystems. First, an anthropocentric framing focuses on nature's instrumental value to human well-being and on the harms to humans when such use is diminished or inequitably distributed. These are the traditional concerns of social justice and formed the basis for the environmental justice movement that took off in the 1980s. These harms to humans might be economic, arising from the fact that humans live from nature, but can also be cultural, because nature is the arena in which our societies and cultures evolve and play out (O'Neill et al., 2008). Thus, to degrade an ecosystem might reduce the economic basis of livelihoods and/or might remove the ability to live the kind of life that a person or community has reason to value. Second, ecocentric framings focus on the intrinsic value of nature and the harms experienced by nature itself.[3] Humans are not viewed as an exceptional species, uniquely deserving of moral concern, but as part of the biotic community. In this framing, the community of justice (those we are morally responsible towards) is extended to include elements of the non-human world, including, for example, sentient animals, or entities such as rivers and mountains. This is the terrain of **ecological justice** and, whilst it has only recently been incorporated into environmental justice thinking, it is highly relevant to biodiversity conservation, which is often motivated by combinations of anthropocentric and ecocentric values. For example, mountain gorillas have economic value as tourism assets, but many think they deserve protection regardless of that.

Anthropocentrism and **ecocentrism** represent polarised worldviews. In the former, what is valued is human well-being, and the protection of nature is instrumental to that; in the latter, nature itself is valued, and its protection is therefore an intrinsic good. The polarisation of this debate has often overlooked alternative worldviews, especially prevalent among Indigenous peoples, in which value lies not in humans or nature but in balanced relationships and in the virtue of being mindful of these (Whiteman, 2009). For this reason, we refer to three ways of valuing nature: instrumental, intrinsic and relational values (Chan et al., 2016). For example, as I write this chapter, I see a news report of the Lummi people of Washington state symbolically dropping a chinook salmon into the ocean, in response to the food shortage faced by orca whales due to human over-fishing and engineering of rivers. The Lummi are certainly not making this gesture of support to the orca out of any simple instrumental concerns for their own well-being (although their well-being is clearly linked to that of the orca). Nor are they doing this out of concern for an objective, external nature whose value is entirely independent of human preferences. As their chief explains to a journalist, it is hard to explain to others what values compel the Lummi to this act of commitment to the orca, but it can be likened to standing by their relatives in a time of need (Pulkkinen, 2019). Justice in such contexts might be likened to a healing process in which the right thing to do is to restore the balance of relationships (McCaslin, 2005). Whilst such deep relational experiences may be largely confined to Indigenous peoples, many more of us feel reciprocal relationships with pets, and it has been found that the great majority of conservation professionals hold worldviews that appear compatible with relational values: in a recent survey of more than 9,000 conservationists from 149 countries, 92% disagreed with the statement that 'humans are separate not part of' nature (Sandbrook et al., 2019).[4]

This three-way typology of how humans value nature—instrumental, intrinsic and relational—is useful for thinking about the justices and injustices of the biodiversity emergency. First, it helps to identify the range of subjects that are deserving of moral concern and commitment (the community of justice). Second, it helps to analyse the kind of harms arising from loss of biodiversity. In terms of the subjects of justice, instrumental framings point to harms to well-being of current and future human individuals and groups. Intrinsic framings expand the range of moral subjects to non-human nature. For example, in 2018, rights equivalent to personhood were granted to the Yamuna and Ganges rivers in India, the Whanganui River in New Zealand and the Atrato in Colombia. Relational framings do not so much point to new subjects of justice, but do introduce more holistic ways of understanding value and justice, as balance within assemblages of subjects. Equally importantly, recent moves to recognise relational framings of nature alert us to the plurality of ways in which people value nature and experience harms from loss and degradation. Indeed, one of the fundamental injustices experienced by those with different ways of knowing nature is that their values are ignored; that is that the kind of harm they experience—when orca whales, for example, are dying—is not recognised or factored into decision-making. Such recognition injustice is a major cause of conservation conflicts: situations where those with greater power assert particular, often singular, ways of valuing nature whilst dismissing values that arise from, for instance, relational perspectives. Such a failure to accept and account for multiple ways of valuing nature is increasingly considered to be a major obstacle, not only to effectively responding to the biodiversity emergency (Díaz et al., 2018, Pascual et al., 2017), but also to conservation justice (Martin, 2017).

This discussion resonates with the well-established typology of injustices explored in the first part of this book. Widely expressed in terms of three dimensions of injustice, it is related to: the unequal **distribution** of environmental goods and harms, the lack of inclusiveness of different groups in environmental decision-making **procedure**, and the systemic lack of **recognition** for some identities and associated ways of knowing and valuing nature.[5] These are different forms of harm that matter differently to different groups, at different times. The conservation profession typically 'sees' harms based on distribution and participation, and has therefore developed technical responses to these; for example, concerns about unfair distribution of the costs and benefits of protected areas has been a reason for introducing a number of mechanisms such as:

- *Alternative livelihood programmes*: where local people are no longer allowed to farm or harvest in the park, alternative ways of making a living are supported, such as beekeeping (see e.g. Roe et al., 2015);
- *Compensation schemes*: where monetary payments cover the costs of wildlife damage, including loss of crops, livestock and even human life (see e.g. Ravenelle and Nyhus, 2017);
- *Benefit-sharing schemes*: where a proportion of revenues from tourism and entry fees are shared with local communities, normally as provision of goods such as wells or health clinics (Cundill et al., 2017);
- *Payments for ecosystem services*: where local people are paid for their contribution to conservation, for example to cover the income they forego by not harvesting in the park.

Communities are often supportive of such distribution-oriented schemes, but they don't always hit the mark, notably when they don't align with the kind of harms people are experiencing. For example, the expansion of populations of large predators throughout much of Europe has increased conflicts with livestock farmers. This is often viewed as a distributional problem, leading to solutions that involve some form of financial compensation. But this type of compensation will not resolve feelings of injustice that arise primarily from relational values, such as a shepherd's sense of a duty of care towards her flock. In France, for example, wolves killed an estimated 10,000 sheep in 2016. Farmers received 3.2 million euros in compensation, but money itself will not remove their sense of injustice.

Protected-area conservation

Protected areas are clearly defined terrestrial or marine spaces that are dedicated to the long-term conservation of nature and associated cultural values (Dudley, 2008). Protecting biodiversity is in principle a universal good that will benefit all humanity, including rural smallholders whose livelihoods are most directly dependent on ecosystem services and who often have much to gain from defending their locales from mining, infrastructure or conversion to plantation crops. However, this potential alliance between rural communities and conservation in the Global South has been blemished by a long history of inequality in which the benefits are mainly enjoyed by wealthier groups whilst the costs are borne locally (Adams et al., 2004). The phrase **'fortress conservation'** became associated with the common practice of protected-area formation in which existing land users were evicted or excluded and then kept out by armed park staff (Brockington, 2002). In the United States, for example, early national parks such as Yellowstone failed to honour territorial claims of native peoples. In East Africa, wildlife had predominantly been a local asset until the 1880s but, following European colonisation, this changed quickly. In Kenya, the North and South game reserves were established in 1896, whilst Africa's first 'national park', the Albert National Park (now Virunga) was created in Belgian Congo in 1925. The expansion of park networks accelerated after the 1950s and the World Database on Protected Areas now includes more than 200,000 protected areas covering about 15% of terrestrial and inland water areas and 4% of marine areas (UNEP-WCMC, 2016).

The dominance of a fortress conservation model of protected areas has been supported by cultural, scientific and moral thinking. Culturally, Western environmentalism has been strongly influenced by belief in the value of pristine wilderness relative to human-shaped landscapes. Scientifically, this valuation of a 'pristine' nature was bolstered by early 20th-century **ecological climax theory**, according to which ecosystems could only achieve their ultimate, climax assemblage of fauna and flora in the absence of human co-habitation. Morally, this case for segregation was sealed through appeal to utilitarian ethics in which the benefit of wilderness for the many justified the imposition of costs on the few who faced exclusion from their territories. This model of conservation has proved remarkably resilient despite challenges to all three of its supporting pillars. Alternative models involving forms of co-existence between humans and wildlife employ alternative cultural traditions (e.g. relational values), alternative science (e.g. social-ecological systems thinking) and alternative moralities (e.g. rights-based ethics).

Despite the resilience of the colonial 'fortress' model, protected-area conservation practices have evolved. One key driver of change has been global agreement on prioritising poverty alleviation, linked to a narrative of **'sustainable development'**. The latter, popularised by the 1987 Brundtland report (WCED, 1987) was a game-changer inasmuch as it altered the terms of debate, away from the view that conservation would inevitably pit global needs against local ones towards a more optimistic view that conservation and local development could be pursued in tandem. The World Parks Congress and other conservation policy forums began to focus on more people-friendly approaches to conservation (Roe, 2008). Greater participation through forms of community conservation became increasingly popular, as did more 'alternative livelihoods'-oriented interventions, known as integrated conservation and development projects (ICDPs). By the late 1990s, it was reported that nearly all donor funding for conservation in Indonesia was for ICDPs (Wells et al., 1999) and this shift in funding priorities became worldwide (Miller et al., 2013). In the early 2000s, market-based (green economy) approaches to integrating conservation and development gained rapid popularity, especially in Latin America. These approaches include payments for ecosystem services schemes, certification and labelling schemes, and biodiversity and carbon offsetting (see e.g. Martin, 2017).

From an environmental justice perspective, the shift from fortress conservation to stronger emphasis on development and poverty alleviation is to be welcomed. But there are some limitations. First, implementation has been patchy, such that exclusionary practices are still commonplace. Second, implementation is often not well done. For example, in the Western Ghats of India, up to 25–30 deaths from elephants occurred annually, and 64% of households near protected areas have reported crop losses (Bal et al., 2011). There is a compensation scheme, but villagers report that it is difficult to access. According to an expert panel report, poor implementation could undermine conservation because the sense of injustice is eroding the long-standing cultural relationship between humans and elephants (Force, 2012). Third, as has been argued in this chapter, these approaches tend to address more 'superficial' issues of economic distribution, without addressing more fundamental concerns related to recognition of alternative ways of knowing and living with nature (another way of putting this is to say that these approaches do little to 'decolonise' conservation practices). Despite the limited scope for delivering conservation justice, economic interventions may still be important in some contexts. For example, during earlier work on mountain gorilla conservation, I found that focusing on economic inequities (through e.g. benefit-sharing schemes) was a necessary condition for tackling more fundamental issues. Addressing economic concerns, such as loss of crops and reduced access to forest resources, helped to develop goodwill and trust that was a first step towards deeper forms of collaboration among stakeholders (Martin et al., 2011).

Towards just conservation

Whilst there have been significant conservation successes, the overall picture is bleak. Reporting in July 2019, the Red List of over 100,000 species compiled by the International Union for Conservation of Nature (IUCN) did not record an improved status for a single species. Current rates of biodiversity loss represent an injustice to future generations and to those current people whose well-being is already harmed by failing productive systems and loss of opportunity to follow culturally significant practices. Furthermore, currently dominant models of conservation too often perpetuate failures of recognition; for example by seeking to resolve conflicts in ways that require assimilation to instrumental ways of valuing nature. In this final section, I consider this current predicament and the growing call for 'transformative' change to conservation. Reformist changes are those that primarily operate within existing framings and therefore do not challenge underlying drivers of injustice in political economies and incumbent conservation regimes. Nor do they seek to decolonise conservation, recognise relational value, reject the growth imperative or challenge the dominant segregationist model for protected areas. By contrast, I refer to transformative changes as those that challenge incumbent framings and regimes and seek to redistribute power in ways that serve environmental justice and sustainability.

Box 11.2   Debates about the future of conservation

Recent debates about future directions for biodiversity conservation highlight two competing perspectives: 'new conservation' and 'traditional protectionism' (Sandbrook et al., 2019; Büscher et al., 2019). These are largely reincarnations of old debates between anthropocentrics and ecocentrics.

In the last twenty years, there has been a major scientific effort to quantify the benefits that humans derive from biodiversity and ecosystem services, including the influential report on The Economics of Ecosystems and Biodiversity (TEEB, 2010). The 'new conservation' position springs from this enhanced knowledge of the instrumental values of nature for human well-being. It argues that conservation should be presented in anthropocentric ways because this is the only realistic way to get enough people to support it. New conservation therefore calls for biodiversity to be aligned with prevailing societal goals such as economic growth and poverty elimination. It calls for conservation to be pursued through partnerships with businesses and by getting the financial incentives right—for the many, not just the few (see e.g. Kareiva and Marvier, 2012).

By contrast, advocates of protectionism argue that an ecocentric ethic (based on intrinsic values of nature) is the only credible basis for sustained care for nature. They argue that market-based values are fickle and favour those species and habitats that are currently most valued by humans. Whilst they see poverty elimination as a worthy objective, they suggest it is better to pursue this separately to biodiversity conservation rather than confusing the two. An example of a protectionist position is the

> 'Half-Earth' call for a massive expansion of protected areas (Wilson, 2016). From a social justice perspective, this mentality is alarming. The idea that half the world should be for humans and half for nature seeks to impose a predominantly Western, segregationist worldview of human relationships with nature onto the rest of the world. Furthermore, the best available evidence to date suggests that extending protected areas without addressing underlying drivers of biodiversity loss would be ineffective (Geldmann et al., 2019).

Box 11.2 summarises two contrasting schools of thinking about the future direction conservation should take. Bram Büscher and Rob Fletcher (2019) succinctly capture why neither of these views represents a sound vision for a transformative and just future for conservation. The 'new conservation' camp has the advantage of rejecting segregationist and elitist approaches, but it fails to challenge the inequalities or unsustainability of current economic systems and priorities. The 'protectionist' camp does challenge current economic systems, but it is essentially an upscaling of a segregationist model of protected-area conservation that is unlikely to be effective and would fail to recognise other ways of knowing and living with nature. I finish this chapter by suggesting two future directions that I think could help a transformation to a more just conservation. The first is, admittedly, more reformist initially but includes practical steps that I think can build towards more transformative change. The second is a brief attempt to exemplify a more radical alternative that has already gained some traction.

Towards justice in protected and conserved areas

The IUCN now refers to 'protected and conserved areas', a phrase that incorporates areas that are conserved by local and Indigenous peoples outside of legally designated protected areas. Such areas, sometimes termed other effective area-based conservation measures (OECMs), provide diverse alternatives to mainstream conservation, incorporating more plural values of nature and often involving co-existence of people and nature rather than segregation. This recognition of alternative conservation practices (and alternative worldviews) is crucial for a more just conservation, and it does not involve any necessary trade-off between social justice and conservation effectiveness: there is growing evidence that such alternatives, including the empowerment of local communities and the granting of territorial rights to Indigenous peoples, is not only ethical but may also be more effective than traditional protected areas (Schleicher et al., 2017).

In addition to diversifying the models of area-based conservation, there are potentially significant plans to incorporate the assessment of social equity into performance monitoring and planning processes. A collaborative process led by the International Institute for Environment and Development has employed the environmental justice typology of distribution, procedure and recognition to develop an equity framework for assessment in protected and conserved areas (Schreckenberg et al., 2016; Franks et al., 2018). Use of this framework has now been adopted as voluntary guidance by the Convention on Biological Diversity and is being promoted by IUCN.

Currently, a tool for implementing equity assessments is being field-tested with the hope that it can be widely used as a means of enhancing the justness of areas-based conservation. Whilst the application of an equity assessment tool is reformist in some respects, because it seeks to tweak the existing system rather than change it, one rationale is that the cumulative educational effect of analysing justice might contribute to transformational change; for example, by mainstreaming multi-stakeholder discussion about recognition into governance processes.

Towards alternative worldviews: towards interconnected social and ecological justice

> what happens when human exceptionalism and the utilitarian individualism of classical political economics become unthinkable. . .? Seriously unthinkable: not available to think with.
> Donna Haraway (2016, p. 57)

Biocultural diversity is one alternative framing of diversity that deliberately integrates human and biological values into a holistic expression of diversity. It reconceives conservation as co-inhabitation or co-existence of humans and non-humans, with the value of diversity in the balance of relationships (Rozzi, 2018). In some ways, it is an antithesis of the 'Half-Earth' way of thinking because it promotes diversity everywhere. For Ricardo Rozzi, there are two key conditions for making progress towards such an alternative way of thinking about and practising conservation. First, we need to break free from some of the mental dispositions that we are currently conditioned to think with. First and foremost, this means ceasing to think with the dominant economic ideology that makes a goal of economic growth, consumerism and individualism. It is this way of thinking that now threatens the destruction of humans and the rest of nature. Second, we need to understand and embrace the many past and current cultures "that promote harmonious forms of co-inhabitation among communities of diverse human and other-than-human beings" (Rozzi, 2018, p. 304).

I am using 'biocultural diversity' here as an umbrella term to try to capture old and new ways of thinking about conservation based around co-habitation and that cut across instrumental, intrinsic and relational ways of valuing nature (I would include e.g. '**Buen Vivir**', 'earth stewardship' and 'conviviality' under this umbrella). The idea of biocultural diversity has become important to Indigenous peoples as a reframing of conservation that distils diverse wisdoms about living well and living with nature. As an example, Rozzi looks at the 2009 constitution of the plurinational state of Bolivia, including the phrase '*Suma Qamaña*'. This translates as 'living well together'. In the Aymara language, it means to inhabit, in the sense of both living in and living with, and it emphasises the relational value of co-habitation. *Suma* means beautiful, but in the sense of perfect and fully formed. It conceives 'well' as an achieved and sufficient state. This Indigenous worldview of living well together provides an alternative to the individualism of neoliberalism, to ideas of living well that rely on never-ending progress and to non-relational ways of knowing and valuing nonhuman nature.[6]

Conclusion

This chapter has shown that biodiversity loss is a matter for justice, but so too is the way in which society chooses to respond to this loss. Both a lack of response and the wrong response can produce and reproduce injustices. These injustices tend to follow existing social chasms and discriminations such that already marginalised groups suffer disproportionately. This includes, for example, poor subsistence farmers who are already bearing the burden of the conjoined biodiversity and climate-change crisis, and it includes Indigenous peoples and local communities worldwide who are politically marginalised and whose cultures and wisdom are too often sidelined.

Understanding and valuing alternative worldviews is in itself an essential basis for environmental justice. It is a question of recognition and fulfils what is increasingly viewed as an obligation to decolonise conservation. One useful exercise is to think about the kind of framing or worldview that is evident in particular notions about biodiversity and conservation. For example, is it mainly based on instrumental, intrinsic or relational values of nature? Is it coming from a framing of living *from* nature, living *with* nature, or perhaps living *in* nature or *as* nature? Perhaps, above all, is it confronting the underlying ideologies and practices that are currently driving us to the brink of ecological disaster?

Follow-up questions

- Do you think that transformational change can be achieved without changes to the current economic system?
- In order to progress towards a more just conservation, how important is it to recognise and promote worldviews and values that are currently marginalised?
- What are the strengths and weaknesses of current models of biodiversity conservation, including the 'new conservation' and 'protectionist' models?

Notes

1. On epistemic justice, see Chapter 7 of this volume; on recognition, see Chapter 5 of this volume; on Indigenous environmental justice, see Chapter 20 of this volume.
2. On environmental justice issues related to food and agriculture, see also Chapter 14 of this volume.
3. On ecocentrism, see also Chapter 21 of this volume.
4. On relational values, see also Chapter 20 of this volume.
5. On distribution, procedure and recognition, see Chapters 3, 4 and 5 of this volume, respectively.
6. On the concept of *Buen Vivir* in Latin America, see Chapter 7 of this volume.

References for further reading

- Díaz, S., Pascual, U., Stenseke, M., Martín-López, B., Watson, R. T., Molnár, Z., Hill, R., Chan, K. M., Baste, I. A. & Brauman, K. A. 2018. Assessing nature's contributions to people. *Science*, 359, 270–272.
- Martin, A. 2017. *Just Conservation: Biodiversity, Wellbeing and Sustainability*. Abingdon: Routledge.
- Schreckenberg, K., Franks, P., Martin, A. & Lang, B. 2016. Unpacking equity for protected area conservation. *Parks*, 22, 11–26.

References used in this chapter

ACBIO. 2014. *The Political Economy of Africa's Burgeoning Chemical Fertiliser Rush*, African Centre for Biosafety. www.acbio.org.za/sites/default/files/2014/12/Fertilizer-report-201409151.pdf

Adams, W. M., Aveling, R., Brockington, D., Dickson, B., Elliott, J., Hutton, J., Roe, D., Vira, B. & Wolmer, W. 2004. Biodiversity conservation and the eradication of poverty. *Science*, 306, 1146–1149.

Bal, P., Nath, C. D., Nanaya, K. M., Kushalappa, C. G. & Garcia, C. J. E. M. 2011. Elephants also like coffee: Trends and drivers of human—Elephant conflicts in coffee agroforestry landscapes of Kodagu, Western Ghats, India. *Environmental Management*, 47, 789–801.

Barrett, M., Belward, A., Bladen, S., Breeze, T., Burgess, N., Butchart, S., Clewclow, H., Cornell, S., Cottam, A. & Croft, S. 2018. *Living Planet Report 2018: Aiming higher*. Gland, Switzerland: WWF International.

BP. 2019. *Statistical Review of World Energy*. www.bp.com/content/dam/bp/business-sites/en/global/corporate/pdfs/energy-economics/statistical-review/bp-stats-review-2019-full-report.pdf

Brockington, D. 2002. *Fortress Conservation: The Preservation of the Mkomazi Game Reserve, Tanzania*. Oxford, UK: James Currey.

Brockington, D. & Igoe, J. 2006. Eviction for conservation: A global overview. *Conservation and Society*, 4.

Büscher, B., Fletcher, R. 2019. Towards convivial conservation. *Conservation & Society*, 17, 283–296.

Chan, K. M., Balvanera, P., Benessaiah, K., Chapman, M., Díaz, S., Gómez-Baggethun, E., Gould, R., Hannahs, N., Jax, K. & Klain, S. 2016. Opinion: Why protect nature? Rethinking values and the environment. *PNAS*, 113, 1462–1465.

Clark, B. and Foster, J.B., 2009. Ecological imperialism and the global metabolic rift: Unequal exchange and the guano/nitrates trade. *International Journal of Comparative Sociology*, 50, 311–334.

Coolsaet, B. 2016. Towards an agroecology of knowledges: Recognition, cognitive justice and farmers' autonomy in France. *Journal of Rural Studies*, 47, 165–171.

Cundill, G., Bezerra, J. C., De Vos, A. & Ntingana, N. 2017. Beyond benefit sharing: Place attachment and the importance of access to protected areas for surrounding communities. *Ecosystem Services*, 28, 140–148.

Díaz, S., Pascual, U., Stenseke, M., Martín-López, B., Watson, R. T., Molnár, Z., Hill, R., Chan, K. M., Baste, I. A. & Brauman, K. A. 2018. Assessing nature's contributions to people. *Science*, 359, 270–272.

Díaz, S., Settele, J., Brondízio, E., Ngo, H., Guèze, M., Agard, J., Arneth, A., Balvanera, P., Brauman, K. & Butchart, S. 2019. *Summary for Policymakers of the Global Assessment Report on Biodiversity and Ecosystem Services*. United Nations: Intergovernmental Science-Policy Platform on Biodiversity and Ecosystem Services.

Dudley, N. 2008. *Guidelines for Applying Protected Area Management Categories. International Union for Conservation of Nature and Natural Resources*. Gland, Switzerland: IUCN Publications Services ISBN.

Estes, J. A., Terborgh, J., Brashares, J. S., Power, M. E., Berger, J., Bond, W. J., Carpenter, S. R., Essington, T. E., Holt, R. D. & Jackson, J. B. 2011. Trophic downgrading of planet Earth. *Science*, 333, 301–306.

Ferrier, P. M., Rucker, R. R., Thurman, W. N. & Burgett, M. 2018. Economic effects and responses to changes in honey bee health. Economic Research Report Number 246, United States Department of Agriculture, March 2018.

Force, K. 2012. *Report of the Karnataka Elephant Task Force*. Bangalore: High Court of Karnataka.

Franks, P., Booker, F. & Roe, D. 2018. *Understanding and Assessing Equity in Protected Area Conservation*. IIED Issue Paper, International Institute of Environment and Development, London, UK.

Geldmann, J., Manica, A., Burgess, N. D., Coad, L. & Balmford, A. 2019. A global-level assessment of the effectiveness of protected areas at resisting anthropogenic pressures. *Proceedings of the National Academy of Sciences*, 116, 23209–23215.

Hallmann, C. A., Sorg, M., Jongejans, E., Siepel, H., Hofland, N., Schwan, H., Stenmans, W., Müller, A., Sumser, H. & Hörren, T. 2017. More than 75 percent decline over 27 years in total flying insect biomass in protected areas. *PloS One*, 12, e0185809.

Haraway, D. J. 2016. *Staying with the Trouble: Making Kin in the Chthulucene*. Duke University Press.

Harvey, D. 2010. *The Enigma of Capital and the Crises of Capitalism*. London: Profile Books. London.

Kareiva, P. & Marvier, M. 2012. What is conservation science? *Bio Science*, 62, 962–969.

Li, T. M. 2014. What is land? Assembling a resource for global investment. *Transactions of the Institute of British Geographers*, 39, 589–602.

Martin, A. 2017. *Just Conservation: Biodiversity, Wellbeing and Sustainability*. Abingdon, UK: Routledge.

Martin, A., Rutagarama, E., Cascao, A., Gray, M. & Chhotray, V. 2011. Understanding the co-existence of conflict and cooperation: Transboundary ecosystem management in the Virunga Massif. *Journal of Peace Research*, 48, 621–635.

McCaslin, W. D. 2005. *Justice as Healing: Indigenous Ways*. St. Paul, MN: Living Justice Press.

Miller, D. C., Agrawal, A. & Roberts, J. 2013. Biodiversity, governance, and the allocation of international aid for conservation. *Conservation Letters*, 6, 12–20.

Mousseau, F. 2019. *The Highest Bidder Takes It All: The World Bank's Scheme to Privatize the Commons*. Oakland, CA: The Oakland Institute.

Muradian, R., Walter, M. & Martinez-Alier, J. 2012. Hegemonic transitions and global shifts in social metabolism: Implications for resource-rich countries. Introduction to the special section. *Global Environmental Change*, 22, 559–567.

Newell, P. & Taylor, O. 2018. Contested landscapes: the global political economy of climate-smart agriculture. *The journal of peasant studies*, 45, 108–129.

Norton, B. G. 2003. *Searching for Sustainability: Interdisciplinary Essays in the Philosophy of Conservation Biology*. Cambridge, UK: Cambridge University Press.

O'Connor, S. & Kenter, J. 2019. Making intrinsic values work; integrating intrinsic values of the more-than-human world through the Life Framework of Values. *Sustainability Science*, 14, 1247–1265.

O'Neill, J., Holland, A. & Light, A. 2008. *Environmental Values*. Abingdon, UK: Routledge.

Pascual, U., Balvanera, P., Díaz, S., Pataki, G., Roth, E., Stenseke, M., Watson, R. T., Dessane, E. B., Islar, M. & Kelemen, E. 2017. Valuing nature's contributions to people: The IPBES approach. *Current Opinion in Environmental Sustainability*, 26, 7–16.

Pretty, J., Benton, T. G., Bharucha, Z. P., Dicks, L. V., Flora, C. B., Godfray, H., Goulson, D., Hartley, S., Lampkin, N. & Morris, C. 2018. Global assessment of agricultural system redesign for sustainable intensification. *Nature Sustainability*, 1, 441.

Pulkkinen, L (2019) A pod of orcas is starving to death. A tribe has a radical plan to feed them. *The Guardian*, available at https://www.theguardian.com/environment/2019/apr/25/orca-starving-washington-feed-salmon-lummi-native-american

Rasmussen, L. V., Coolsaet, B., Martin, A., Mertz, O., Pascual, U., Corbera, E., Dawson, N., Fisher, J. A., Franks, P. & Ryan, C. M. 2018. Social-ecological outcomes of agricultural intensification. *Nature Sustainability*, 1, 275.

Ravenelle, J. & Nyhus, P. J. 2017. Global patterns and trends in human—wildlife conflict compensation. *Conservation Biology*, 31, 1247–1256.

Redford, K. 1992. The empty forest. *BioScience*, 42, 412–422.

Redford, K. & Mace, G. M. 2019. Conserving and contesting biodiversity in the homogocene. In: Lele, S., Brondizio, E. S., Byrne, J., Mace, G. M. & Martinez-Alier, J. (eds.) *Rethinking Environmentalism: Linking Justice, Sustainability, and Diversity*. Cambridge, MA: The MIT Press.

Roe, D. 2008. The origins and evolution of the conservation-poverty debate: A review of key literature, events and policy processes. *Oryx*, 42, 491–503.

Roe, D., Booker, F., Day, M., Zhou, W., Allebone-Webb, S., Hill, N. A., Kumpel, N., Petrokofsky, G., Redford, K. & Russell, D. 2015. Are alternative livelihood projects effective at reducing local threats to specified elements of biodiversity and/or improving or maintaining the conservation status of those elements? *Environmental Evidence*, 4, 22.

Rozzi, R. 2018. Biocultural conservation and biocultural ethics. In *From Biocultural Homogenization to Biocultural Conservation*. New York: Springer.

Sánchez-Bayo, F. & Wyckhuys, K. 2019. Worldwide decline of the entomofauna: A review of its drivers. *Biological conservation*, 232, 8–27.

Sandbrook, C., Fisher, J. A., Holmes, G., Luque-Lora, R. & Keane, A. 2019. The global conservation movement is diverse but not divided. *Nature Sustainability*, 2, 316.

Schleicher, J., Peres, C. A., Amano, T., Llactayo, W. & Leader-Williams, N. 2017. Conservation performance of different conservation governance regimes in the Peruvian Amazon. *Scientific reports*, 7, 11318.

Schreckenberg, K., Franks, P., Martin, A. & Lang, B. 2016. Unpacking equity for protected area conservation. *Parks*, 22, 11–26.

Sikor, T. (ed.) 2013. *The Justices and Injustices of Ecosystem Services*. London: Earthscan.

TEEB. 2010. *The Economics of Ecosystems and Biodiversity: Mainstreaming the Economics of Nature: A Synthesis of the Approach, Conclusions and Recommendations of TEEB*. United Nations.

UNEP-WCMC, I. 2016. Protected Planet Report 2016 How Protected Areas Contribute to Achieving Global Targets for Biodiversity. Cambridge, UK: World Conservation Monitoring Centre.

Vivanco, D. F., Kemp, R. & Van Der Voet, E. 2016. How to deal with the rebound effect? A policy-oriented approach. *Energy Policy*, 94, 114–125.

WCED. 1987. *Our Common Future*. World Commission on Environment and Development. Oxford: Oxford University Press.

Wells, M., Guggenheim, S., Khan, A., Wardojo, W. & Jepson, P. 1999. *Investing in Biodiversity: A Review of Indonesia's Integrated Conservation and Development Projects*. Washington, DC: The World Bank.

Whiteman, G. 2009. All my relations: Understanding perceptions of justice and conflict between companies and indigenous peoples. *Organization Studies*, 30, 101–120.

Wilson, E. O. 2016. *Half-Earth: Our Planet's Fight for Life*. New York: WW Norton & Company.

12 Climate justice

Gareth A.S. Edwards

Learning outcomes

- Recognise why climate change can be understood as fundamentally and intrinsically a justice dilemma.
- Develop an understanding of key debates within climate ethics, and key questions raised by the climate justice movement.
- Appreciate the limitations of framing climate justice in purely international, statist terms.
- Understand the basic contours of climate ethics as an academic pursuit and climate justice as a political pursuit, and the relationships between them.

Introduction

When I teach Masters students about climate change, I often start by quoting Principle 1 from Article 3 of the United Nations Framework Convention on Climate Change (UNF-CCC). This principle says:

> The Parties should protect the climate system for the benefit of present and future generations of humankind, on the basis of equity and in accordance with their common but differentiated responsibilities and respective capabilities.
>
> (UNFCCC, 1992)

What I ask my students to notice is that no fewer than four justice-related concepts make an appearance in just this one principle: **intergenerational justice**, equity, common but differentiated responsibilities, and capabilities. So it is not surprising that the philosopher Simon Caney (2010) argues that both climate change mitigation and adaptation should be understood as moral duties: **mitigation** is the duty to cut back on climate change-producing activities; adaptation is the duty to help those who will be affected by unavoidable climate change. There is widespread acceptance that a key challenge of climate change is how to make nation-states do their duty. This is done by assigning *responsibilities* to some (broadly, the richer countries of the Global North) and *rights* to others (broadly, the poorer countries of the Global South). Indeed, the whole UNFCCC process—known as the 'international climate regime'—is geared towards this goal. When countries are setting their emissions targets, they are negotiating about their responsibility for climate change mitigation compared to others. When they are proposing adaptation measures or finance mechanisms, they are negotiating what rights to assistance they or other countries have as a result of unavoidable climate change.

Responding to climate change raises both *ethical* and *political* questions. The relationship between the ethical and the political will reverberate throughout this chapter, which will focus on '**climate justice**', an idea that emerged simultaneously in academia and civil society as climate change rose to prominence in the 1990s. I have arranged this chapter in two sections. In the first, I introduce climate justice as an ethical question, focusing on the principles of justice which scholars have sought to apply to guide climate policy. In the second, I introduce climate justice as a political question, focusing on the emergence and claims of the 'climate justice movement'. At first glance, it might appear that this second section is much more clearly situated in the environmental justice tradition and relevant to a text on environmental justice. But by the end of the chapter, it should be clear that knowledge of the ethical debate about climate change is essential, because this debate both shapes and is shaped by the political claim-making of the climate justice movement. The ethics of climate change is intrinsically interconnected with the politics of climate change, so pursuing climate justice is always a process of dialogue, mediation and compromise between them. Box 12.1 proposes one way of bringing them together.

Climate justice as an ethical question

One of the earliest treatments of the ethical implications of climate change was Edith Brown Weiss' (1989) book *In Fairness to Future Generations*, which explicitly set out to provide a theory of intergenerational justice which accounted for climate change, and was published three years before the 1992 Rio Earth Summit at which the UNFCCC was established. Early contributors to the debate sought to establish the key ethical questions raised by climate action. For instance, in his influential paper 'Subsistence Emissions and Luxury Emissions', Shue (1993, p. 51) raised four key questions which subsequent work has continued to grapple with:

1 Allocating the costs of prevention [i.e. who has *responsibility* to pay for mitigation];
2 Allocating the costs of coping [i.e. who has *responsibility* to pay for adaptation];
3 The background allocation of resources and fair bargaining [i.e. how can procedural fairness and participation be ensured];
4 Allocating emissions: transition and goal [i.e. to whom should ongoing emissions *rights* be allocated].

Contributions like these laid the groundwork for subsequent academic work on climate ethics, which has sought to devise principles for responding to climate change justly. The focus of most of the literature has been on the international and intergenerational justice implications of climate change. Despite Shue's early reference to adaptation—picked up by Paavola and Adger (2006), who develop four analogous questions framed specifically in terms of **climate change adaptation**—most of the literature has focused on climate change mitigation, with Gardiner's (2004, pp. 578–579) prescription that "[the] core ethical issue concerning global warming is that of how to allocate the costs and benefits of greenhouse gas emissions and abatement" resonating through the literature.

Most of the climate ethics literature sits within the liberal philosophical tradition. While this is not the place to give an expansive introduction to liberal political philosophy, it is important to understand this because it explains why this literature—in contrast to some of the social movements to be discussed in the next section—does not fundamentally question the structure of society, the status of humans, or the broader socio-economic system. John Rawls' (1958) developed an egalitarian approach to justice known as 'justice as **fairness**', which he expanded and codified in his 1972 book *A Theory of Justice*. Rawls sets out two basic principles for a just society:

1. equality in the assignment of basic rights and duties; and
2. social and economic inequalities, for example inequalities of wealth and authority, are just only if they result in compensating benefits for everyone, and in particular for the least advantaged members of society.

(Rawls, 1999 [1972], p. 13)

These principles have been the dominant reference points for liberal philosophers ever since, and underpin the two fundamental principles to allocating the costs of climate change mitigation discussed in the climate ethics literature. The first is known as the '**polluter pays principle**', and holds that it is just that "those who have caused a problem (such as pollution) should foot the bill" (Caney, 2005, p. 752). The second is known as the 'ability to pay' principle and holds that "it is reasonable to ask [the most advantaged] (rather than the needy) to bear this burden since they can bear such burdens more easily" (Caney, 2005, p. 769). However, there are clear ethical and practical problems associated with applying either principle in isolation, which Caney explores in more depth. As a result, he advocates a 'hybrid' approach which is similar to the principle of 'common but differentiated responsibility' in the 1992 Rio Declaration which reads:

> In view of the different contributions to global environmental degradation, States have **common but differentiated responsibilities** [emphasis added]. The developed countries acknowledge the responsibility that they bear in the international pursuit of sustainable development in view of the pressures their societies place on the global environment and of the technologies and financial resources they command.
>
> Principle 7, Rio Declaration, 1992
> (given force in Article 3[1] of the UNFCCC—see Section 1)

From the Rio Declaration, action (in this case mitigation) is *everyone's* responsibility, but developed countries should bear the greater portion of this burden because: (a) they benefited most from the causes of climate change; and (b) they are most able to pay by virtue of their wealth. The Rio Declaration is a political statement, but it effectively points to what is ultimately a fairly simple principle at the heart of climate ethics: 'those with the most responsibility for climate change should act first'. Climate justice, in this formulation, is a matter of establishing responsibilities and rights, and then allocating the burdens of climate change mitigation and adaptation accordingly. However, in reality, it is incredibly complex even to establish who 'those with the most responsibility' are and what 'acting first' would mean. For instance, if we limit ourselves to an international perspective on the question of who should pay for mitigation, there is considerable acceptance that the developed world bears greater responsibility for causing climate change, due to historical patterns of development and industrialisation. However, it is also true that until the late 20th century, much of this was done in ignorance of the effects. This ignorance means this 'pollution' is ethically different than emissions since the effects of climate change have been more widely understood (Page, 2008), a growing proportion of which is coming from developing countries. At the same time, the current emissions of these developing countries are often directly related to their attempts to escape the cycle of poverty that colonialism bequeathed them. Assigning responsibility at the international scale very quickly becomes extremely complex.

Adaptation further complicates this already difficult situation, because whereas mitigation generally only involves 'fairly sharing the burden', adaptation involves the creation of both burdens and benefits. The beneficiaries (whose 'rights' we would seek to protect) could validly be seen as developing countries, those most at risk of climate change impacts, the poor, or even those who are otherwise disadvantaged, for instance by gender, colour, race, ethnicity or religion, or various combinations of all of these. It is likely that adaptation measures will also end up benefiting people other than these beneficiaries, however they are defined. This raises additional questions about who a valid beneficiary of adaptation is, and who should pay for these benefits.

On top of this, Paul Harris (2010) has argued that the focus on nation-states is itself fundamentally flawed, because it ignores the vastly different responsibilities and rights of people *within* those states. To give just one example, surely it would not be acceptable if the growing number of Indian billionaires were able to completely avoid any responsibility to assist with climate change mitigation and adaptation? For Harris, the solution is a cosmopolitan approach to justice in which the responsibilities and rights associated with climate change accrue to individual people rather than nation-states, decoupling justice from the state. But this further complicates the already difficult questions raised of who is responsible for mitigation or adaptation action, and how can they be compelled to do their duty. Moreover, it adds an additional layer of complexity in terms of deciding on what terms actors at different levels should be included in or excluded from climate change decision-making. For instance, we might be able to agree that 'luxury' emissions (such as air travel for holidays) should be seen as ethically different from 'subsistence' emissions (such as burning cooking fuel), but it is less clear who should be held responsible for emissions from land use change over time, or how we should account for varying emissions profiles due to different available energy sources or different heating or cooling needs in different parts of the world (see Füssel, 2010).

It is incredibly complex to devise a universal ethical system to guide action on climate change, even if we limit ourselves to thinking within the western philosophical tradition. But there are at least two further issues with framing climate justice as a matter of rights and responsibilities. The first was actually prefaced by Shue (1993), which is that any allocation of costs or burdens will only be just if procedural fairness and participation of relevant parties can be ensured. We will come to this in the next section, because one of the core contentions of the climate justice movement is that this has *not* been the case. The second is that, because the scale of the climate change problem is global, the state cannot be the arbiter of justice as it normally would in liberal justice theory. Lacking this natural 'judge', both mitigation and adaptation action always rely on the creation and effective operation of institutions which can reach agreements and then hold the relevant parties to their commitments. The history of the UNFCCC demonstrates just how difficult this is in practice. Indeed, the ethicist Stephen Gardiner argues that on multiple levels:

> The peculiar features of the climate change problem pose substantial obstacles to our ability to make the hard choices necessary to address it. Climate change is a perfect moral storm. One consequence of this is that, even if the difficult ethical questions

could be answered, we might still find it difficult to act. For the storm makes us extremely vulnerable to moral corruption.

(Gardiner, 2006, p. 398)

As a result, there have been suggestions from some quarters in recent years that a focus on justice actually *detracts from*—rather than enhances—climate policy (see Gardiner and Weisbach, 2016; Klinsky et al., 2017). My argument, which I develop in the following section, is that we should not abandon a justice focus, but that we must understand climate justice as both an ethical *and* a political question, and that these characteristics inform and shape each other.

Box 12.1 Calling for climate justice

In their report 'Greenhouse Gangsters vs. Climate Justice', Bruno et al. (1999, p. 3) outline seven things that they think climate justice means:

1. "**Climate Justice means**, first of all, **removing the causes of global warming**";
2. "Climate Justice means **opposing destruction wreaked by the Greenhouse Gangsters at every step of the production and distribution process**" including "a moratorium on new oil exploration" and significant reductions in emissions, particularly in the transport sector;
3. "Climate Justice in the United States means . . . fostering a *just transition*" for "low income communities, communities of color, or the workers employed by the fossil fuel industry";
4. "Climate Justice means **providing assistance to communities threatened or impacted by climate change**";
5. "Climate Justice means that while all countries should participate in the drastic reduction of greenhouse gas emissions, **the industrialized nations, which historically and currently are most responsible for global warming, should lead the transformation**";
6. "Climate Justice for developing nations means that international institutions such as **the World Bank and World Trade Organization should halt their funding and promotion of corporate-led fossil fuel-based globalization**" and base development on "clean energy technologies";
7. "Ultimately, Climate Justice means **holding fossil fuel corporations accountable for the central role they play in contributing to global warming**."

Notice that these definitions are written as demands, and that each demand is targeted at a particular actor in society. Demand 1 is posed to everyone; Demands 2 and 7 to civil society and government; Demands 3 and 4 to government and corporations; Demand 5 to the governments of the Global North; and Demand 6 to international institutions.

In 2018, in the run-up to COP24 in Katowice, Poland, a broad coalition of social movements published the 'People's Demands for Climate Justice' (2018) and invited organisations and individuals to endorse them. The demands were as follows:

1   Keep fossil fuels in the ground.
2   Reject false solutions that are displacing real, people-first solutions to the climate crisis.
3   Advance real solutions that are just, feasible, and essential.
4   Honor climate finance obligations to developing countries.
5   End corporate interference in and capture of the climate talks.
6   Ensure developed countries honor their 'Fair Shares' for largely fueling this crisis.

Exercise: bridging ethics and politics

Any political demands are always informed by and built on ethical principles, for instance about what is considered fair and just. Simon Caney has proposed that:

> an adequate theory of justice in relation to climate change must explain in what ways global climate change affects persons' entitlements and it must do so in a way that (i) is sensitive to the particularities of the environment; (ii) explores the issues that arise from applying principles at the global rather than the domestic level; and (iii) explores the intergenerational dimensions of global climate change.
> (Caney, 2005, p. 750)

Consider the two sets of climate justice demands introduced in this box, and ask yourself the following questions:

1   What ethical principles underlie each set of demands?
2   What assumptions do the demands make about responsibility, rights, recognition, and representation in the context of climate change?
3   Are the demands compatible with Caney's proposal for a just response to climate change?
4   Do the demands call for anything that Caney's proposal neglects which would be important for a just response to climate change?

Climate justice as a political question

Though the climate ethics literature is clearly engaged with what we would now call 'climate justice', the first printed use of the term was not in the academic literature, but in a November 1999 self-published report by the San Francisco-based organisation Corporate Watch (CorpWatch) which carried the title 'Greenhouse Gangsters vs. Climate Justice' (Bruno et al., 1999; see also Chatterton et al., 2013; Tokar, 2014). Focused on the disproportionate power of the oil industry in the context of globalisation, the report contains the first attempt at a political declaration in favour of climate justice (Box 12.1). Tokar (2014) notes that the authors had been influenced by the US environmental justice movement, and their definition of climate justice reflects established environmental justice principles, combined with a radical critique of globalised neoliberal capitalism and an aspiration for a new political settlement for a more equal and more sustainable world.

Driven in part by those involved in authoring this report, the usage of the term 'climate justice' in political activism has exploded since then, leading to the development of a considerable and diverse climate justice movement. In 2002 a collective of non-governmental organisations, led by Corpwatch, gathered in Bali and endorsed the 'Bali Principles of Climate Justice', which they deliberately modelled on the 'Environmental Justice Principles' developed at the 1991 **People of Color Environmental Justice Leadership Summit** in Washington DC (CorpWatch et al., 2002; First National People of Color Environmental Leadership Summit, 1991). The Bali Principles are important because they actually prefigure debates that would subsequently emerge in the climate ethics literature. For instance, eight years before Paul Harris' influential paper discussed in the previous section, the Bali Principles observed that "unsustainable consumption" was at the heart of the climate crisis and "exists primarily in the North, but also among elites within the South" (CorpWatch et al., 2002).

The Bali Principles were theoretically innovative, too. Almost a third of the principles (eight, in total) concerned the recognition and representation of groups largely marginalised in the international negotiations, including Indigenous peoples, women, and children. Principle 19 captures the spirit of many of these, stating that "climate justice demands that public policy be based on mutual respect and justice for all peoples, free from any form of discrimination or bias" (CorpWatch et al., 2002). In other words, no matter how we apportion responsibilities or rights with respect to climate change between countries, justice will never be achieved unless we also act to overcome the patterns of representation, interpretation, and communication by which some people's voices are systematically silenced and others are amplified, since these also produce and reinforce injustice. Since the 1990s, civil society has played a significant role in putting this 'who' question onto the agenda, pointing to what Nancy Fraser calls the **'recognition'** dimension of justice (Fraser, 1997; also Honneth, 2004), which environmental justice scholars have consistently advocated be elevated to at least equal footing with distributive, procedural and participatory dimensions of justice (Schlosberg, 2007).[1] Assigning responsibilities and rights is important, but any such allocative paradigm is a political process shot through with structural inequalities in power and resources, and so there must be mechanisms in place to ensure that marginalised groups are not excluded by virtue of being invisible to policymakers.

A number of scholars and scholar-activists have tracked the development of climate justice as a political proposition over the decade after 2002, including Brand et al. (2009), Bond (2012), and Tokar (2014). Chatterton et al. (2013) provide a good entry to this work, as well as to the diversity of political demands being pursued under the banner of climate justice. Those interested in the development of the climate justice movement would do well to start reading here. Three calls for climate justice from within the movement, all from 2015 (the year of the influential COP21 meeting in Paris), illustrate this diversity.

1 "To build resilience, including of marginalized people and vulnerable groups, we need to promote climate justice and maximize resources for investment in low-carbon development paths through adequate and appropriate financing, technology transfer and

capacity building for poorer countries" (Wael Hmaidan, Director, CAN International, 2015: www.climatenetwork.org/node/5234).

2 "The world is facing two related challenges that threaten the lives and livelihoods of billions of people: climate change and the global energy crisis. . . . Key to the solution is energy sovereignty: the right of communities to choose their sustainable energy sources and to develop healthy consumption patterns that will lead to sustainable societies. This, combined with the need for greenhouse gas emissions reduction and for all people to share an equitable amount of resources within ecological limits, is essential to achieving climate justice" (Friends of the Earth International 2015: www.foei.org/what-we-do/climate-justice-and-energy-explained).

3 "The heart of climate justice is the understanding that the urgent action needed to prevent climate change must be based on community-led solutions and the well-being of local communities, Indigenous Peoples and the global poor, as well as biodiversity and intact ecosystems. Climate justice is the understanding that we will not be able to stop climate change if we don't change the neoliberal, corporate-based economy which stops us from achieving sustainable societies. It is the understanding that corporate globalization must be stopped" (Global Justice Ecology Project, 2015: http://globaljusticeecology.org/climate-justice/).

All these approaches to climate justice acknowledge the interrelatedness of climate action and poverty and see structural factors as the key barriers to action. However, the demands reflect different notions of the key points of intervention and modes of engagement. In the first example, climate justice is a means to building resilience, almost on an equal footing with assistance from the developed world (coming in the form of finance, technology, and capacity-building assistance). It is clearly seeking to influence the international climate debate, and, as such, it adopts a liberal egalitarian notion of justice, focuses on incremental rather than transformational change, and constrains itself to a mild critique of mainstream understandings of development. It is a political proposition which would sit quite comfortably with the ethical principles advocated by liberal political philosophers (see the previous section of this chapter).

The second and third approaches are more radical. Both of them emphasise the role of multinational corporations and multilateral aid lenders, and both focus more directly on fossil fuel extraction and use. The second focuses on how the climate crisis intersects with the energy crisis and sees community **energy sovereignty** as key to achieving climate justice. This implies a shift away from seeing the state as the arbiter of justice, and a much more cosmopolitan approach to justice. It also articulates a much clearer sense of the role of ecological limits in setting the parameters for climate justice than the first approach. The third approach goes one step further, by placing responsibility squarely on the global economic system, which it argues earlier "encourages banks and corporations to ignore ethical and moral considerations and gamble with the Earth, peoples' lives, and our collective futures in the service of higher profits". The use of any market-based mechanisms to address the climate crisis is seen from this view as fundamentally unjust, because it is this system which is deemed the cause of the problem.

The point to highlight here is that these political demands are not made in a normative or ethical vacuum. Rather, each political demand reflects the outcome of a process of deliberation which is informed by both ethical principles and practical and strategic calculations. Sometimes, the purpose of the demand might be to shift the political space for action towards alternatives that were previously not under consideration. Other times, the demands reflect an open dialogue on the meaning of climate justice itself, since "like climate change debates, the term climate justice itself is a terrain of contestation" (Chatterton et al.,

2013, p. 607). This goes some way to explaining the diversity of claims being made under the banner of climate justice:

> Briefly defined, climate justice refers to principles of democratic accountability and participation, ecological sustainability and social justice and their combined ability to provide solutions to climate change. Such a notion focuses on the interrelationships between, and addresses the roots causes of, the social injustice, ecological destruction and economic domination perpetrated by the underlying logics of pro-growth capitalism. In particular, climate justice articulates a rejection of capitalist solutions to climate change (eg carbon markets) and foregrounds the uneven and persistent patterns of eco-imperialism and **"ecological debt"** as a result of the historical legacy of uneven use of fossil fuels and exploitation of raw materials, offshoring, and export of waste.
>
> (Chatterton et al., 2013, p. 606)

Put differently, it is important to understand both the ethical principles which underpin these demands, but it is critical to remember that as political statements, they cannot simply be taken at face value, since the formulation of each demand involves a process of calculation in which the proponent assesses both what they think is demanded ethically (we might call this their 'normative orientation') and also what they think will have purchase with their target audience. The ethics of climate change is intrinsically interconnected with the politics of climate change, so pursuing 'climate justice' is always a process of dialogue, mediation and compromise between them. Box 12.2 proposes one way of bringing them together.

Box 12.2 Conceptualising climate justice as a 'pyramid'

Is there a way to bring the literature on climate ethics into conversation with that from an environmental justice perspective? One way is that, rather than seeing it as a matter of balancing the responsibilities and rights of different actors through distributional or procedural mechanisms, climate justice must be seen as *multifaceted*. Distributions, procedures, rights, and responsibilities are all vital facets of justice, but recognition underpins all of them, because without recognition any attempt at redistribution (i.e., balancing the responsibilities and rights of different actors through distributional or procedural mechanisms) will ultimately be unjust because it will always reproduce structural injustices within society.

One way to do this is to see climate justice as a transparent square pyramid, whereby each of these five facets of justice forms one face (Figure 12.1a). Viewed from above, the pyramid (Figure 12a) highlights the four facets of justice which have been the focus of the international debates. But viewed from the side, the facet of recognition comes into view (Figure 12.1b). This highlights the fact that all these facets of justice are bound to each other, interlinked and interdependent. The pyramid captures several important things about climate justice. First, it highlights the fact that recognition undergirds all the other facets of justice. Second, it is a visual reminder that, though climate justice is multifaceted, the different facets are all intrinsically interconnected, and it is the relationships between them that produce justice.

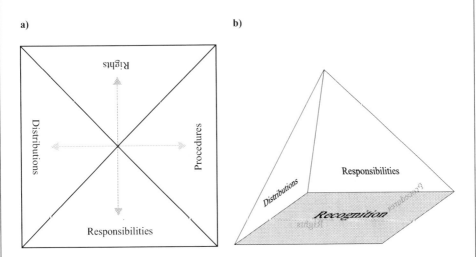

Figure 12.1 The three-dimensional climate justice pyramid, as viewed from a) directly above; and b) obliquely below.

Source: Reproduced from Bulkeley et al., 2014, p. 34.

The climate justice pyramid can be used both as a conceptual framework to help understand climate justice itself and as an analytical tool to help shape the design of new forms of climate change intervention. In introductory high school physics classes, a common exercise is to shine a light through a glass prism to show how white light is made up of the colours of the visible light spectrum. Similarly, the climate justice pyramid helps us understand how justice is made up of many components, and how each facet of justice must be understood in relation to the others. Just as changing the position of the light source or glass prism changes the dispersal of the colour spectrum in a high-school physics exercise, changing the orientation of the pyramid or which face of the climate justice pyramid we look through first affects what kind of justice issues become most visible in any given climate change intervention or policy. Furthermore, because any one perspective is inevitably refracted through the prism formed by the interrelation of different elements of justice, the pyramid forces us to always consider the perspective of multiple facets of justice, and the interaction between them.

For instance, consider an initiative to improve the energy efficiency of a city neighbourhood to reduce greenhouse gas emissions or promote more comfortable living spaces. The perspective of *recognition* will force us to examine who might be excluded from such action, which has the effect of confirming their *rights* to benefit from such action, but also creates a *responsibility* on government to ensure these rights. Government might then use a combination of *distributive* and *procedural* mechanisms to guarantee those rights.

Both academic work on climate justice and its application in practice tends to focus on how specific facets of justice—notably rights and responsibilities—relate to questions of distribution. The use of other entry points into questions of justice—from the perspective of procedural justice or the notion of recognition—has been more limited. But these facets of climate justice cannot be ignored, because climate justice

> demands both ethical coherence and political salience. The climate justice pyramid is just one tool to help those involved in the design or implementation of climate change responses to more effectively design policy and interventions which robustly consider and promote climate justice.
>
> Note: the above text is adapted from Bulkeley et al. (2014)

Conclusion

This chapter has examined climate justice as an ethical question and as a political question. But while we can broadly separate these out analytically, the discussion has highlighted that the ethical principles being devised—mainly, though not exclusively, by academics—are not developed in a political vacuum, and neither are the political demands of the climate justice movement developed in an ethical vacuum. Climate ethics tends to focus on international and intergenerational justice, the responsibility and rights of nation-states, and issues of distributional and procedural justice in the context of the international climate regime. The climate justice movement tends to focus on structural inequality within society at multiple scales and draws more on radical justice concepts, such as recognition, to frame its political demands. But both develop in conversation with and reference to each other. Climate change is fundamentally and intrinsically a justice dilemma, and understanding climate justice means examining both the ethical principles which can inform our response to the problem of climate change and the ways in which scholar-activists and the movement for climate justice are framing their political demands for climate justice.

> **Follow-up questions**
>
> - What is the relationship between the ethics of climate change and the politics of climate change?
> - What are the key debates within climate ethics? How does the climate justice movement seek to influence them?
> - What implications would adopting an environmental justice approach have for how we approach climate change?

Note

1 On distribution, procedure and recognition, see Chapters 3, 4 and 5 of this volume, respectively.

References for further reading

Chatterton P, Featherstone D and Routledge P (2013) Articulating climate justice in Copenhagen: Antagonism, the commons, and solidarity. *Antipode* 45(3): 602–620.

CorpWatch, Friends of the Earth International, Global Resistance, Greenpeace International, groundwork, Indigenous Environmental Network, Indigenous Information Network, National Alliance of People's Movements, National Fishworkers Forum, OilWatch Africa, OilWatch International,

Southwest Network for Environmental and Economic Justice, Third World Network and World Rainforest Movement (2002) *Bali Principles of Climate Justice*. Adopted 29 October 2002. Available from: www.ejnet.org/ej/bali.pdf.

Dietz M and Garrelts H (eds) (2014) *Routledge Handbook of the Climate Change Movement*. London: Routledge.

Gardiner SM and Weisbach D (2016) *Debating Climate Ethics*. Oxford: Oxford University Press.

Harris PG (2010) *World Ethics and Climate Change: From International to Global Justice*. Edinburgh Studies in World Ethics. Edinburgh: Edinburgh University Press.

Heyward C and Roser D (eds) (2016) *Climate Justice in a Non-Ideal World*. Oxford: Oxford University Press.

References used in this chapter

Bond P (2012) *Politics of Climate Justice: Paralysis Above, Movement Below*. Scottsville: University of KwaZulu-Natal Press.

Brand U, Bullard N, Lander E and Mueller T (2009) *Contours of Climate Justice: Ideas for Shaping New Climate and Energy Politics*. Critical Currents No. 6, October 2009. Uppsala: Dag Hammarskjöld Foundation.

Brown Weiss F. (1989) *In Fairness to Future Generations: International Law, Common Patrimony, and Intergenerational Equity*. Doobs Ferry, NY: Transnational Publishers Inc.

Bruno K, Karliner J and Brotsky C (1999) *Greenhouse Gangsters vs. Climate Justice*. San Francisco: TRAC—Transnational Resource & Action Center.

Bulkeley H, Edwards GAS and Fuller S (2014) Contesting climate justice in the city: Examining politics and practice in urban climate change experiments. *Global Environmental Change* 25: 31–40.

Caney S (2005) Cosmopolitan justice, responsibility, and global climate change. *Leiden Journal of International Law* 18(4): 747–775.

Caney S (2010) Climate change and the duties of the advantaged. *Critical Review of International Social and Political Philosophy* 13(1): 203–228.

Chatterton P, Featherstone D and Routledge P (2013) Articulating climate justice in Copenhagen: Antagonism, the commons, and solidarity. *Antipode* 45(3): 602–620.

CorpWatch, Friends of the Earth International, Global Resistance, Greenpeace International, groundwork, Indigenous Environmental Network, Indigenous Information Network, National Alliance of People's Movements, National Fishworkers Forum, OilWatch Africa, OilWatch International, Southwest Network for Environmental and Economic Justice, Third World Network and World Rainforest Movement (2002) *Bali Principles of Climate Justice*. Adopted 29 October 2002. Available from: www.ejnet.org/ej/bali.pdf (accessed 14 November 2019).

First National People of Color Environmental Leadership Summit (1991) *Principles of Environmental Justice*. Summit held 24–27 October 1991. Available from: www.ejnet.org/ej/principles.html (accessed 14 November 2019).

Fraser N (1997) *Justice Interruptus: Critical Reflections on the 'Postsocialist' Condition*. New York: Routledge.

Füssel H-M (2010) How inequitable is the global distribution of responsibility, capability, and vulnerability to climate change: A comprehensive indicator-based assessment. *Global Environmental Change* 20(4): 597–611.

Gardiner SM (2004) Ethics and global climate change. *Ethics* 114(3): 555–600.

Gardiner SM (2006) A perfect moral storm: Climate change, intergenerational ethics and the problem of moral corruption. *Environmental Values* 15(3): 397–413.

Global Justice Ecology Project (2015) Available from: http://globaljusticeecology.org/climate-justice/.

Gardiner SM and Weisbach D (2016) *Debating Climate Ethics*. Oxford: Oxford University Press.

Harris PG (2010) *World Ethics and Climate Change: From International to Global Justice*. Edinburgh Studies in World Ethics. Edinburgh: Edinburgh University Press.

Honneth A (2004) Recognition and Justice: Outline of a plural theory of justice. *Acta Sociologica* 47(4): 351–364.

Klinsky S, Roberts T, Huq S, Okereke C, Newell P, Dauvergne P, O'Brien K, Schroeder H, Tschakert P, Clapp J, Keck M, Biermann F, Liverman D, Gupta J, Rahman A, Messner D, Pellow D, and Bauer S (2017) Why equity is fundamental in climate change policy research. *Global Environmental Change* 44(Supplement C): 170–173.

Paavola J and Adger WN (2006) Fair adaptation to climate change. *Ecological Economics* 56(4): 594–609.

Page EA (2008) Distributing the burdens of climate change. *Environmental Politics* 17(4): 556–575.

People's Demands for Climate Justice (2018) *The People's Demands for Climate Justice*. Available from: https://www.peoplesdemands.org/#read-the-demands-section.

Rawls J (1958) Justice as Fairness. *The Philosophical Review* 67(2): 164–194.

Rawls J (1999 [1972]) *A Theory of Justice*. Revised. Cambridge MA: Harvard University Press.

Schlosberg D (2007) *Defining Environmental Justice: Theories, Movements, and Nature*. Oxford: Oxford University Press.

Shue H (1993) Subsistence Emissions and Luxury Emissions. *Law & Policy* 15(1): 39–60.

Tokar B (2014) *Toward Climate Justice: Perspectives on the Climate Crisis and Social Change*. Porsgrunn: New Compass Press.

UNFCCC (1992) *United Nations Framework Convention on Climate Change*. Available from: https://unfccc.int/sites/default/files/convention_text_with_annexes_english_for_posting.pdf.

13 Energy justice

Rosie Day

Learning outcomes

- Think about how wider environmental and social justice theories can and should be applied to energy systems.
- Consider how injustices arise through different modes of energy production.
- Gain an understanding of energy poverty and injustices in energy consumption.
- Think about how energy justice connects with climate justice.
- Consider how new energy developments including renewable energy continue to raise justice issues that need careful addressing.

Introduction

Energy justice has for some years now been gaining a life of its own as a concept, but it is nevertheless strongly rooted in environmental justice and social justice. **Energy justice** is broadly concerned with social difference in access to the benefits of energy systems, and in suffering negative consequences from the operation of energy systems. Energy is, of course, inextricably linked with the environment, in that energy in the forms that we use it arises from the metabolism of environmental resources, such as wood, coal, minerals, or from the harnessing of the energy of natural systems—sun, wind, water flows. Indeed, energy is inseparable from the environment. Our use and generation of energy also modifies the environment around us through creating structures, landscape changes, and residues. The effects of our use of some forms of energy are modifying the earth's systems in far-reaching ways, as the consequent release of greenhouse gases brings about climate change, and thus energy justice needs also to be thought about in the context of climate justice.

Theoretically, energy justice draws on the same reference points as environmental justice more widely. The most used framings of justice, which will already be familiar from other chapters in this volume, are distributional justice, procedural justice, and justice as recognition. With **distributional justice**,[1] we are concerned with the social and spatial distribution of the costs and benefits of energy production, and of access to useful energy, and in terms of deciding what is fair or just, we draw upon theorists such as John Rawls (1971). Procedural energy justice is concerned with decision-making in relation to energy access and energy system design, at different scales. The Aarhus Convention (UNECE, 1998) provides a good reference point for defining **procedural environmental justice**,[2] setting out three aspects—information, decision-making, and legal redress. The notion of justice as **recognition**,[3] taken from theorists such as Nancy Fraser (1995), is drawn on in energy justice to remind us to pay attention to the needs of diverse groups of people and to make sure that they are treated with due respect in energy-related matters. Other theories of justice are also brought into play as academic work on energy justice develops, notably the **capability approach**[4] of Amartya Sen (1999) and Martha Nussbaum (2011).

In this chapter, I examine how energy justice and injustice have been explored, highlighted, and sometimes addressed, mainly through academic research but also touching on social campaigns and on policies. I start with the energy supply and production side of things, then move on to thinking about justice in energy consumption, while also concluding that we need to think about justice in energy systems in a joined-up way. I end with some thoughts on where some major current and future energy justice concerns lie, but also emphasise that because our use of energy is constant, so too are energy justice questions always present in our everyday lives.

Justice in energy production

Although the language of justice may not always have been used, recorded data and research have long reflected the differential impacts of systems of energy production—or what Healy and Barry (2017) refer to as the externalities of supply-side energy policy. Coal mining has historically been one of the most dangerous jobs in the world, blighting working-class communities with fatalities through accidents and chronic illnesses (see e.g. McIvor and Johnston, 2007; Coalmining History Resource Centre, 2018, with regard to the UK). It remains a worldwide killer of hundreds if not thousands annually, mostly in developing countries (Dhillon, 2010), although safety is improving year on year. More often framed as an issue of occupational health and addressed by labour movements, along with other issues of safety in working-class occupations,[5] it is nevertheless clearly an issue of energy injustice.

Given rather more attention as an incidence of energy injustice has been the effects of uranium mining for the nuclear industry on Indigenous populations, especially those in the Southwestern US. Mining sites on Navajo and other tribal lands have been overlooked by regulators and left unremediated for a long time, with consequent damage to community health, on top of the impacts on those directly engaged in mining. In contrast, the benefits of nuclear generation flow to the electricity consumers, often located in different states (Hoffman, 2001; Sharpe, 2008). Similar concerns have mobilised justice discourses regarding the effects of the more recent tar sands exploitation on Indigenous communities in Canada (Black et al., 2014)[6]. The multi-dimensional negative impacts of oil exploration and extraction activities in the Amazon regions of Ecuador and Peru on Indigenous people have also drawn deserved attention (Orta-Martinez and Finer, 2010; San-Sebastián et al., 2001), if insufficient redress. Clearly understandable as distributional injustices against already disadvantaged groups, such activities by powerful interests also thrive on procedural injustices in terms of lack of informed consent and lack of restorative justice, as marginalised communities struggle to get access to legal redress and compensation.

The scale at which injustice operates is increasingly international, as flows of resources become more global and energy developments more open to international investors. Along with the mining and oil extraction corporations, it is high energy consumers in wealthier countries who benefit most from the extraction of fossil fuels in regions such as Amazonia and southern Africa. Regarding waste products, the injustice can also be intergenerational, especially (but not only) where radioactivity is concerned. Nuclear waste disposal distributes risk inequitably not only within generations, but also across generations, and it poses a challenge to procedural justice, as any decision-making procedures struggle to account fairly for future generations who will be left with the risk (Shrader-Frechette, 1994).

Even greater is the threat of future climate change—increasingly felt in the present, as rising temperature, sea level rise, and disrupted weather patterns kick in—that comes about largely through the burning of fossil fuels in power plants, as well as for mobility. This is where energy justice meets **climate justice**, as discussed in Chapter 12 of this volume. Whilst energy justice does tend to be more concerned with the more immediate relationships between people and energy in current generations, the two concepts are clearly connected in important ways. Some authors address this by including the **sustainability**[7] of energy sources within concepts of energy justice (e.g. Sovacool and Dworkin, 2015), although addressing global and climate sustainability does require a different scale of thinking from many other energy justice questions—and hence, climate justice is often a more helpful framing for thinking about the wider climate-related impacts of fossil fuel burning.

Either way, for reasons of climate change and also other forms of pollution, such as air pollution, which also has its own justice dimensions (see e.g. Miranda et al., 2011; Wheeler and Ben-Shlomo, 2005), moving away from the consumption of fossil fuels to generate electricity and instead towards more low-carbon forms of energy production is an imperative. Nevertheless, this does not mean the end of energy justice issues; on the contrary, although it might be good for climate justice, low-carbon energy production has raised a series of energy justice concerns in different contexts across the world.

Justice in renewable energy production

The first interest in the justice aspects of renewable energy development mostly came about through studies of the social acceptability of renewable energy developments, especially wind farms. An observation repeated in several countries was that whilst there was relatively strong public support for renewable energy, including wind power, in more general terms, specific developments were encountering local opposition. Initially, planners and developers tended to dismiss this as a form of the so-called **NIMBY** (not in my back yard) syndrome, but social scientists were quick to refute this and to uncover more complex and less negative explanations for these resistances. A consistent finding has been that perceived justice is an important factor in securing acceptance and support for such developments. Both distributional and procedural justice aspects have been found to be important. In terms of distributional justice, one issue is who receives the profits from selling the energy that is generated. For the most part, this will be private-sector developers, often large companies. Another set of profits comes in the form of payments to landowners whose land may be bought or rented for the development. A further important element that can come into play are 'community benefit packages': funds or equivalent benefits that developers pay to a community—usually to a community organisation rather than to individuals—to compensate them for any impacts of hosting the development. Community benefits may also be in other forms, such as improved infrastructure. Justice concerns and grievances around some or all of these aspects have come to light in, for example, Australia (Gross, 2007; Hall et al., 2013), Austria (Scherhaufer et al., 2017), Germany (Langer et al., 2016), Scotland (Macdonald et al., 2017), Canada (Walker and Baxter, 2017a), and across cities of different countries (Khorsand et al., 2015).

Procedural justice concerns often go hand in hand with these distributional justice concerns and, if anything, can be more important (Liebe et al., 2017; Wolsink, 2007; Walker and Baxter, 2017a), although, as Gross (2007) points out, the concerns can vary for different people in any given community. Procedural justice aspects raised in these studies generally include the availability of timely and good quality information, the perceived trustworthiness of developers and planners, transparent processes, and, especially, the opportunity to have meaningful input into decision-making about siting and/or about any community benefits (see e.g. Gross, 2007; Walker and Baxter, 2017b; Macdonald et al., 2017).

With regard to wind farm siting especially, a number of studies have also found concerns with the effect on local landscapes and ecology to be a significant factor in local opposition (Wolsink, 2007; Hall et al., 2013). Mason and Milbourne (2014) actually suggest that **'landscape justice'** should be conceived as a further aspect of justice, encapsulating issues around acknowledging the value of landscapes to local and also to more distant people. An alternative way of addressing this aspect could be within procedural justice efforts, by making sure that landscape-related concerns were clearly heard and attended to in decisions about siting.

Community energy schemes are those whereby the means of energy generation—such as wind turbines or solar panels, and sometimes local distribution networks as well—are owned by local people, rather than by a larger commercial organisation such as an energy developer (although sometimes more geographically distant individuals can also invest). This means that any profits generated by selling the energy tend to be kept local and, whilst some of the profits will go to the people who invested financially in the scheme to pay for the physical infrastructure, sometimes part of the money goes to a fund for wider community benefits. Various community energy arrangements and business models exist, but for the reasons just stated, they are generally assumed to be better in terms of energy justice. Still, they are not immune to justice concerns. Simcock's (2016) study of a community windpower project in South Yorkshire (UK), for example, revealed dissatisfactions with the perceived fairness of decision-making: in part because different individuals had different experiences and positions, and in part because people had conflicting notions of what constitutes fair or just processes. Community energy schemes are also not fully inclusive, as investors need to have spare money to invest, and so the benefits do not necessarily flow to all. Denmark's success with wind farms is often attributed to its high level of community co-ownership, but, as Johansen and Emborg (2018) argue, not everyone is equally able to invest, and those that are against wind farms are unlikely to. In Nova Scotia, Canada, since 1999 windpower developments have also generally offered opportunities for 'local' citizens to invest, as well as to be involved in decisions, which has helped to contribute to a greater level of acceptance than in other regions of Canada. Nevertheless, there have still been concerns about the availability of information about the opportunities and on the criteria for being counted as a 'local' community investor (Walker and Baxter, 2017b), meaning that both the distributive and procedural justice aspects have room for improvement.

At a smaller scale still, individual households can become energy producers by installing some renewable energy generation equipment at their homes, typically solar panels. They may do this purely to generate electricity or to heat water for their own consumption, but in many countries they can sell excess energy back to the grid at a guaranteed price set by the government, called a 'feed-in tariff'. As with larger-scale developers, this price is usually guaranteed for a number of years so that the householder can be confident of covering the costs of installing the equipment and eventually making a profit. Such a system represents poor distributional justice because money to pay the feed-in tariff is drawn by taxes on all householders or bill-payers, whilst the profits flow to people who are able to make the up-front investment in equipment such as solar panels. Measures such as subsidies on the equipment have the potential to improve justice, although are unlikely to eradicate it. Recent research in the US found that, despite various policies designed to include lower-income groups, higher-income and home-owning households were more likely to have solar installations. Also, non-white households were less likely to have them, even after controlling for income and home ownership (Sunter et al., 2019). Nevertheless, a survey in Australia found support for renewable energy subsidies to be high, notwithstanding distributional justice concerns, as long as certain procedural justice criteria were in place (Simpson and Clifton, 2016).

Renewable energy is 'clean' at the point of production and better for climate justice, but we should not overlook the fact that the components such as turbines and solar panels have to be manufactured and eventually disposed of. Box 13.1 further examines some justice-relevant issues surrounding solar panel production and waste.

Box 13.1 Justice concerns in solar panel production and disposal

Solar panels are basically large electronic components. Their structures vary due to technological developments over time, but the newer generation of 'thin film' cells require semiconductor materials which contain toxic minerals such as cadmium, tellurium, indium, and selenium. At present, these minerals are produced often as by-products of mining for widely used minerals such as aluminium and copper, but, as demand increases, mining for such minerals will also need to increase. This raises concerns about environmental and social impacts, especially as most such **mining** is in the Global South. Similarly, the batteries often deployed alongside solar panels contain minerals such as lithium and cobalt, with much of the latter currently coming from unregulated and unsafe mines in the Democratic Republic of Congo. Beyond mining, the manufacture of solar panels involves workers handling highly toxic chemicals, subjecting them to occupational health risks. Innovation in manufacturing has led to solar panels becoming much cheaper, but associated contract and labour conditions in Asian factories have been called into question.

Solar panels only last for a few years on average, and what happens to broken and discarded panels has rapidly become a major concern, as solar panels make up an increasing proportion of global e-waste. The industry consensus is that recycling is the answer, but the specialist recycling infrastructure is not yet developed enough and, indeed, in some parts of the world barely exists. Programmes to increase producer responsibility often require producers to pay (often distant) governments for the disposal of the panels they sell in those governments' countries, but not to provide waste management strategies or facilities. As such, especially in less developed regions, solar panel waste often enters mainstream waste streams or accumulates in informal dumps. Meanwhile, the regulation that focuses on recycling at specialist sites does not support localised small-scale repair and repurposing operations and risks undermining this important, existing, and sustainability-positive sector of many local economies.

Sources: Mulvaney (2013); Cross and Murray (2018); Bazilian (2018); UNEP (2019)

Renewable energy production in the Global South

So far, I have been discussing **renewable energy** development mostly with reference to more developed countries, but, in recent years, many countries in the Global South have sought to support their development through large-scale renewable energy deployment, including solar, hydropower, and biofuel. India, for example, has a very ambitious National Solar Mission (Government of India, 2019), and renewable energy development in China, including solar, wind, and hydropower, accounted for 45% of global renewable energy investment in 2017 (UNEP, 2018). Often, such development is through partnership arrangements between governments and private developers, whereby the government facilitates the development by providing land and guaranteeing to buy electricity, whilst the developer builds the infrastructure, runs operations, and ultimately receives the profits. Whilst the scaling up of renewable energy is undoubtedly good for climate justice as already discussed, this rapid and large-scale development has brought up some severe justice issues more specific to Global South contexts.

A major concern is that large-scale developments often take place in poor, rural areas—where there is enough space—on land which was previously not in private ownership, instead being either explicitly or de facto a **common property resource** and used by local communities for subsistence activities such as grazing animals and harvesting plants. In the absence of private land titles, governments can choose to 'enclose' these commons to allow developments to go ahead, without needing to pay compensation to communities. At other times, land might be bought at low rates from uninformed communities. The benefits of this land therefore become transferred from the public to the already wealthy private developers, who profit from being able to build on it, or in the case of biofuels, to plant on it. Under-developed rural communities can lose their livelihoods as a result. This is a form of distributional injustice which can also be termed a spatial injustice, whereby land or space is accumulated for profit by some, at the cost of other, already disadvantaged, people (see Baka, 2017; Yenneti et al., 2016; also Mulvaney, 2013 on enclosure of desert land in the Southwestern US).

Procedural injustices have been implicated in facilitating such large-scale energy developments and '**energy dispossessions**', to use Baka's term. For example, information given to communities has in some cases been inadequate and in unsuitable formats, in contexts where many people may be unable to read or may speak only local or tribal languages. Informed consent is therefore unlikely in such cases. Poorer and lower-status communities are also seldom given a meaningful voice in decision-making. As a result of this, they are unable to express their concerns about impacts, or to negotiate meaningful compensation or community benefit packages, meaning that opportunities to lessen the distributional injustice are missed. Komali Yenneti's case studies of solar development in India (Yenneti and Day, 2015, 2016) give a useful analysis of alleged procedural injustices involved and of distributional injustice outcomes, although it should be noted that India has since updated its land acquisition laws and, on paper at least, affected communities now have stronger rights to compensation, including in the form of land.

A further dimension is that, in Global South contexts, such communities relying on common property resources may be Indigenous communities with specific cultures and identities that have been threatened and oppressed for centuries by colonial and neo-colonial powers. Developments can pose particular threats to the way of life and cultural viability of Indigenous groups, such as the Rabari nomadic herders in Gujarat, India, deprived of access to grazing land by a mega-solar development (Yenneti and Day, 2015, 2016), and the Binnizá and Ikoots Indigenous communities of the Tehuantepec peninsula, Mexico, threatened and displaced by large-scale windpower developments (Avila-Calero, 2017). For such communities, which may be lacking status and respect even within their own regions and countries, these energy justice struggles are closely tied to struggles for recognition and anti- or de-colonisation, as discussed in Chapters 7 and 20 of this volume. In many ways, these are not new issues, as they reflect many countries' histories of displacement of lower-status communities for major infrastructure projects such as dams (see e.g. McDowell, 1996), but the important point is that, although positive in many ways, renewable energy developments without proper consideration of multi-dimensional justice are at risk of continuing these oppressions.

Not all renewable energy development in the Global South is at this large utility scale; smaller mini-grid systems powered by solar or hydro are increasingly used in development projects as a way to provide electricity to rural communities which are more distant from the main grid infrastructure. Whilst such systems have good potential in terms of relieving energy poverty (see next section), issues around maintenance and waste disposal, in particular (as explored in Box 13.1) need attention.

Justice and energy consumption

Energy justice in consumption terms relates largely to uneven access to energy and energy services. **Fuel poverty** was first recognised as a problem in the UK, academically and in policy, where it was defined as an inability to be able to afford enough energy to heat the home adequately, at the same time as providing for other energy needs (Boardman, 1991; DTI, 2001). The focus on heating as the main energy use of concern came about due to the relatively cold and damp climate of the UK, which means that heating represents the main component of the average annual energy (electricity plus gas) spending, coupled with the evident health impacts of living in a cold home. Poor affordability of energy comes about through a combination of all or some of the three factors of low income, high energy costs, and houses having poor energy efficiency (i.e. easily losing heat). In 2011 in the UK, fuel poverty was estimated to affect 14.6% of the population; following a technical redefinition of fuel poverty, this dropped to 10.4% in 2012 and has remained at a similar level since (DBEIS, 2019). Officially, older people, young children and people who are long-term sick or with disabilities are considered 'vulnerable' in fuel poverty policy in England and Wales, because of their greater propensity to be affected by cold.

Whilst the UK initiated the concept of fuel poverty, researchers, and to some extent policy communities in other countries, followed in recognising versions of the same problem—for example, France (Dubois, 2012), New Zealand (Howden-Chapman et al., 2012), Ireland (McAvoy, 2007), and across the EU (Thomson and Snell, 2013). These have tended to reproduce the same notion of the difficulty in heating homes being the main problem, although problems with other energy services (see Box 13.2), such as cooling and lighting, have also been noted (Brunner et al., 2012; Thomson et al., 2019).

Box 13.2 Energy services and their relationship with everyday needs

Energy services can be defined as "those functions performed using energy which are means to obtain or facilitate desired end services or states" (Fell, 2017, p.137). Access to various energy services is necessary for the fulfilment of everyday needs in most contexts. Energy services that are generally considered essential at household level are listed following. How much of these services we need will vary according to the context in which we live; and how much is essential rather than luxurious is a matter of debate, and the topic of some more recent work on energy poverty.

> **Cooking**: essential for the preparation of hot, readily digestible, and culturally acceptable food.
> **Water heating**: separate from cooking, hot water is needed for personal hygiene, laundry, and other cleaning, and sometimes for domestic industries such as textile production.
> **Space heating**: necessary to maintain a warm, dry, and healthy living environment for at least part of the year in many parts of the world.
> **Space cooling**: necessary to maintain a comfortable and healthy living environment for at least part of the year in many regions, especially in more modern building designs.
> **Refrigeration**: needed for the preservation of food, to avoid illness and wastage; also sometimes for the storage of medicines.
> **Lighting**: needed for safety, for the ability to carry out all kinds of tasks after dark, including schoolwork for children, and often for productive work relating to generating an income.
> **Information and communication**: needed for accessing information, keeping in touch with family and friends, and often for purposes related to education and paid work.
> **Mechanical and appliance power**: appliances help with a multitude of everyday tasks, such as cooking and ironing, and small machines might be used for productive activities such as sewing or cultivating produce.
> **Mobility**: a form of mechanical power and not always considered essential at household level, but many people rely on being able to power a motorbike, car, or van for personal mobility and transporting small amounts of goods, e.g. to market.

(See e.g. Practical Action, 2010; Bouzarovski and Petrova, 2015; Simcock et al., 2016; Walker et al., 2016.)

Walker and Day (2012) were the first to write explicitly about fuel poverty in the UK in terms of injustice. According to them, it can be viewed as:

- a distributional justice concern, in terms of distribution of incomes, energy prices, and quality housing, and the resultant distribution of access to energy services;
- a procedural justice concern, in terms of access to information, to participation in energy and other associated policy, and the extent of legal rights; and
- a recognition concern, in terms of the recognition of differential vulnerabilities and requirements and the importance assigned to the needs of some groups.

They argue that these three aspects are mutually reinforcing in the constitution of fuel poverty. Their analysis also offers some ways forward in terms of identifying areas of intervention to improve the different aspects of justice and arguments to justify these interventions.

Energy poverty and its effects in the Global South

In other, less developed parts of the world, the related term '**energy poverty**' has been more often used to refer to situations of people suffering from energy resource constraint. Although affordability of energy can definitely be an issue in such contexts, often the concern is more over the availability of energy, and especially of so-called 'modern' energy. 'Modern' energy is that which is of high quality and intensity and creates less pollution to the user—electricity, natural gas, and LPG (liquefied petroleum gas) count as modern energy—whilst its opposite, 'traditional' energy, includes biomass such as wood, agricultural residues and animal dung, and charcoal. These 'traditional' fuels give low energy output, and they also produce a lot of pollution at the point of use. Some other fuels such as kerosene are between the two in terms of pollution and output quality and are sometimes classed as 'transitional' fuels.

There is clearly a distributional inequality globally in access to modern energy (World Bank and IEA, 2017). Without access to modern energy, people's development is limited because they can only achieve a low level of essential energy services, such as cooking, lighting, and communication (see Box 13.2). In addition, if they are using fuels such as biomass and animal dung, they will be exposed to high levels of pollution from smoke, with severe health effects—the WHO (2018) estimates that 3.8 million people die prematurely each year due to cooking with solid fuels or kerosene. This affects women more than men, because women are generally responsible for cooking, and also young children, because they tend to stay close to their mothers. Collecting biomass for burning, such as firewood, is a time-consuming activity and also one which is often done by women or (female) children, taking time away from other pursuits, such as education, income-generating work, or socialising. Thus, there is a strong gender dimension to the impacts of the use of traditional fuels (Parikh, 2011). Nevertheless, their use often persists, even where alternatives such as LPG for cooking are available. One reason for this is that biomass collected from the local environment is often free, but another contributing factor can be that in some (not all) contexts women lack economic decision-making power, and so their needs might not be prioritised in household spending (Pachauri and Rao, 2013).

Improving access to clean energy, including electricity (through grid expansion or mini-grids powered by renewable energy) and cleaner cooking fuels such as LPG, is crucial, although not without justice issues, as noted in the previous section. Even without such access, the use of biomass can be made more efficient and less polluting through the use of better designed stoves, rather than traditional open fireplaces, while solar powered appliances, such as lanterns, and battery chargers are also possible. Often, the rollout of such technologies is supported by aid organisations and NGOs. However, we need to be wary of simplistic 'technological fixes' and assumptions that simply introducing new technologies will transform people's lives. Technologies are always taken up in a social context and within sets of social relations. Questions about the distribution of the costs and benefits of transition away from biomass use need to be examined, and participation of community members is essential in order to understand their needs, concerns, and preferences (see e.g. Munro et al., 2017).

Joined-up approaches and demand management

Recently, there have been attempts to introduce more unified conceptualisations of fuel poverty or energy poverty that can work across more and less developed society contexts. Bouzarovski and Petrova (2015) have suggested a definition of energy poverty based on sufficiency of access to necessary energy services. This does not privilege any specific service, such as heating, over another, so it can be used flexibly in different contexts, and it takes into account affordability, but does not prescribe how much income should be spent. Day et al. (2016) proposed a conceptualisation of energy poverty using the capabilities approach. They argue that the point of consuming energy and energy services is not for their own sake but to support important capabilities and functionings, and therefore, that energy poverty should be seen in terms of an inability to realise important capabilities due to a lack of affordable, reliable, and safe energy, and in the absence of alternative means of realising these capabilities. This last part of the definition offers a way forward to think about the injustice of energy poverty, whilst also being concerned about the injustice of climate change—in that it encourages us to think about how we might support our needs in less energy-consuming ways, and whether there are alternatives to all households having access to a certain level of services.

Almost all work on justice in energy consumption has been concerned with situations of energy poverty, where some people are arguably not using enough energy to meet their needs, and where addressing the problem may well mean increasing their energy consumption. However, increasing the consumption of energy in the form of biomass and fossil fuels stands to increase both climate change and air pollution, issues that, as we know, have severe impacts on millions of people and, as noted earlier, themselves have strong (in)justice dimensions, also involving **future generations** and other species. In order to address energy poverty in the context of climate change, two kinds of solution are generally proposed.

The first is to increase the **energy efficiency** of buildings, such as through large-scale insulation programmes—this has been the favoured approach in the UK where older-style housing tends to be prone to heat loss, leading to greater energy consumption for heating. Who pays for and who benefits from such programmes is, of course, a justice question. The other generally proposed solution is the stepping up of renewable energy generation. As this chapter has already discussed, however, large-scale renewable energy production does not come without environmental and social costs, nor must we forget to pay attention to the justice aspects of its deployment.

A more neglected issue in the energy justice and consumption literature is that of over-consumption. Yet, if our ability to generate energy is limited for environmental and atmospheric reasons, then justice requires that we should look at the distribution of the opportunities to consume energy. The definition of energy needs and of rights to energy cannot be concerned only with establishing the lower limit, but also needs to consider where the upper limits lie. This brings up a host of questions about context, vulnerability, recognition of specific needs, and the fairness of procedures by which any limits are established.

New developments in energy demand management are being ushered in for a combination of reasons, including climate change imperatives and rising prices of fossil fuels, but also because the electricity generating and distribution systems of many countries, such as the UK, are stretched to capacity at peak times, and a larger share of generation coming from renewables such as wind and solar means that capacity is more variable. Coming down the line, therefore, are initiatives such as flexible pricing, to encourage consumers to use more energy when the grid has greater capacity—for example, on windy nights—and less at peak demand times. Such developments bring their own justice questions, such as: who can benefit from such flexibility, and whose consumption may be more time constrained? These will need further investigation.

Challenges and future directions

For the sake of organising this chapter, I addressed energy justice in production and consumption arenas separately, and indeed, most of the literature tends to be focused on issues related to either one or the other. However, it is important to have awareness of how justice questions arise and need consideration across a whole energy system. We need to be aware of how systems of production and consumption connect, and to remember to think about systems of distribution, such as power lines, grids, and storage, which have to date been paid less attention in energy justice research, but which are increasingly important sites of development, especially as the share of electricity generated from renewable energy increases. It is not always possible or appropriate in individual studies to take a whole systems approach, but it is important to remember that any given energy situation is connected—often literally—to sets of issues in other parts of the system, where the justice questions may be quite different.

One of the challenges of energy justice is that of addressing scale, as justice questions arise at a number of scales. They may even be in tension, in that addressing injustices at one scale may lead to injustices arising at another scale, as for example when large renewable energy installations have local effects on specific communities, or when relieving energy poverty for households in one region increases overall consumption and adds to global climate pressures. Awareness of the whole system and of the connections between energy justice, climate justice and other environmental and social injustices (such as labour conditions addressed in Chapter 19 of this volume) is essential to help to resolve these scalar tensions.

In terms of future directions, new developments in energy systems throw up new situations for paying attention to justice all the time. On the distribution and consumption side, I have already mentioned moves towards flexible pricing and smart grids, which are likely to benefit some more than others. On the production side, some of the major current concerns are around the extraction of unconventional hydrocarbons through fracking and the like, the increasing demand for minerals from under-developed regions for renewable energy technologies, and the e-waste streams from solar production. The effects on not only future generations but on **non-humans**[8] probably do not receive enough attention within energy justice specifically, and in this context, connecting with developments in environmental justice would help. However, the more established questions of energy justice have not gone away and continue to need close attention, so that we work towards adequate energy for all, through means that are as low impact as possible, and whose externalities are fairly shared.

Follow-up questions

- What policies and regulations do we need to put in place to ensure that renewable energy development does not create unnecessary injustices?
- What are the best technologies and arrangements through which to provide energy services to millions of people who live in energy poverty?

- How much energy consumption is too much, and how do we decide this? How do we discourage people in wealthier regions from using excessive energy?
- How do we ensure that addressing energy justice at one scale or in one place does not lead to other injustices at other scales or in other places?

Notes

1 See Chapter 3 of this volume for more on distributional, or distributive, justice.
2 Chapter 4 of this volume goes into more depth on procedural justice.
3 Explored in Chapter 5 of this volume.
4 See Chapter 6 of this volume for more explanation of the capabilities approach.
5 On labour movements and environmental justice, see Chapter 19 of this volume.
6 On Indigenous environmental justice, see also Chapter 20 of this volume.
7 On sustainability and environmental justice, see Chapter 9 of this volume.
8 On non-human justice, see Chapter 21 of this volume.

References for further reading

- Day, R., Walker, G., Simcock, N. (2016). Conceptualising energy use and energy poverty using a capabilities framework. *Energy Policy*, 93, pp. 255–264.
- Jenkins, K., McCauley, D., Heffron, R., Stephan, H., Rehner, R. (2016) Energy Justice: a conceptual review. *Energy Research and Social Science*, 11, pp. 174–182.
- Walker, C., Baxter, J. (2017). 'It's easy to throw stones at a corporation': wind energy development and distributive justice in Canada. *Journal of Environmental Policy and Planning*, 19(6), pp. 754–768.
- Yenneti, K., Day, R., Golubchikov, O. (2016) Spatial justice and the land politics of renewables: dispossessing vulnerable communities through solar energy mega-projects. *Geoforum*, 76, pp. 90–99.

References used in this chapter

Avila-Calero, S. (2017) Contesting energy transitions: wind power and conflicts in the Isthmus of Tehuantepec. *Journal of Political Ecology*, 24, pp. 992–1012.

Baka, J. (2017) Making space for energy: wasteland development, enclosures, and energy disposessions. *Antipode*, 49(4), pp. 977–996.

Bazilian, M. (2018) The mineral foundation of the energy transition. *The Extractive Industries and Society*, 5, pp. 93–97.

Black, T., D'Arcy, S., Weis, T., Russell, S.K. (2014) *A Line in the Tar Sands: Struggles for Environmental Justice*. Between the Lines Books, Toronto.

Boardman, B. (1991) *From Cold Homes to Affordable Warmth*. Belhaven Press, London.

Bouzarovski, S., Petrova, P. (2015) A global perspective on domestic energy deprivation: overcoming the energy poverty-fuel poverty binary. *Energy Research and Social Science*, 10, pp. 31–40.

Brunner, K.-M., Spitzer, M., Christanell, A., (2012) Experiencing fuel poverty: coping strategies of low-income households in Vienna, Austria. Energy Policy, 49, pp. 53–59.

Coalmining History Resource Centre (2018) Coalmining Accidents and Deaths. Available at https://web.archive.org/web/20160305031248/www.cmhrc.co.uk/site/disasters/index.html, accessed 11/04/19.

Cross, J., Murray, D. (2018) The afterlives of solar power: waste and repair off the grid in Kenya. *Energy Research and Social Science*, 44, pp. 100–109.

Day, R., Walker, G., Simcock N., (2016) Conceptualising energy use and energy poverty using a capabilities framework. *Energy Policy*, 93, pp. 255–264.

Department for Business, Energy and Industrial Strategy (UK) (2019) Fuel poverty statistics. Available at https://www.gov.uk/government/collections/fuel-poverty-statistics, accessed 03/04/20.

Department of Trade and Industry (UK) (2001) *The UK Fuel Poverty Strategy*. London: DTI.

Dhillon, B.S. (2010) *Mine Safety: A Modern Approach*. Springer, London.

Dubois, U. (2012) From targeting to implementation: the role of identification of fuel poor households. *Energy Policy*, 49, pp. 107–115.

Fell, M.J. (2017) Energy services: a conceptual review. *Energy Research and Social Science*, 27, pp. 129–140.

Fraser, N. (1995) From redistribution to recognition? Dilemmas of justice in a 'post-socialist' age. *New Left Review*, I/212, pp. 68–93.

Government of India (2019) Scheme/documents (National Solar Mission). Available at https://mnre.gov.in/scheme-documents, accessed 28/11/19.

Gross, C. (2007) Community perspectives of wind energy in Australia: the application of a justice and community fairness framework to increase social acceptance. *Energy Policy*, 35(5), pp. 2727–2736.

Hall, N., Ashworth, P., Devine-Wright, P. (2013) Societal acceptance of wind farms: analysis of four common themes across Australian case studies. *Energy Policy*, 58, pp. 200–208.

Healy, N., Barry, J. (2017) Politicizing energy justice and energy system transitions: Fossil fuel divestment and a 'just transition'. *Energy Policy*, 108, pp. 451–459.

Hoffman, S.M. (2001) Negotiating eternity: energy policy, environmental justice, and the politics of nuclear waste. *Bulletin of Science, Technology and Society*, 21(6), pp. 456–472.

Howden-Chapman, P., Viggers, H., Chapman, R., O'Sullivan, K., Barnard, L.T., Lloyd, B. (2012) Tackling cold housing and fuel poverty in New Zealand: a review of policies, research, and health impacts. *Energy Policy*, 49, pp. 134–142.

Johansen, K., Emborg, J. (2018) Wind farm acceptance for sale? Evidence from the Danish wind farm co-ownership scheme. *Energy Policy*, 117, pp. 413–422.

Khorsand, I., Kormos, C., Macdonald, E.G., Crawford, C. (2015) Wind energy in the city: an inter-urban comparison of social acceptance of wind energy projects. *Energy Research and Social Science*, 8, pp 66–77.

Langer, K., Decker, T., Roosen, J., Menrad, K. (2016) A qualitative analysis to understand the acceptance of wind energy in Bavaria. *Renewable and Sustainable Energy Reviews*, 64, pp. 248–259.

Liebe, U., Bartczak, A., Meyerhoff, J. (2017) A turbine is not only a turbine: the role of social context and fairness characteristics for the local acceptance of wind power. *Energy Policy*, 107, pp. 300–308.

Macdonald, C., Glass, J., Creamer, E. (2017) What is the benefit of community benefits? Exploring local perceptions of the provision of community benefits from a commercial wind energy project. *Scottish Geographical Journal*, 133(3–4), pp. 172–191.

Mason, K., Milbourne, P. (2014). Constructing a 'landscape justice' for windfarm development: the case of Nant Y Moch, Wales. *Geoforum*, 53, pp. 104–115.

McAvoy, H. (2007) *All-Ireland Policy Paper on Fuel Poverty and Health*. Institute of Public Health in Ireland, Dublin.

McDowell, C. (ed.) (1996) *Understanding Impoverishments: The Consequences of Development Induced Displacement*. Berghahn, Oxford.

McIvor, A., Johnston, R. (2007). *Miner's Lung: A History of Dust Disease in British Coal Mining*. Routledge, London.

Miranda, M.L., Edwards, S.E., Keating, M.H., Paul, C.J. (2011) Making the environmental justice grade: the relative burden of air pollution exposure in the United States. *International Journal of Environmental Research in Public Health*, 8(6), pp. 1755–1771.

Mulvaney, D. (2013) Opening the black box of solar energy technologies: exploring tensions between innovation and environmental justice. *Science as Culture*, 22(2), pp. 230–237.

Munro, P., van der Horst, G., Healy, S. (2017) Energy justice for all? Rethinking sustainable development goal 7 through struggles over traditional energy practices in Sierra Leone. *Energy Policy*, 105, pp. 635–641.

Nussbaum, M.C. (2011) *Creating Capabilities: The Human Development Approach.* Harvard University Press, Cambridge, MA.

Orta-Martínez, M., Finer, M. (2010) Oil frontiers and indigenous resistance in the Peruvian Amazon. *Ecological Economics*, 70, pp. 207–218.

Pachauri, S., Rao, N.D. (2013) Gender impacts and determinants of energy poverty: are we asking the right questions? *Current Opinion in Environmental Sustainability*, 5(2), pp. 205–215.

Parikh, J. (2011) Hardships and health impacts on women due to traditional cooking fuels: a case study of Himachal Pradesh, India. *Energy Policy*, 39(12), pp. 7587–7594.

Practical Action (2010) Poor People's Energy Outlook 2010. Available at https://policy.practicalaction.org/resources/publications/item/poor-people-s-energy-outlook-2010, accessed 27/11/19.

Rawls, J. (1971) *A Theory of Justice.* Belknap Press, Cambridge, MA.

San-Sebastián, M., Armstrong, M., Cordoba, J.A., Stephens, C. (2001) Exposures and cancer incidence near oil fields in the Amazon basin of Ecuador. *Occupational and Environmental Medicine*, 58, pp. 517–522.

Scherhaufer, P., Holtinger, S., Salak, B., Schauppenlehner, T., Schmidt, J. (2017) Patterns of acceptance within energy landscapes: a case study on wind energy expansion in Austria. *Energy Policy*, 109, pp. 863–870.

Sen, A. (1999) *Development as Freedom.* Oxford University Press, Oxford.

Sharpe, V. (2008) 'Clean' nuclear energy? global warming, public health and justice. *Hastings Center Report*, 38(4), pp. 16–18.

Shrader-Frechette, K. (1994) Equity and nuclear waste disposal. *Journal of Agricultural and Environmental Ethics*, 7(2), pp. 133–156.

Simcock, N. (2016) Procedural justice and the implementation of community wind energy projects: a case study from South Yorkshire, UK. *Land Use Policy*, 59, pp. 467–477.

Simcock, N., Walker, G., Day, R. (2016) Fuel poverty in the UK: beyond heating? *People, Place and Policy*, 10(1), pp. 25–41.

Simpson, G., Clifton J. (2016) Subsidies for residential solar photovoltaic energy systems in Western Australia: distributional, procedural and outcome justice. *Renewable and Sustainable Energy Reviews*, 65, pp. 262–273.

Sovacool, B., Dworkin, M.H. (2015) Energy justice: conceptual insights and practical applications. *Applied Energy*, 142, pp. 435–444.

Sunter, D.A., Castellanos, S., Kammen, D.M. (2019) Disparities in rooftop photovoltaics deployment in the United States by race and ethnicity. *Nature Sustainability*, 2, pp. 71–76.

Thomson, H., Simcock, N., Bouzarovski, S., Petrova, S. (2019) Energy poverty and indoor cooling: an overlooked issue in Europe. *Energy and Buildings*, 196, pp. 21–29.

Thomson, H., Snell, C. (2013) Quantifying the prevalence of fuel poverty across the European Union. *Energy Policy*, 52, pp. 563–572.

United Nations Economic Convention for Europe (1998) Convention on Access to Information, Public Participation in Decision-Making and Access to Justice in Environmental Matters. Available at www.unece.org/fileadmin/DAM/env/pp/documents/cep43e.pdf, accessed 12/04/19.

United Nations Environment Programme (2018) Global Trends in Renewable Energy Investment 2018. Available at https://europa.eu/capacity4dev/unep/documents/global-trends-renewable-energy-investment-2018, accessed 12/04/19.

United Nations Environment Programme (2019) Global Resources Outlook 2019. Available at https://wedocs.unep.org/bitstream/handle/20.500.11822/27518/GRO_2019_SPM_EN.pdf?sequence=1&isAllowed=y, accessed 27/11/19.

Walker, C., Baxter, J. (2017a) 'It's easy to throw stones at a corporation': wind energy development and distributive justice in Canada. *Journal of Environmental Policy and Planning*, 19(6), pp. 754–768.

Walker, C., Baxter, J. (2017b). Procedural justice on Canadian wind energy development: a comparison of community-based and technocratic siting processes. *Energy Research and Social Science*, 29, pp. 160–169.

Walker, G., Day, R. (2012) Fuel poverty as injustice: integrating distribution, recognition and procedure in the struggle for affordable warmth. *Energy Policy*, 49, pp. 69–75.

Walker, G., Simcock, N., Day, R. (2016) Necessary energy uses and a minimum standard of living in the United Kingdom: energy justice or escalating expectations? *Energy Research and Social Science*, 18, pp. 129–138.

Wheeler, B., Ben-Shlomo, Y. (2005) Environmental equity, air quality, socioeconomic status, and respiratory health: a linkage analysis of routine data from the Health Survey for England. *Journal of Epidemiology and Community Health*, 59(11), pp. 948–954.

Wolsink, M. (2007) Wind power implementation: the nature of public attitudes: equity and fairness instead of 'backyard motives'. *Renewable and Sustainable Energy Reviews*, 11(6), pp. 1188–1207.

World Bank and International Energy Agency (2017) Sustainable Energy for All Global Tracking Framework: Progress Toward Sustainable Energy. Available at http://trackingsdg7.esmap.org/data/files/download-documents/eegp17-01_gtf_full_report_for_web_0516.pdf, accessed 12/04/19.

World Health Organisation (2018) Household Air Pollution and Health. Available at www.who.int/news-room/fact-sheets/detail/household-air-pollution-and-health, accessed 18/03/2019.

Yenneti, K., Day, R. (2015) Procedural (in)justice in the implementation of solar energy: the case of Charanaka solar park, Gujarat, India. *Energy Policy*, 86, pp. 664–673.

Yenneti, K., Day, R. (2016) Distributional justice in solar energy implementation in India: the case of Charanka solar park. *Journal of Rural Studies*, 46, pp. 35–46.

Yenneti, K., Day, R., Golubchikov, O. (2016) Spatial justice and the land politics of renewables: dispossessing vulnerable communities through solar energy mega-projects. *Geoforum*, 76, pp. 90–99.

Plate 1 The government of the Maldives holds an underwater cabinet meeting to raise awareness about the threats of climate change for Small Island Developing States (SIDS). While being among the world's smallest emitters of greenhouse gases, the SIDS are the most affected by climate change.

Source: © Mohamed Seeneen / @mohamedseeneen / @seeneen.photography

Plate 2 An estimated 400,000 people took the streets demanding climate justice during the 2014 People's Climate March in New York City, USA. See Chapter 12 of this volume.

Source: © Robert van Waarden / Survival Media Agency

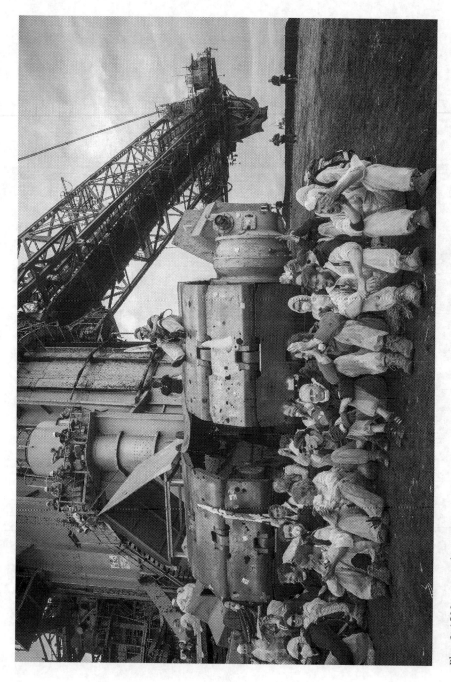

Plate 3 1500 anti-coal activists shutting down the Tagebau Garzweiler surface mine during the 2015 *Ende Gelände* ("here and no further") protests in North-Rhine Westphalia, Germany. In 2018, mining was found to be the deadliest environmental cause around the world, according the non-governmental organization Global Witness. See Chapter 8 of this volume.

Source: © Paul Lovis Wagner

Plate 4 Squatters erecting a barricade in a 'Zone-to-Defend' (ZAD) in Notre-Dame-des-Landes, France, against a boondoggle airport construction project. Starting in the 1970s, opposition to the project is said to have been France's longest running struggle for land. The project was canceled in 2018.

Source: © Reuters / Stephane Mahe

Plate 5 A resident in Flint, Michigan, USA collects water from a distribution center during the lead-contaminated water crisis, which Paul Mohai called 'the most egregious example of environmental injustice'. The crisis started in 2014. See Chapters 3, 10 and 16 of this volume.

Source: © Jake May / MLive.com

Plate 6 A boy scavenging for metals at the world's largest e-dump, Agbogbloshie, Ghana, where most of the locals die from cancer while in their 20s. Unjust exposure to toxic hazards is a pivotal and long-standing concern within the environmental justice movement. See Chapter 10 of this volume.

Source: © Michael Ciaglo

Plate 7 A woman with a "Water is Life" headscarf protesting the construction of the Dakota Access Pipeline (DAPL) near the Standing Rock Indian Reservation, Northern USA. Beyond the ecological impacts, the #NoDAPL movement laid bare the common roots of carbon-intensive industries and settler colonialism in the USA. See Chapters 5, 7 and 20 of this volume.

Source: © Avery White

Plate 8 In 2017, Mount Taranaki on North Island, New Zealand was given "legal personality, in it:s own right" by the government. Eight local Māori tribes and the government share guardianship of the sacred mountain. See Chapter 21 of this volume.

Source: Donna O'Donoghue (public domain)

14 Food, agriculture, and environmental justice
Perspectives on scholarship and activism in the field

Kristin Reynolds

Learning outcomes

- Explore environmental justice issues related to food and agriculture.
- Develop familiarity with the evolution of food justice and food sovereignty, as concepts and movements, in the late 20th and early 21st centuries.
- Become aware of some debates in the field.
- Consider reflections on emerging areas of focus among scholars and social movements addressing food, agriculture, and environmental justice.

Introduction

Environmental justice encompasses interest in and action on a variety of environmental issues. One of those is food, and the **food system**—the social, cultural, economic, political, and environmental aspects of food, including its production, harvest, gathering, distribution, processing, consumption, compost, and waste.

Inherent in the food system are issues of environmental inequity and environmental justice. For example, when pesticides and herbicides harmful to human health are used in agriculture, some members of global society, often low-income people and/or international migrants, experience greater or more damaging exposures to these chemicals through work in agricultural fields and/or consumption of foods that are produced with these products. When farmed land is converted to non-agricultural uses, particularly in agrarian societies, agricultural livelihoods are diminished or lost. In urban as well as rural environments, healthful and culturally relevant foods are not available to all; in both Global North and Global South settings, this often cuts along economic class, gender/gender identity, ethnic, and/or racial lines—and, while it is common to divide rural/urban food systems issues along production and consumption lines (that is, seeing agriculture production as an exclusively rural activity and consumption as a predominantly urban practice), considering food and agriculture through a justice lens, as we do in this chapter, compels us to recognize both the links between rural and urban areas and the important practices that exist beyond these divides—rural inhabitants also consume food, while urban agriculture is a growing practice!

Activists, non-governmental organizations, policy makers, academics, students, and frontline communities are addressing these issues, in the interest of global social justice as it relates to agriculture, food, and the environment. While these actors may not always do so through an environmental justice lens, environmental justice has explicitly informed some such initiatives and intersects to varying degrees with others, including food justice and food sovereignty. This chapter examines these intersections, and the evolution of key food and justice concepts and movements.

Food, agriculture, and environmental justice

Foundational environmental justice research and documentation have traced how parts of the environmental justice movement have addressed issues related to agriculture and food, most notably with respect to farmworkers and pesticides, but also interrelated issues of land, water, and civil rights. One of the most oft-cited, historical examples of this is the farmworker movement in the US. Indeed, Cole and Foster (2001, p. 27) note that the farmworker labour movement of the 1960s was "perhaps the first nationally known effort by people of colour to address an environmental issue" (see also Gottlieb, 2005; Moses, 1993; Taylor, 2011.)

Agricultural labour in the US has, since colonization and the foundation of the settler-colonial nation, been performed largely by people of colour working under exploitative conditions including: enslavement of Indigenous peoples and Native Americans,[1] and Africans and African Americans (prior to the end of legalized slavery in the 1860s); inequitable sharecropping arrangements; and successive reliance on ethnic migrants from China, the Philippines, India, Japan, and Mexico, as well as other parts of Central and South America, the latter of which continue to this day (Moses, 1993; Bullard, 1993; Mitchell, 2013; Minkoff-Zern, 2018, 2019) (white labourers also experienced exploitation in US agriculture, particularly following the Dust Bowl crisis in the 1930s; see Mitchell, 2013). In the mid-20th century, the Bracero Program, which resulted from a series of agreements between Mexico and the US, brought nearly 5 million Mexican men into the US to work on a temporary basis in agriculture. Though the programme was designed as a wartime effort to alleviate labour shortages during World War II, it was in existence from 1942–1964 (Mitchell, 2013), and set the stage for the current context in which the majority of farmworkers in the US are migrants from Mexico or Central America.

Shortly after the termination of the Bracero Program, in 1965, the UFW Organizing Committee (today the United Farm Workers), led by César Chavez and Dolores Huerta, coalesced, fighting for better working conditions, including a number of initiatives focused on pesticides, notably: workers' right to know about pesticide use in fields, inclusion of pesticide safety and the dangers of health risks in worker contracts, and strategic collaborations with mainstream environmental groups that resulted in the eventual banning of the use of the carcinogenic pesticide DDT in the US (Moses, 1993; Gottlieb, 2005; Cole and Foster, 2001; see also Minkoff-Zern, 2018, 2019). As noted earlier, this movement is often cited as the first national initiative led by people of colour to address an environmental issue and is one that clearly connected environment and the food system.

Box 14.1   Concept in action: solidarity activism for farmworker livelihoods

In many countries of the Global North, immigrants are the most highly represented group in agricultural labour. As a case in point, farm labour in the US is performed primarily by immigrants. An analysis of US Department of Labor's Agricultural Workers

Survey by the nonprofit group Farmworker Justice noted that, in 2015–2016, approximately 75% of farmworkers were immigrants, most of whom hailed from Mexico. Among these, according to the Farmworker Justice report, approximately 49% lacked work authorization. Agricultural workers have been excluded from worker labour protections since the 1938 Fair Labor Standards Act, which established minimum wages, maximum number of work hours, and prohibited child labour (Farmworker Justice, 2018). While the farmworker labour movement during the 1960s accomplished gains for farmworkers, including those related to pesticide use and exposure, subpar working conditions for farmworkers, along with poverty wages and other labour abuses, remain.

The organizing tactics of the 1960s included strikes and consumer boycotts (Gray, 2013; Minkoff-Zern, 2017, 2019), but consumer action to improve labour conditions in agriculture waned, beginning in the 1980s as the sustainable agriculture movement gained momentum and neoliberalism took hold (see Allen et al., 2003 for an analysis). Beginning in the late 1990s and early 2000s, a new wave of consumer concern about working conditions for farmworkers gave rise to a variety of initiatives to support changes in this system, including domestic fair-trade labels, as well as farmer-led strategies, notably by the Coalition of Immokalee Workers, to raise farmworker wages and address forms of exploitation, including 'modern day slavery' (CIW, n.d.; Minkoff-Zern, 2017). These efforts have won some gains for workers, such as increased wages and improved working conditions (CIW). In conjunction with related organizing, led by coalitions such as the organization Food Chain Workers' Alliance and the Student-Farmworker Alliance, have enacted solidarity campaigns that move beyond the common '**vote-with-your-fork**' mentality in which consumers are presumed only to have power through purchase (Minkoff-Zern, 2017; Lo and Koening, 2017). Such strategies can be read as a contemporary iteration of infusing the environmental justice tenet of **participation** and representation—in this case among farmworkers—in addressing justice issues in the food system, and they are also inspirational in terms of the possibilities for non-farmers or non-farmworkers to support such initiatives.

Indigenous peoples' struggles around land and territory in many parts of the world also constitute work that is central to food. To cite another US example, the forced removal of Indigenous peoples from ancestral lands, the conquest of Indigenous territories (first by European colonists and powers and, following the establishment of the US, by US government, military, and individual settler-colonists), and a variety of land and resource-use policies—among other horrific realities of these conquests—have since denied Native Americans access to traditional foodways and foods, with ramifications for cultural and bodily survival (Norgaard et al., 2011; Mihesuah and Hoover, 2019; see also Dunbar-Ortiz, 2014, and Chapters 7 and 20 of this volume.) In the early 1990s, the Indigenous Peoples' Movement coalesced, alongside the environmental justice movement, initially addressing toxic waste siting,[2] which can have direct negative impacts on food procurement for Native American peoples, particularly with respect to water contamination (Moses, 1993; Cole and Foster, 2001, pp. 134–150). As noted following, the Indigenous Environmental Network, established in 1990 within the US by Indigenous peoples and individuals to deal with environmental and economic justice matters, addresses issues surrounding food and sovereignty more explicitly (www.ienearth.org/ Accessed 6/5/19).

As discussed in Chapter 2 of this volume, the US Civil Rights Movement of the 1950s, 1960s, and 1970s was a main tributary to the environmental justice movement (Cole and Foster, 2001, pp. 20–21). In our consideration of food and environmental justice, it is important to recognize that the Black Power Movement also addressed specific intersections of food, agriculture, and cultural survival in the context of the ongoing oppression of Black people and communities in the US. A notable example of this is the Free Breakfast Program of the Black Panther Party, which, in the late 1960s and 1970s provided free breakfasts to Black children as a means for both bodily and cultural survival (Heynen, 2009; White, 2018; Penniman, 2018). Indeed, included in the Black Party for Self-Defense Ten-Point Platform (cf. Bloom and Martin, 2016, pp. 70–73) was a demand for "land, bread, housing, education, clothing, justice and peace". This further underscores the connections that those fighting for social justice have made between food and environmental justice, well before the concept of environmental justice or the environmental justice movement coalesced per se (See Taylor, 2011).

Food and agriculture in the environmental justice movement

The previous examples support the contentions by contemporary scholars and activists both that food and agriculture concerns are central to the environmental justice concerns, and that activists and social movements were already addressing these concerns in the decades (and centuries, as Taylor, 2011, argues) preceding the early 1990s. As the environmental justice movement coalesced, the *Principles of Environmental Justice*, a founding document of the environmental justice movement, specifically called for the protection from environmental hazards that "threaten the fundamental right to clean air, land, water, and *food*" (1991, emphasis added). Still, some scholars have argued that food was not addressed systematically in the environmental justice movement until the National People of Color Leadership Summit II (held in 2002 as a follow-up to the first National People of Color Leadership Summit in 1991) (Mares and Peña, 2011). At that time, a resource paper commissioned for the second summit discussed connections between environmental justice, food, and sustainable agriculture, proposing action plans for the environmental justice movement to address a range of interrelated issues including: farmworker health (as discussed previously); several facets of racial discrimination on the part of lending agencies and the US Department of Agriculture that negatively affected the ability of African American, Latinx, and Native American farmers to remain in agriculture; and the loss of access to common or ancestral lands on which Native Americans would grow, hunt, and/or gather food (Peña, 2002). Separately, a 1996 academic paper argued for a link between environmental justice and the burgeoning community food security movement (discussed further ahead), which evolved throughout the mid-1990s to address hunger and food insecurity as a community-level, rather than an individual, problem (Gottlieb and Fisher, 1996).

These papers, among others, highlighted growing connections between food and agricultural issues in the environmental justice movement and scholarship. Since then, food systems issues have gained more prominence in the environmental justice frame. Today some activists, scholars, and supportive policymakers more explicitly connect a broader array of food systems issues to environmental justice, either by addressing food and agricultural issues through environmental justice claims, or by articulating what we might understand as environmental justice concerns within distinct, but sympathetic, analyses (Agyeman et al., 2016). Two movements and fields of scholarship—food justice and food sovereignty—are key examples of this latter trajectory and are discussed next.

> **Box 14.2 Black farmers in the US**
>
> Black and African American farmers were almost literally the backbone of the development of the US economy prior to and during the nation's establishment, beginning in the 17th century. As most readers will be aware, this role was long a forced one: enslavement and post-slavery exploitation by white people sanctioned by the national or state governments. Post-Reconstruction, Black farmland ownership reached its peak in 1910, when 14% of farmland was owned by Blacks; the subsequent decline was due largely to Jim Crow laws and racist loan and government practices (cf. Penniman, 2018; White, 2018). Despite these historical facts, this is not the entire picture of Black farming in the US.
>
> As environmental justice scholar Monica White has shown, Black and African American farmers and agricultural scholars also used agriculture as a source of strength and **resilience** (White, 2018). For example, Black farmers and agricultural scholars invented many of the strategies that are today described as 'alternative' food initiatives, including agricultural cooperatives and cooperative land management, and Black and African Americans, including Ms. Fannie Lou Hamer, used their farms or university-based agricultural research to support the struggle for civil rights (White, 2018; Penniman, 2018; McCutcheon, 2019). Despite this ingenuity, there has been a general lack of recognition both of these and other ways in which Black farmers in the US used agriculture as a source of liberation and the essential contributions that Black farmers have made to agriculture as we know it (ibid.).
>
> In recent years, a renewed movement of Black and African American farmers and scholar-activists is making strides in both helping African American and Black people in the US to 'heal' their relationship with the land and training a next generation of Black diasporic farmers. Groups like Black Urban Growers, which organizes and hosts an annual farming conference, Soul Fire Farm, a working farm that also runs agricultural training programmes for Black, Indigenous, and people of colour, and a number of additional local groups and national coalitions are reinvigorating the traditions of African American and Black agriculture in the US (see White, 2018; Penniman, 2018). Broadening understanding about African Americans and agriculture is connected to the environmental justice principles of recognition and representation.

Food justice: a movement and field of practice

Food justice is a concept and related movement that considers the social and political roots of inequities in the food system and holds that these structural issues must be addressed to solve problems such as disparate access to healthy food and exploitative or unfair labour practices. Similarly to the environmental justice movement, many food justice activists and scholars insist that people of colour and working-class people—often those most marginalized from food system decision-making and mainstream economies—should lead initiatives to realize food justice in their own communities (Alkon and Agyeman, 2011; Gottlieb and Joshi, 2010; Redmond, 2013; Yakini, 2013).

Prior to the emergence of the food justice movement, in the US in particular, a number of initiatives and other movements were already connecting environmental and some small-scale farmers' and consumers' concerns to the food system. What is often discussed as the **alternative food movement** (with its attendant alternative food initiatives, or AFIs) has brought together an array of critiques of the dominant industrialized food system to seek different modes of production and distribution of food, with goals including confronting corporate agriculture and addressing the difficulties that small-scale farm owners and operators face as they strive to remain financially viable in an increasingly consolidated and neoliberal global marketplace (Allen et al., 2003; Allen and Guthman, 2006). Examples of initiatives often discussed as being part of this movement are farmers markets, community-supported agriculture, and other forms of direct marketing; urban agriculture, and contemporary homesteading (in rural or urban spaces) through which people aim to be more self-sufficient in food production, deflecting their purchasing power from the mainstream food system; and sustainable agriculture, a now-broad rubric that is in its most general sense understood to be more environmentally benign than the most ecologically damaging forms of industrial agriculture relying heavily on chemical inputs, large-scale machinery, and digital technology.

Of particular relevance to our consideration of food and environmental justice is the community food security movement, a subset of the alternative food movement, which saw its origins in the mid-1990s. The origins of this movement are often traced to a 1996 paper (Gottlieb and Fisher, 1996) in which authors Robert Gottlieb and Andrew Fisher proposed that sustainable agriculture and environmental justice movements should work together to tackle inequitable food access, the latter of which was at the time most commonly addressed through anti-hunger strategies ranging from food banks to federal government subsidies that provide food to eligible low-income recipients. While their proposal did not gain traction in the way they may have originally envisioned it (sustainable agriculture and environmental justice movements remain separate, by and large), the community food security movement that coalesced subsequently (and in which Fisher played a key role) articulated a vision for achieving food access for low-income people that consciously stepped away from prevailing notions that food insecurity is an individual or household problem, resulting from individual decisions and habits (i.e., 'blame the victim' approaches). Community food security, as a concept and movement in the US, held that, although personal choices are important drivers of consumption, structural issues at the community level and beyond were equally—if not more—important causes of inequitable food access.

Community food security was thus conceptualized (primarily in the US) as "the ability of all persons obtaining, at all times, a culturally acceptable, nutritionally adequate diet through local, non-emergency [such as government programmes] sources" (Gottlieb and Fisher, 1996, p. 24).

In contrast, at the global scale, and as defined by the United Nations (UN), the original (1970s) definition of **food security,** focused rather narrowly on adequate food supply for the global population, without regard to individual or household-level access to that food, much less cultural preferences (as discussed in Patel, 2009). The UN definition was used then, as it is now, by governments and global bodies, including most notably the UN Food and Agriculture Organization (FAO), to shape food and agricultural aid programmes. Community food security thus expanded upon the concept of food security, as it existed in the early 1990s at the global scale.[3] For community food security, the solutions to food *in*security lay not simply in increasing aid or entitlement programmes (such as government subsidies to purchase food), or even in educating low-income people about healthy food (i.e., intervention at the individual or household levels), but also addressing community-level and policy drivers of food insecurity. Questions such as "Is there a grocery store with affordable and fresh food within reasonable distance?" and "What policies prevent or discourage grocery stores from locating in low-income communities?" exemplify this approach.

The community food movement of the 1990s was successful in many regards in helping to shift thinking about the causes and potential solutions to food insecurity and hunger and gave rise to consideration of these issues at the neighbourhood or community levels. By the early 2000s, however, scholars and activists were becoming more vocal about the lack of focus, within alternative and community food movements, on social justice issues such as economic equity and, particularly, fair labour practices *in the food system* (Allen et al., 2003; Allen, 2008). Relatedly, in some ways parallel to critiques of the mainstream environmental movement as one that is dominated by middle-class or elite whites, some began to critique the alternative food movement as being dominated by white cultural values and white people (Slocum, 2006; Guthman, 2008; Reynolds, 2015). It is in the context of these critiques that the food justice movement began to coalesce, particularly as activists wove together social justice, food, and, at times, environmental concerns (Alkon and Agyeman, 2011; People's Grocery, 2009; Levkoe, 2006; Just Food, n.d.; Coalition of Immokalee Workers, n.d.).

Box 14.3   The 'food desert' debate

The term **'food desert'** was first used in the early 2000s to describe places in which healthy food was difficult to procure (Wrigley, 2002; Walker et al., 2010). The term responded in important ways to what was at the time a common analysis about food security, or community food security, in many policy and public health arenas: that people's inability to access healthy food, with a focus on supermarkets as the main source, was rooted in their own decision-making and/or economic circumstance, *rather than* surrounding structures or environments. Food desert terminology has since gathered recognition in food systems literature, as well as policy-making arenas. For example, the US Department of Agriculture uses its own definition of a food desert— "Low-income census tracts with a substantial number or share of residents with low levels of access to retail outlets selling healthy and affordable foods are defined as food deserts"—in order to map areas with these characteristics and communicate this to the public (www.ers.usda.gov/amber-waves/2011/december/data-feature-mapping-food-deserts-in-the-us/ Accessed 11/29/2019).

The food desert frame was a step toward broader public awareness that lack of ability to access healthy foods is not only an individual problem, but instead one rooted in broader inequalities such as economic inequality or racist urban-planning practices (see Kwate et al., 2009). Nonetheless, scholars and activists have since argued that the term is misleading (at best), as it: has tended to ignore smaller stores (Short et al., 2007); suggests that lack of access to fresh and healthy food is a 'natural' phenomenon (like a desert, coupled with 'food swamps' in which much *un*healthy food is available: Taylor and Ard, 2015); and, relatedly, implies that this phenomenon is a result of laissez-faire market systems, whereas, according to some activists, inequitable access to healthy food is seen as intentional, as a way to further weaken the position of people of colour, particularly in white-dominant systems (Penniman, 2018; Reese, 2019). Responding to the latter of these arguments, activists have proposed the term 'food apartheid' to describe the system in which people have differential access to healthy food based not only on economic status (which is itself racialized), but also on race or ethnicity (ibid.).

Like the environmental justice movement, the food justice movement and related scholarship coalesced around, and is built upon, previous efforts and historical legacies of food systems activism, including those discussed at the beginning of this chapter (cf. Alkon and Agyeman, 2011; Sbicca, 2018). And, as those in the environmental justice movement have already done, food justice activists, scholars, and supportive policymakers address a broad array of issues, some of which are directly connected environmental claims, and some less so. These include farmworker exposure to agricultural chemicals, but also a broader set of concerns such as labour conditions and wages for food workers, social equity in business sectors, and food gentrification, along with food access inequities that divide along ethno-racial and class lines.

What makes food justice distinct from related initiatives and movements, notably alternative food and community food security movements discussed previously, is its emphasis on the grounding of food systems inequities in structural racism and class-based disparities, and its explicit focus on dismantling these structures as a means to make improvements in specific food and justice issues (Alkon and Agyeman, 2011; Gottlieb and Joshi, 2010; Sbicca, 2018). To cite a classic example of this distinction, access to healthy and affordable food can plausibly be improved by opening a large-scale supermarket in a given community that is in need of this basic necessity. However, in itself, such an initiative may do little to address the working conditions, racialized poverty, or community self-determination that underlie food inequities. Moreover, the food justice frame critically considers the paradoxical, unintended effects of some alternative food systems issues—such as the **'ecogentrification'** associated with farmers markets or urban agriculture—in which initiatives that may make food and/or green spaces more accessible also drive up property values and result in longtime residents being forced to leave the neighbourhoods if they cannot afford increased property taxes or rents (Joassart-Marcelli and Bosco, 2018; Anguelovski, 2016; McClintock, 2018a). Food justice claims would argue that all of these must be tackled to achieve a truly just system (Reynolds and Cohen, 2016; McClintock, 2018b; Horst, McClintock, and Hoey, 2017).

The food justice movement and related scholarship saw their origins in the US and Canada (e.g., Alkon and Agyeman, 2011; Levkoe, 2006). However, more recently, scholars in other regions have integrated the food justice frame into their work (e.g., in France: Paddeu, 2016; Hochedez and Le Gall, 2016.) And, though some scholars situate the origins of food justice in urban movements, both the legacy of farmworker organizing around agricultural issues, discussed previously, and contemporary farmworker organizing that began to gain momentum in the early 1990s (for example, the Coalition of Immokalee Workers [CIW], which began organizing for farmworker rights and labour improvements in 1993) are examples of longer-standing rural-oriented food justice work (Coalition of Immokalee Workers, n.d.; Minkoff-Zern, 2017) (see also McEntee, 2011 for a discussion of rural food justice concerns). Moreover, as Taylor and Ard (2015) note, for some activists, food justice is rooted in environmental justice principles, but they may also incorporate the concept of food sovereignty into their work, as discussed next.

Food sovereignty: from movement to policy imperative

While we have traced the coalescence of the food justice movement to the early 2000s, the food sovereignty movement took roots a decade earlier and has been led from its onset by grassroots activists—many of them peasant (or subsistence) farmers—in the Global South. A driving force in the movement is the global grassroots network La Vía Campesina, at whose Second International Conference, held in Mexico in 1996, the idea of food sovereignty was first discussed (Wittman et al. 2010). Participants believed that there was a need for a framework more relevant to addressing the food and agricultural realities of peasants and other low-wealth peoples in the Global South, within the context of a neoliberalizing global economy, than the Neo-Malthusian concept of global food security discussed previously (Wittman et al. 2010). Food sovereignty, as a concept, addresses broader structures of sociopolitical power.

La Vía Campesina (2003) first articulated a vision for food sovereignty in 2003, and the concept has since evolved, notably during the 2007 Forum for Food Sovereignty in Nyéléni, Mali. The Forum produced the Nyéléni Declaration, which defines **food sovereignty** as "the right of peoples to healthy and culturally appropriate food produced through ecologically sound and sustainable methods, and their right to define their own food and agriculture systems" (Nyéléni Declaration, 2007). While Patel (2009) argues that food sovereignty has been 'over-defined,' this is perhaps the most expedient and commonly used conceptualization of the movement's goals. Food sovereignty is distinct from environmental justice as a framing concept per se, though elements of environmental justice appear, including an explicit focus on self-determination, as well as environmental and ecological sustainability in agriculture. Meanwhile, some environmental justice scholars have argued that food sovereignty is among the issues that *global* environmental justice activists and movements are addressing today (Agyeman et al., 2016).

Intersections between movements

With a nearly 25-year history, food scholars and activists at times debate where food sovereignty "sits" as a movement, what its successes have been to date, and how it relates to the food justice concept and movement discussed previously. One area of discussion is the question of political scale and context. Given that much of the food sovereignty movement's work centres on peasant agriculture in the Global South, observers have wondered how far the food sovereignty frame can be extended to understand and drive change in context of the Global North, where the majority of people do not engage in farming, fishing, hunting, or gathering for subsistence, and where the agrarian transition (e.g., the integration of agriculture into capitalist economies) has already taken place. To address this question, studies have explored the relevance of food sovereignty in the Global North (e.g. Fairbairn, 2012) and ways that farmers and activists have used the global food sovereignty framing to advocate for the rights of small-scale farmers in regional policy arenas in the US (Kurtz, 2015); for example, the Indigenous Environmental Network and distinct Indigenous groups within the US engage in work for food sovereignty for Indigenous communities (Indigenous Environmental Network, n.d.; Mihesuah and Hoover, 2019). These studies and initiatives suggest that the food sovereignty frame is also relevant in Global North contexts and at various political scales.

A related topic of debate is whether and to what extent the food sovereignty frame does, or should, intersect with the food justice frame (e.g., Alkon and Mares, 2012). As discussed previously, food justice has largely evolved in US and Canadian contexts and has recently gained salience in some other Global North regions, almost always with a strong focus on racial and/or class difference—within advanced capitalist societies. Meanwhile, though food sovereignty evolved in the Global South, some Global North activists have begun articulating their work through the food sovereignty lens, as the examples cited here suggest. As these two movements continue to evolve, activists and scholars have argued and/or acted upon the idea that food sovereignty and food justice can *both* be important and relevant in diverse contexts, even though they are distinct ways of approaching struggles for social justice in the food system with different meanings and different sets of social or political goals (White, 2011; Taylor and Ard, 2015; Taylor, 2018; Reynolds and Cohen, 2016).

Box 14.4 Urban agriculture as activism

Urban agriculture, the growing of food and non-food crops, and the raising of livestock in and around cities, is an activity with multiple *possible* benefits, including increasing food access, community economic development, and urban greening (Smit et al., 1996). Types of urban agriculture include community gardens, institutional—school or hospital—gardens or farms, community farms, and commercial farms (Cohen et al., 2012). Recent developments in the field include high-tech indoor agriculture, which, in several global cities, is gaining ground (Reynolds and Darly, 2018).

While the potential benefits of urban agriculture are often elided by those advocating for the practice (that is, advocates may suggest that *all* environmental, public health, and social justice benefits will be realized once a farm or garden is successfully established), in reality, the outcomes may paradoxically reinforce existing systems and inequities. For example, inequities such as race or class disparities in food access or policy-making may be deepened by a given urban agricultural initiative; gentrification may ensue in neighbourhoods surrounding farms and gardens as property values rise along with urban greening (McClintock, 2014; Reynolds, 2015).

Despite these potential effects of projects that do not explicitly seek to improve food and environmental equity, some farmers and gardeners use urban agriculture to directly address such injustices. They may do so through on-farm education about the roots of food and environmental inequities; political coalition-building in gardens; supporting self-determination by encouraging the establishment of locally owned food businesses in economically marginalized communities; and other means (Reynolds and Cohen, 2016). Such 'urban agriculture activists' and the '**scholar-activists**' with whom they may work may or may not discuss their work in terms of food justice, environmental justice, or food sovereignty (Reynolds and Cohen, 2016). However, their aims align in important ways with the goals of these respective movements in their focus on building equity, recognition, and self-determination (Reynolds and Cohen, 2016). In this way, such initiatives align with environmental justice principles of representation and participation.

A third important question surrounding food sovereignty is to what extent this powerful grassroots movement is or can be effective in policy arenas, beyond its success in building a global social movement (Claeys, 2014). This is particularly relevant as the food sovereignty movement calls for changes, including agrarian or land reform and a global food trade that would require political action or concession on the part of national governments (see Wittman et al., 2010). Two brief examples suggest the potential of the movement to impact policies and address food sovereignty at global or local scales.

- Researchers have examined how food sovereignty policy priorities have been enacted, including in the case of Belo Horizonte, Brazil, whose Secretariat for Food and Nutrition Security (SMASAN) programme, founded in 1993, has successfully addressed food insecurity through a suite of interconnected programmes including low-cost 'popular' restaurants and support for farmers to sell their products at a low cost at urban farmers markets (Chappell, 2018; IPES-Food, 2017). While the programme does not explicitly address food sovereignty, Chappell (2018) has argued that it enacts many of the focal points of food sovereignty, as well as the food justice frame.
- At the scale of global governance, in 2018, after more than a decade of advocacy led by LVC, the UN Human Rights Council adopted the UN Declaration on the Rights of Peasants and Other People Working in Rural Areas (UNDROP). This Declaration includes a provision on food sovereignty as follows:

Peasants and other people working in rural areas have the right to determine their own food and agriculture systems, recognized by many States and regions as the right to food sovereignty. This includes the right to participate in decision-making processes on food and agriculture policy and the right to healthy and adequate food produced through ecologically sound and sustainable methods that respect their cultures.

(UN Declaration, 2018, Article 15, pt 5)

Analyses of the decade-long process leading up to the adoption of UNDROP raise important points about the concessions the food sovereignty movement made in order to advance this convention (Clayes and Edelman, 2019), but it nonetheless represents an important step toward political recognition of food sovereignty concerns on the global stage.

The questions and examples presented underscore the salience of food sovereignty as a global concept and movement that will doubtless continue to be relevant, particularly in the context of climate change and its differential impact on agriculture and agricultural livelihoods, in addition to the continued march of the genetic-engineering and biotechnology revolution. The next section offers perspectives on emerging debates.

Evolving areas of focus in food justice and food sovereignty

As discussed in the first part of this volume, in addition to distributional aspects of justice, environmental justice focuses both on the defining of environment and environmental problems, and on leadership and **recognition**[4] by people who are disproportionately burdened by environmental degradation and have less access to environmental 'amenities' or rights (such as clean water) for developing strategies to confront these problems (Schlosberg, 2013). If approaches to achieving food and social justice follow this line of thinking, we are compelled to continually pay attention to who defines food justice or food sovereignty, and who are seen as leaders in related activist efforts, scholarship, and policy-making (Bradley and Herrera, 2016; Redmond, 2013; Reynolds et al., 2018).

While much food justice and food sovereignty work has been led by people of colour, low-income and working-class people, peasant and subsistence farmers, and fisher folk throughout the world, there is at times a tendency for members of privileged communities to dominate narratives or solution-seeking actions. This has particularly been documented by food systems scholars and activists in the US, where race is a fundamental differentiator in terms of social and environmental equity, and where seeing social (in)justice in the context of racism has a long legacy (e.g., Slocum, 2006; Guthman, 2008; Reynolds, 2015; Yakini, 2013). Yet, while inequity and injustices based on ethno-racial identity may be less openly discussed in some places outside of the US (particularly in the Global North), histories of colonialism and its legacies render a focus on problem definition and leadership that is salient in many, if not most, parts of the world. Addressing these discrepancies is an important current and future focus for the movements and attendant scholarship.

Leadership in the food sovereignty movement, at least in its global iteration, is usually attributed to peasant farmers in the Global South, who have, from the origins of the movement, defined a vision for the food-sovereign communities they aim to realize. Recent work has begun to weave these visions together with the realities of small-scale farmers in the Global North, forging another new area of research that may also inform change (Coolsaet, 2016). As for food justice, recent scholarship suggests that this, too, is growing in its integrity as pertains to **self-determination** and recognition of leadership. Important recent work on Black agrarianism and Black food geographies, coupled with movement-building to reinvigorate the relationships between agriculture, food, and racial justice, is forging essential directions in food justice scholarship and activism (see White, 2018; Penniman, 2018; Reese, 2019; Smith, 2019). Emerging food justice and food sovereignty issues thus include:

- The imperatives of climate change and its effects on agricultural production and food security, with the expectation that those global citizens who are already materially worse off, will experience the most ravaging effects and negative impacts on food security and agricultural livelihoods (see Agyeman et al., 2016; Clapp et al., 2018).
- **Land-grabbing**, or 'large-scale land acquisitions' in which big firms or governments purchase very large tracts of land, often in other countries, for financial speculation and/or agricultural development and ensuring future national-level food security (see Borras et al., 2011).
- The connections between food and gentrification, such as increased property values in gentrifying neighbourhoods driving out traditional food stores, thus making healthy food less economically accessible to longtime residents, many of whom can no longer afford to live in the neighbourhood because of increased rents (see Jones, 2019; Sbicca, 2019).
- Black Agrarianism, its historical importance and contemporary renaissance, in the US in particular, and, for students, scholars, and outside supporters, recognition and prioritization of Black leadership in this movement (see McCutcheon, 2019; Smith, 2019; Touzeau, 2019).

Conclusion

This chapter has explored the intersections of food, agriculture, and environmental justice, including initiatives to address food and social justice issues related to the environment prior to the emergence of the environmental justice movement and ways in which some activists and scholars have argued that environmental justice movement should address food or agriculture more explicitly. It has examined food justice and food sovereignty as concepts and movements, and offered perspectives on how and when these movements have either been grounded in, or more loosely connected to, environmental justice movement analyses and activism. It has introduced some of the contemporary debates in the fields of food justice and food sovereignty; provided examples of ways in which grassroots movements, scholars, and activists have engaged with these concepts; and offered insights into emerging directions in the field.

Key takeaways from this discussion are that the intersection of food systems and environmental justice concerns have history that pre-dates the emergence of the environmental justice movement in the early 1990s, even if these are often not recognized as such. Likewise, movements and scholarship addressing these intersections continue to evolve in line with, or in response to, contemporary social, political, and environmental phenomena, such as is evidenced by the emergence of the food sovereignty movement during the neoliberal revolution of the 1980s and 1990s, the increased focus on climate change as it relates to food and environmental justice, and the renaissance of Black agrarianism and attendant scholarship. In the future, we will surely come to see new ways that these frames, alongside or in conjunction with the environmental justice movement, are useful in realizing a more just food system, from local to global scales.

Follow-up questions

- What are some examples of a food justice or food sovereignty issue where you live, work, or spend leisure time?
- Do you think that food sovereignty, as a concept and movement led by, and focused primarily on, the concerns of subsistence farmers and fisher folk in the Global South, is relevant in Global North settings? What brings you to this opinion?
- How might consumers, or students, support food and social justice initiatives? Is using 'buying power' an adequate strategy? What other types of actions might be possible and effective?

Notes

1 Following Dunbar-Ortiz (2014), the term 'Indigenous peoples' in the context of the US is used interchangeably with the term 'Native Americans' in this chapter, though it is noted that Indigenous peoples exist throughout the world. Proper nouns (such as the Indigenous Peoples' Movement) retain their original terminology.
2 On the toxic legacies of siting practices, see also Chapter 10 of this volume.
3 In 1996, FAO revised its definition of food security, such that today it reflects more closely the ideas articulated in community food security (as discussed in Patel 2009).
4 On justice as recognition, see Chapter 5 of this volume.

References for further reading

Alkon, A.H. and Agyeman, J. eds., 2011. *Cultivating food justice: Race, class, and sustainability*. Cambridge, MA: MIT Press.

Chappell, M.J., 2018. *Beginning to end hunger: Food and the environment in Belo Horizonte, Brazil, and beyond*. Berkeley, CA: University of California Press.

Gottlieb, R. and Joshi, A., 2010. *Food justice*. Cambridge, MA: MIT Press.

Reese, A.M., 2019. *Black food geographies: Race, self-reliance, and food access in Washington, D.C.* Chapel Hill, NC: University of North Carolina Press.

Reynolds, K. and Cohen, N., 2016. *Beyond the kale: Urban agriculture and social justice activism in New York City* (Vol. 28). Athens, GA: University of Georgia Press.

Sbicca, J., 2018. *Food justice now!: Deepening the roots of social struggle*. Minneapolis, MN: University of Minnesota Press.

White, M.M., 2018. *Freedom farmers: Agricultural resistance and the Black freedom movement*. Chapel Hill, NC: UNC Press Books.

Wittman, H., Desmarais, A. and Wiebe, N. eds., 2010. *Food sovereignty: Reconnecting food, nature, and community*. Nova Scotia, Canada: Fernwood Publishing.

References used in this chapter

Agyeman, J., Schlosberg, D., Craven, L. and Matthews, C., 2016. Trends and directions in environmental justice: from inequity to everyday life, community, and just sustainabilities. *Annual Review of Environment and Resources*, 41, pp. 321–340.

Alkon, A.H. and Agyeman, J. eds., 2011. *Cultivating food justice: Race, class, and sustainability*. Cambridge, MA: MIT Press.

Alkon, A.H. and Mares, T.M., 2012. Food sovereignty in US food movements: Radical visions and neoliberal constraints. *Agriculture and Human Values*, 29(3), pp. 347–359.

Allen, P., 2008. Mining for justice in the food system: Perceptions, practices, and possibilities. *Agriculture and Human Values*, 25(2), pp. 157–161.

Allen, P., FitzSimmons, M., Goodman, M. and Warner, K., 2003. Shifting plates in the agrifood landscape: The tectonics of alternative agrifood initiatives in California. *Journal of Rural Studies*, 19(1), pp. 61–75.

Allen, P. and Guthman, J., 2006. From 'old school' to 'farm-to-school': Neoliberalization from the ground up. *Agriculture and Human Values*, 23(4), pp. 401–415.

Anguelovski, I. 2016. Healthy food stores, greenlining and food gentrification: Contesting new forms of privilege, displacement and locally unwanted land uses in racially mixed neighborhoods. *International Journal of Urban and Regional Research*, 39(6), pp. 1209–1230.

Bloom, J. and Martin, W.E., 2016. *Black against empire: The history and politics of the Black Panther Party*. Berkeley, CA: University of California Press.

Borras Jr, S.M., Hall, R., Scoones, I., White, B. and Wolford, W., 2011. Towards a better understanding of global land grabbing: An editorial introduction. *The Journal of Peasant Studies*, 38(2), pp. 209–216.

Bradley, K. and Herrera, H., 2016. Decolonizing food justice: Naming, resisting, and researching colonizing forces in the movement. *Antipode*, 48(1), pp. 97–114.

Bullard, R.D. ed., 1993. *Confronting environmental racism: Voices from the grassroots*. Boston, MA: South End Press.

Chappell, M.J., 2018. *Beginning to end hunger: Food and the environment in Belo Horizonte, Brazil, and beyond*. Berkeley, CA: University of California Press.

Claeys, P. 2014. Food sovereignty and the recognition of new rights for peasants at the UN: A critical overview of La Vía Campesina's rights claims over the last 20 years. *Globalizations*, 12(4), pp. 452–465.

Claeys, P. and Edelman, M., 2019. The United Nations declaration on the rights of peasants and other people working in rural areas. *The Journal of Peasant Studies*, pp. 1–68.

Clapp, J., Newell, P. and Brent, Z.W., 2018. The global political economy of climate change, agriculture and food systems. *The Journal of Peasant Studies*, 45(1), pp. 80–88.

Coalition of Immokalee Workers, n.d. http://ciw-online.org/. Accessed 6/6/19.

Cohen, N., Reynolds, K. and Sanghvi, R., 2012. *Five borough farm: Seeding the future of urban agriculture in New York City*. New York: Design Trust for Public Space.

Cole, L.W. and Foster, S.R., 2001. *From the ground up: Environmental racism and the rise of the environmental justice movement*. New York: NYU Press.

Coolsaet, B., 2016. Towards an agroecology of knowledges: Recognition, cognitive justice and farmers' autonomy in France. *Journal of Rural Studies, 47*, pp. 165–171.

Dunbar-Ortiz, R., 2014. *An Indigenous peoples' history of the United States* (Vol. 3). Beacon Press.

Fairbairn, M., 2012. Framing transformation: The counter-hegemonic potential of food sovereignty in the US context. *Agriculture and Human Values, 29*(2), 217–230.

Farmworker Justice, 2018. *Selected statistics on farmworkers (2015–16 Data)*. (Analysis of U.S. Department of Labor's National Agricultural Workers Survey (NAWS).

Gottlieb, R., 2005. *Forcing the spring: The transformation of the American environmental movement*. Washington, DC: Island Press.

Gottlieb, R. and Fisher, A., 1996. Community food security and environmental justice: Searching for a common discourse. *Agriculture and human values, 13*(3), pp. 23–32.

Gottlieb, R. and Joshi, A., 2010. *Food justice*. Cambridge, MA: MIT Press.

Gray, M., 2013. *Labor and the locavore: The making of a comprehensive food ethic*. Berkeley, CA: University of California Press.

Guthman, J., 2008. 'If they only knew': Color blindness and universalism in California alternative food institutions. *The Professional Geographer, 60*(3), pp. 387–397.

Heynen, N., 2009. Bending the bars of empire from every ghetto for survival: The black Panther party's radical antihunger politics of social reproduction and scale. *Annals of the Association of American Geographers, 99*(2), pp. 406–422.

Hochedez, C. and Le Gall, J. 2016. Justice alimentaire et agriculture: Introduction. *Justice Spatiale-Spatial Justice, 9*.

Horst, M., McClintock, N. and Hoey, L., 2017. Urban agriculture and food justice: A review of the literature and implications for planning. *Journal of American Planning Association, 83*(3), 277–295.

Indigenous Environmental Network, n.d. www.ienearth.org/category/food-sovereignty/. Accessed 6/8/19.

IPES-Food, 2017. *What makes urban food policy happen? Insights from five case studies*. International Panel of Experts on Sustainable Food Systems.

Joassart-Marcelli, P. and Bosco, F., 2018. Alternative food and gentrification of farmers' markets, community gardens and the transformation of urban neighborhoods. In W. Curran and T. Hamilton (Eds.), *Just green enough: Urban development and environmental gentrification* (pp. 80–94). London: Routledge.

Jones, N., 2019. (Re) Visiting the corner store: Black youth, gentrification, and food sovereignty. In *Race in the marketplace* (pp. 55–72). Cham: Palgrave Macmillan.

Just Food, n.d. www.justfood.org.

Kurtz, H.E., 2015. Scaling food sovereignty: Biopolitics and the struggle for local control of farm food in rural Maine. *Annals of the Association of American Geographers, 105*(4), pp. 859–873.

Kwate, N.O.A., Yau, C.Y., Loh, J.M. and Williams, D., 2009. Inequality in obesogenic environments: Fast food density in New York City. *Health & Place, 15*(1), pp. 364–373.

La Vía Campesina, 2003. Food sovereignty. https://viacampesina.org/en/food-sovereignty/. Accessed 6/6/19.

Levkoe, C.Z., 2006. Learning democracy through food justice movements. *Agriculture and Human Values, 23*(1), pp. 89–98.

Lo, J. and Koenig, B., 2017. Food workers and consumers organizing together for food justice. In A. Alkon and J. Guthman (Eds.), *The new food activism: Opposition, cooperation, and collective action*. Berkeley, CA: University of California Press.

Mares, T. and Peña, D., 2011. Environmental justice and food justice: Toward local, slow, and deep food systems. In A. Alkon and J. Agyeman (Eds.), *Cultivating food justice: Race, class, and sustainability*. Cambridge, MA: MIT Press.

McClintock, N., 2014. Radical, reformist, and garden-variety neoliberal: Coming to terms with urban agriculture's contradictions. *Local Environment*, 19(2), pp. 147–171.

McClintock, N., 2018a Cultivating (a) sustainability capital: Urban agriculture, ecogentrification, and the uneven valorization of social reproduction. *Annals of the American Association of Geographers*, 108(2), 579–590.

McClintock, N., 2018b. Urban agriculture, racial capitalism, and resistance in the settler-colonial city. *Geography Compass*, 12(6), e12373.

McCutcheon, P., 2019. Fannie Lou Hamer's freedom farms and black agrarian geographies. *Antipode*, 51(1), pp. 207–224.

McEntee, J., 2011. Realizing rural food justice: Divergent locals in the Northeastern United States. In A.H. Alkon and J. Agyeman (Eds.), *Cultivating food justice: Race, class, and sustainability*. Cambridge, MA: MIT Press.

Mihesuah, D. and Hoover, E. eds., 2019. *Indigenous food sovereignty in the United States: Restoring cultural knowledge, protecting environments, and regaining health*. Norman, OK: University of Oklahoma Press.

Minkoff-Zern, L.A., 2017. Farmworker-Led food movements then and now. In A. Alkon and J. Guthman (Eds.), *The new food activism: Opposition, cooperation, and collective action*. Berkeley, CA: University of California Press.

Minkoff-Zern, L.A., 2018. Race, immigration and the agrarian question: Farmworkers becoming farmers in the United States. *The Journal of Peasant Studies*, 45(2), pp. 389–408.

Minkoff-Zern, L.A., 2019. *The new American farmer: Immigration, race, and the struggle for sustainability*. Cambridge, MA: MIT Press.

Mitchell, D., 2013. 'The issue is basically one of race': Braceros, the labor process, and the making of the agro-industrial landscape of mid-twentieth-century California. In R. Slocum and A. Saldanha (Eds.), *Geographies of race and food: Fields, bodies, markets*. Abingdon, UK: Routledge.

Moses, M., 1993. Farmworkers and pesticides. In R.D. Bullard (Ed.), *Confronting environmental racism: Voices from the grassroots*. Boston, MA: South End Press.

Norgaard, K.M., Reed, R. and Van Horn, C., 2011. A continuing legacy: Institutional racism, hunger, and nutritional justice on the Klamath. In A. Alkon and J. Agyeman (Eds.), *Cultivating food justice: Race, class, and sustainability*. Cambridge, MA: MIT Press.

Nyéléni Declaration, 2007. https://nyeleni.org/spip.php?article290. Accessed 6/6/19.

Paddeu, F., 2016. From one movement to another? Comparing environmental justice activism and food justice alternative practices. *Justice Spatiale-Spatial Justice*, 9.

Patel, R., 2009. Food sovereignty. *The Journal of Peasant Studies*, 36(3), pp. 663–706.

Penniman, L., 2018. *Farming while black: Soul Fire Farm's practical guide to liberation on the land*. White River Junction, VT: Chelsea Green Publishing.

Peoples Grocery, 2009. www.peoplesgrocery.org.

Peña, D. 2002. *Environmental justice and sustainable agriculture: Linking Ecological and Social Sides of Sustainability*. Resource paper series, October 23, 2002. Second National People of Color Environmental Leadership Summit – Summit II.

Principles of Environmental Justice, 1991. https://www.ejnet.org/ej/principles.html. Accessed 3/31/2020.

Redmond, L. 2013. 'Food + Justice = Democracy' (presentation at TEDxMan-hattan). www.cfjn.org/.

Reese, A.M., 2019. *Black food geographies: Race, self-reliance, and food access in Washington, D.C.* Chapel Hill, NC: University of North Carolina Press.

Reynolds, K., 2015. Disparity despite diversity: Social injustice in New York City's urban agriculture system. *Antipode*, 47(1), pp. 240–259.

Reynolds, K., Block, D. and Bradley, K., 2018. Food justice scholar-activism and activist-scholarship: Working beyond dichotomies to deepen social justice praxis. *ACME: An International E-Journal for Critical Geographies*, 17(4), pp. 988–998.

Reynolds, K. and Cohen, N., 2016. *Beyond the kale: Urban agriculture and social justice activism in New York city* (Vol. 28). Athens, GA: University of Georgia Press.

Reynolds, K. and Darly, S. (Equal authorship.) 2018. Commercial urban agriculture in the global city: Perspectives from New York city and Métropole du Grand Paris. In *Food policy monitor*. City University of New York Urban Food Policy Institute.

Sbicca, J., 2018. *Food justice now!: Deepening the roots of social struggle*. Minneapolis, MN: University of Minnesota Press.

Sbicca, J., 2019. Urban agriculture, revalorization, and green gentrification in Denver, Colorado. In *The politics of land*. Emerald Publishing Limited. *Research in Political Sociology*, 26, pp. 149–170.

Schlosberg, D., 2013. Theorising environmental justice: The expanding sphere of a discourse. *Environmental Politics*, 22(1), pp. 37–55.

Short, A., Guthman, J. and Raskin, S., 2007. Food deserts, oases, or mirages? Small markets and community food security in the San Francisco Bay area. *Journal of Planning Education and Research*, 26(3), pp. 352–364.

Slocum, R., 2006. Anti-racist practice and the work of community food organizations. *Antipode*, 38(2), pp. 327–349.

Smit, J., Nasr, J. and Ratta, A., 1996. Urban agriculture: food, jobs and sustainable cities. *New York, USA*, 2, pp. 35–37.

Smith, B.J., 2019. Food justice, intersectional agriculture, and the triple food movement. *Agriculture and Human Values*, pp. 1–11.

Taylor, D.E., 2011. Introduction: The evolution of environmental justice activism, research, and scholarship. *Environmental Practice*, 13, pp. 280–301

Taylor, D.E., 2018. Black farmers in the USA and Michigan: Longevity, empowerment, and food sovereignty. *Journal of African American Studies*, 22(1), pp. 49–76.

Taylor, D.E. and Ard, K.J., 2015. Food availability and the food desert frame in Detroit: An overview of the city's food system. *Environmental Practice*, 17(2), pp. 102–133.

Touzeau, L. (2019). 'Being Stewards of Land is Our Legacy': Exploring the lived experiences of young black farmers. *Journal of Agriculture, Food Systems, and Community Development*, 8(4), 1–16.

UN Declaration on the Rights of Peasants and Other People Working in Rural Areas, 2018. Article 15, pt 5. Geneva, UN.

Walker, R.E., Keane, C.R. and Burke, J.G., 2010. Disparities and access to healthy food in the United States: A review of food deserts literature. *Health & place*, 16(5), pp. 876–884.

White, M.M., 2011. Sisters of the soil: Urban gardening as resistance in Detroit. *Race/Ethnicity: Multidisciplinary Global Contexts*, 5(1), pp. 13–28.

White, M.M., 2018. *Freedom farmers: Agricultural resistance and the Black freedom movement*. Chapel Hill, NC: University of North Carolina Press.

Wittman, H., Desmarais, A. and Wiebe, N. (Eds.), 2010. *Food sovereignty: Reconnecting food, nature, and community*. Nova Scotia, Canada: Fernwood Publishing.

Wrigley, N., 2002. 'Food deserts' in British cities: Policy context and research priorities. *Urban Studies*, 39(11), pp. 2029–2040.

Yakini, M., 2013. Building a racially just food movement. *Be Black and Green (blog)*. www.beblackandgreen.com.

15 Urbanisation

Towns and cities as sites of environmental (in)justice

Jason Byrne

Learning outcomes

- Become aware that towns and cities are characterised by environmental inequalities.
- Understand how the process of urbanisation socially and spatially distributes environmental benefits and harms.
- Identify urban environmental justice issues such as waste management; green space accessibility; energy, heat, food and water security; and opportunities to participate in decision-making.
- Learn how urban policies can have (un)intended consequences, locking some communities into patterns of 'path dependence'.
- Consider how cities in the Global North and South are also sites of resistance and hope in remedying environmental inequalities.

Introduction

During the opening decades of the 21st century, the Earth has become a planet of cities, with very real consequences for people and ecosystems (Turner and Kaplan, 2019). While cities have existed for hundreds, even thousands of years—across Asia, Europe, the Americas and Oceania—the scale and rate of urban change over the past century is unprecedented (Derickson, 2018). It is estimated that in 1900, the proportion of the global population living in cities was only 16%; by 2018, it had soared to 55% (Ritchie and Roser, 2019; United Nations, 2018). Much of that growth occurred during the latter half of the 20th century. In 1950, the global urban population was just 751 million; by 2018, it had grown to 4.2 billion (United Nations, 2018). In the near future, the proportion of city dwellers is expected to comprise somewhere between two-thirds and three-quarters of the world's population (Lederbogen et al., 2011). This expected future growth will predominantly occur in Asia and Africa, and in small to mid-sized cities (Fragkias and Boone, 2013).

The process of a place becoming more urban is termed 'urbanisation'. **Urbanisation** has different forms including urban expansion (i.e. sprawl), suburbanisation, densification and contraction (shrinking cities) (Martinez-Fernandez et al., 2012). Many social and environmental impacts accompany urbanisation. Some impacts are beneficial (e.g., improved health care, employment opportunities and technological innovation), whereas others are harmful (e.g., pollution, species extinction, exposure to natural hazards and unhealthy environments). The differential impact of cities on people's health and well-being can be attributed to: (i) the characteristics of cities (e.g., where land uses are located); (ii) social and economic processes that undergird the creation and maintenance of cities; and (iii) the uneven process of urbanisation.

Much contemporary urban growth (especially in the developing world) is occurring in places ill-suited to intensive human habitation—on steep hillsides, floodplains, arid areas and in regions prone to natural hazards (e.g. earthquakes, tsunamis, severe drought). Most cities have marginalised and disadvantaged residents (e.g., the poor, recent immigrants, homeless and certain ethno-racial groups) who shoulder a disproportionate burden of environmental harm—a situation termed 'environmental injustice'. Efforts to 'right the wrongs' of environmental disparities, achieve more inclusive decision-making and recognise that resources should be redistributed to those most in need can collectively be described as environmental justice.

This chapter examines environmental injustices that stem from the process of urbanisation. We note how these issues manifest similarly—and differently—in cities in the developing and developed world. To paraphrase Sharma-Wallace (2016, p. 176), the chapter seeks to explain the vectors of environmental disparities, moving beyond describing outcomes to considering practices and processes. In other words, we consider how socio-spatially differentiated environmental impacts and unfair decision-making can be traced to specific urban policies, procedures, structures and institutions (also see Haughton, 1999). We begin by assessing the historical foundations for urban environmental injustice, recognising that, although the concept of environmental (in)justice is relatively recent, many urban environmental disparities have long histories. What we now call the environmental justice movement had its roots in the Civil Rights Movement in the United States (US). Beginning in the late 1970s and early 1980s, activists and scholars in the US began to identify cases where so-called minorities (e.g. African Americans and Latino/as – hereafter Latinx) were disproportionately exposed to environmental harms such as toxic waste landfills, polluting factories and agricultural pesticides (Bullard, 1990; see Chapter 2 of this volume). Resistance to such environmental inequalities quickly grew into a national, then international, movement for justice and reparation. Key questions we consider here are *why* and *how* cities distribute environmental benefits and harms differently among urban populations.

Tracing the socially and spatially uneven impacts of urbanisation

Cities are social, economic and environmental entities. From their origins, they have been characterised by power relations, reflecting status, wealth and privilege—in turn shaping access to environmental benefits and exposure to environmental harms (Stanley et al., 2016). Indeed, for towns and cities to exist requires a particular division of labour—agricultural surpluses must be generated by farmers, enabling non-farmers to be fed, freeing up their labour for other activities. These surpluses provide a basis for trade in goods and services, thus supporting the development of social elites, such as religious and political leaders and city administrators (Sjoberg, 1960). Historically, as urban populations have become increasingly specialised, they have also stratified, reflecting the social and economic division of labour and social values and norms (e.g., patterns of prejudice and discrimination). Social 'classes' have included the elites (e.g., religious classes, civic leaders and bureaucrats) who governed and administered cities, merchants who organised trade, workers (e.g. labourers, clerical workers, soldiers, artisans) and the underclass (e.g. slaves, untouchables and the like) (op. cit.).

The social stratification of pre-modern cities was mirrored by differentiated living conditions and life chances, oftentimes expressed spatially. Put slightly differently, specific locations within pre-modern cities were occupied by different groups of people, who experienced environmental conditions unevenly. This **residential differentiation** reflects the political and economic conditions and processes that produce and configure urban space (Harvey, 1973). Pre-modern cities, like their contemporary counterparts, tended to be divided into 'haves' and 'have-nots'. The elites, due to their social and economic power, typically inhabited the centres of cities, usually with good access to urban services and better-quality housing, and experienced freedom from most environmental hazards (though not fire or communicable diseases), whereas the underclasses (e.g. landless peasants) were relegated to the undesirable and often unhealthy and/or unsafe locations (Sjoberg, 1960; Stanley et al., 2016). Such locales were usually outside the city walls, close to tanneries, slaughter yards and refuse sites (thus experiencing air and water pollution), and in locations that were physically dangerous (e.g. prone to flooding and poorly defended against attackers) (Sjoberg, 1960; Stanley et al., 2016). This spatial expression—then as now—mirrored social (e.g. religious) and economic power and systems of governance, which shaped opportunities for social advancement, and in turn configured the capacity of ordinary people to secure the necessities of life through their everyday livelihoods and access to urban services.

Following the industrial revolution, which began in Europe in the 18th century, a different type and scale of urbanisation emerged, produced through new forms of energy, transport and communication, and new modes of social organisation (Sjoberg, 1960). Industrialisation and the mechanisation of agriculture, as well as changing social structures (e.g., bringing about the end of feudalism and rise of capitalism) acted as forces that pushed people away from farmland and hamlets. Changes to agricultural production meant unneeded agricultural labour became 'surplus', while mechanisation created new ways of making and distributing goods and increased demand for labour in cities. Cities thus became magnets for displaced people. As growing numbers of people swelled into the new industrial cities, and new industrial modes of production arose in increasingly specialised urban spaces, different environmental conditions emerged (see Box 15.1). New industrial cities were organised around a specific social, spatial and economic logic (Hall, 1996). Initially, workers needed to live close to industries as there was no form of mass transit, and industries needed to locate proximately to sources of raw materials or in the vicinity of harbours and canals, to which these materials were transported. While wealthier individuals could live further away from noisy mills and noxious industries (e.g. tanneries), the working classes usually lived nearby the ports, factories and waste disposal areas where they worked. The poorest lived in informal settlements (e.g. slums)—areas with substandard or self-built housing (oftentimes dilapidated)—and in places characterised by air and water pollution, and associated high rates of disease (Goodman, 1971).

With the advent of the coal-fired steam engine, and later internal combustion engines, the spatial ordering of industrial cities changed. Factories could locate on cheaper land, close to railway lines that linked them to ports or distribution centres. Factories in cities made products such as textiles, soap, paints, steel and cement, their workers often exposed to noxious and dangerous substances (e.g., lead, mercury, animal wastes, solvents) with little protection. Working classes typically had the most dangerous jobs (exposed to heat in steel mills and noxious gases and cave-ins in coal mines). Many industrialists constructed housing for workers close to factories, so even away from workplaces, the working class were still exposed to noise and dangerous emissions. On the other hand, middle-class families were able to move to new suburban areas, away from the crowded and polluted city centres, with access to their new suburban housing provided by mass transit. Such patterns of disproportionate exposure to harm can still be found in many cities. From this cursory history, we can see that underlying economic and social processes in cities can structure access to environmental benefits and configure exposure to environmental harms in very specific ways (see Box 15.2).

Box 15.1 Environmental impacts of urbanisation

The process of urbanisation creates a broad array of environmental impacts. The clearing of vegetation that precedes urban development displaces and/or kills plants and animals, but can also harm people. Accompanying urbanisation come soil compaction and erosion; altered soil and water chemistry; soil, water and air pollution; drier soils; reduced water infiltration; weed invasion; and changed patterns of fire (Byrne and Houston, 2020). Hard surfaces can comprise up to two-thirds of urban land cover; green areas can fall as low as 16%. Rates of stormwater runoff increase substantially when stream and river catchments are paved, thus increasing downstream flooding. Streams and rivers may carry higher silt loads, decreasing the abundance of some insect, fish, amphibian and bird species—with flow-on effects for livelihoods because impoverished groups often supplement their diets though hunting and foraging.

Many environmental changes associated with urbanisation have uneven social impacts—heat is an example. With lower vegetation cover, temperatures rise—termed the urban heat island effect. Higher temperatures (between 6–8 degrees ambient and 20 degrees Celsius on some surfaces) can lead to increased rates of illness and death, especially among young children and older adults, but also outdoor workers (Byrne et al., 2016). Another effect, termed the wealth effect, occurs where fewer insects, soil fungi and other micro-scale lifeforms (and their beneficial 'ecosystem functions and services') are found in poorer neighbourhoods. Researchers attribute this to reduced levels of vegetation cover in less well-off areas—wealthy residents usually have more money to purchase plants and more time and resources to maintain gardens or employ gardeners. Fewer studies have considered the environmental inequalities that arise from reduced access to urban nature, and altered urban ecologies, but emerging research suggests there could be many.

Moreover, lower levels of mental health have been found in areas with reduced vegetation cover. Child development may be harmed if children have reduced access to greenspace (e.g., fewer places to play and socialise); school performance appears to be impacted when schools have less greenspace; and levels of domestic violence appear to be higher in neighbourhoods with fewer parks and gardens. Rates of recovery from disease and hospitalisation take longer when there is less greenspace accessible. Places with fewer street trees are less walkable (higher traffic speeds and thermal discomfort from sun, wind and rain). People's gut and skin microbes may also be different in parts of the city with fewer natural areas. Health effects include digestion problems, increased stress and anxiety and even reduced immunity (Wolch et al., 2014). While **urban greening** (e.g., community gardens) can ameliorate some of these problems, it can worsen others and/or create new ones. For example, planting the wrong trees in the wrong places can damage infrastructure (e.g., powerlines, sewerage pipes), heighten bushfire risks, increase the risk of trees falling on houses during storms or even increase allergies and asthma, as well as bring people into conflict with wildlife (e.g., colonies of bats roosting in trees can produce noxious odours and spread disease). Some forms of urban greening can also increase property values, thereby displacing poorer and vulnerable residents (i.e. environmental gentrification).

Patterns of exposure and risk in contemporary cities

In today's cities, as in the past, where you live affects your quality of life and who you are strongly shapes where you live (Slater, 2013; Stanley et al., 2016). Particular neighbourhoods have better access than others to education, health care, affordable housing, public transportation, green space, recreation opportunities and employment opportunities. Conversely, some neighbourhoods have higher levels of exposure to environmental pollution, natural hazards and locally unwanted land uses. Such residential differentiation is not coincidental; it reflects entrenched social and economic processes and relations (Currie and Sorensen, 2019). Environmental justice scholars have observed how social stratification affects urban residents in very specific ways. Discrimination, for example, on the basis of race, ethnicity, gender, class/income, (dis)ability, age, sexual preference or religious beliefs can strongly configure urban dwellers' life chances.[1] Furthermore, such axes of difference do not operate independently and can reinforce each other (Malin and Ryder, 2018).[2]

For example, growing up as an African American or Latinx in places like South Los Angeles or inner Charlotte, North Carolina, often means experiencing discrimination across the course of life. For instance, discrimination in access to housing can mean exposure to lead-paint poisoning during childhood (i.e. from paint or water pipes in poorly maintained older houses). Attending school in an impoverished neighbourhood with fewer resources can mean less chance of attending university. Being exposed to air pollution from freeways and factories as a school kid is related to higher probabilities of childhood asthma and, thus, reduced school attendance. As a consequence, higher absenteeism is associated with reduced likelihood of securing a higher-paying job. Residing within, and attending school within, neighbourhoods with reduced access to parks and greenspaces is associated with physical inactivity and co-morbidities (e.g., obesity, diabetes, some cancers) and, ultimately, reduced life expectancy (Pulido et al., 1996; Currie and Sorensen, 2019). In other words, systematic discrimination is etched into the life-worlds of many people of colour living in cities—a situation termed **environmental racism**.[3]

Contemporary patterns of risk and exposure in cities reflect entrenched power relations. So-called minority and disadvantaged communities are typically disproportionately exposed to environmental harms and have reduced access to environmental benefits. Axes of difference such as those described above mean that Indigenous people, migrants, working poor, homeless, displaced peasants and farmers, welfare-dependent and other vulnerable populations are especially prone to experiencing unhealthy and dangerous environments. Wealthy elites can escape environmental problems due to their privileged socio-economic status. In many instances, these same people may have played a major role in the creation of environmental disparities (see Box 15.2). These patterns scale up from individual cities to encompass nation-states. In the developing world, for instance, dangerous manufacturing and waste disposal (e.g., battery recycling) are often exported from cities in wealthier nations to those in poorer ones (e.g., unregulated electronics disposal in Africa and Asia, and poorly regulated electronics manufacturing in the maquiladoras of Mexico).

Such disparities are especially pronounced in slums (informal settlements), which are home to more than one-fifth of the world's population; over a billion city dwellers live in slums (Friesen et al., 2019). In cities prone to flooding, such as Jakarta (Indonesia) and Manila (Philippines), it is the poor who are most at risk of disease and harm. In the impoverished informal settlements near Cape Town (South Africa), Rio de Janeiro (Brazil) and Caracas (Venezuela), disease rates are very high due to poor sanitation, irregular refuse collection and lack of access to safe drinking water. Fire and landslides are a constant threat in many informal settlements, and indoor and outdoor air pollution can regularly exceed World Health Organization standards. Insecure tenure (because residents can't finance housing construction on land they don't own) and limited public transport can trap residents in situations of concentrated disadvantage. Many informal settlements (shantytowns and slums) are also built in the shadows of large industrial areas, meaning that their residents are often impacted by a hazardous cocktail of poorly regulated chemicals and wastes (Young, 2013).

Box 15.2   Drivers of urban environmental inequality

One of the vexing questions often raised in environmental justice research is: 'Which came first, the people or the pollution?' Scholars have identified five different drivers of injustice, including: economic or market forces, discrimination, histories of neglect and oppression, administrative or political negligence, and exclusion from decision-making. Some scholars contend that market mechanisms are primarily responsible for many urban environmental inequalities. Land values are oftentimes lower in areas near industries or locally unwanted land uses (e.g., freeways, landfill sites) because of the increased noise, light, odours, vibrations, truck traffic and air and water pollution, etc., and thus, housing is more affordable. Cheaper housing can spatially concentrate marginalised and disadvantaged people in undesirable localities because they cannot afford to live in places closer to environmental amenities (e.g., the seaside, forests) (Been, 1994).

Other scholars assert that, although land values may play a role, there is strong evidence (e.g., from archival research) showing patterns of elites intentionally targeting marginalised and disadvantaged communities for locally unwanted land uses (e.g., Pastor et al., 2001). An influential study by Pulido et al. (1996) found that in Los Angeles, for instance, racist practices of property developers and land use planners designated parts of the city as 'sacrifice zones' for industrial development, away from comparatively more affluent suburbs and closer to those occupied by people of colour. Some researchers (e.g., Callewaert, 2002) have observed an interaction effect, whereby historical patterns of discrimination create a type of **path dependence** that amplifies contemporary patterns and processes of exposure to harm.

Racist policies and practices—such as apartheid in South Africa (1948–1994), the White Australia Policy in Australia (1901–1973) and Jim Crow in the United States (1865–1968)—have spatially concentrated people of colour in the past, preventing them from living in parts of cities occupied by whites and/or elites. Limited mobility can reinforce inequitable housing markets, even after racially restrictive practices have ceased. The end of racially restrictive housing covenants and practices of so-called red-lining by mortgage lenders in the US from the late 1940s to the late 1960s is an example. Even after these practices ended, their effects have lingered for generations because residents were locked into place—unable to afford to move elsewhere (Currie and Sorensen, 2019). Lower levels of service provision are also typical in such communities (e.g., inadequate public transport, reduced infrastructure maintenance and fewer neighbourhood parks).

Institutional discrimination, unfair policies and biased decision-making—due to disinterest, malign neglect or even corruption on the part of decision-makers and bureaucrats—can exacerbate and entrench poverty and limit mobility (Koehler and Wrightson, 1987). For example, in the US in the 1960s, New York's chief planner Robert Moses targeted African American communities for the routing of the city's new freeways in the belief that they would be less likely to resist (Goodman, 1971; Hall, 1996). Scholar Mike Davis has previously noted how disaster funding is regularly provided to residents of the elite communities of Thousand Oaks and Malibu in greater Los Angeles following wildfires, whereas inner-city apartment dwellers receive no such assistance following fires in buildings resulting from 'slumlord' practices. A growing body of evidence shows that public transport (including age and quality of the vehicles, emissions and level of service) is often better in wealthier and white neighbourhoods, a situation termed 'transit racism'. Other municipal services (e.g. maintenance of public roads, provision of municipal pools, rubbish collection) are similarly differentiated. And environmental laws are oftentimes enforced more quickly and more thoroughly in wealthier locales (Bullard, 1990).

Evidence also shows that disadvantaged populations can become concentrated in places that are hazardous (e.g., flood-prone land) or unhealthy (e.g., next to waste incinerators or oil refineries) due to the interaction of historical racist or elitist practices, contemporary property markets and global economics. In some cases, entire nations have outsourced their unwanted industries to other countries with cheaper labour markets and lax environmental regulations, reproducing inequalities across a regional or even global scale (e.g., Mexico's maquiladoras on the border with the US). When public policy interventions to 'clean up' contaminated neighbourhoods occur—such as 'urban renewal', brownfield remediation or via neighbourhood greening programmes—property prices can increase substantially (both housing markets and commercial property markets), displacing marginalised and disadvantaged residents, their businesses and places of employment to less desirable or harder-to-access parts of the city (Lim et al., 2013; Wolch et al., 2014).

A substantial literature over the past three decades has shown that multiple and sometimes interrelated processes are responsible for the socio-spatial distribution of environmental risk and harm in cities (see Box 15.2). Some impacts are worth noting in more detail. Energy and **fuel insecurity**[4] is a major cause of premature death in cities—especially in informal settlements in India and Africa, where residents rely on fires to cook food and are thus exposed to indoor air pollution—but also in developed world cities where older and often poorer residents can experience extremes of cold or heat—unable to afford heating or air-conditioning (Bolin et al., 1993). In addition, many urban residents in the developing and developed world suffer from poor nutrition. In some cities in the US, researchers have observed a phenomenon called **food deserts**[5] where fast-food chains, convenience stores and liquor stores outnumber fresh-food vendors. These food deserts are typically concentrated in neighbourhoods occupied by people of colour and low-income earners.

Many urban dwellers are also exposed to vector-borne and water-borne disease, especially in the developing world, where typhoid, cholera, malaria, dengue fever, gastro enteritis and parasitic diseases reduce life expectancies, particularly in children and infants. In both the developing and developed worlds, impoverished people are also exposed to waste. Cancer clusters in cities mirror toxic waste dumps located adjacent, or even beneath housing, and concentrations of toxic industries within and adjoining marginalised and impoverished neighbourhoods.[6] Furthermore, noise and light pollution can disrupt sleep, resulting in higher prevalence of mental illness such as anxiety and depression (Lederbogen et al., 2011). Many residents of marginalised and vulnerable neighbourhoods also experience limited access to public transportation or rely on old and polluting vehicles—transport inequity—which reduces access to employment and education opportunities. The most impoverished residents may work as rag-pickers in landfills or as casual labourers, manually breaking down electronic components or sorting through plastic waste in backyard recycling facilities that contaminate soil, water and air.

Box 15.3   Rapid urbanisation in Asia and Africa

Much of the world's urban growth during the 21st century will occur in Asia and Africa. According to the United Nations, 90% of the urban growth over the next 30 years will take place within these two continents (UN, 2018). Many countries in Asia have already experienced explosive urban growth—especially China, which has an annual urban growth rate of around 3%. Here the scale of urbanisation over the past four decades is, arguably, unprecedented in human history. According to the United Nations, in sub-Saharan Africa, almost two-thirds of the urban population consists of slum-dwellers. Africa, which is still rapidly urbanising, is home to emerging megacities such as Lagos in Nigeria.

Global population projections indicate that population growth in Africa will remain high. The populations of many Asian countries, in contrast, are stable or even declining in some countries (e.g., Japan), bringing other challenges (e.g., loneliness among older people, economic decline). Many Asian megacities are found in places that are highly vulnerable to natural hazards such as cyclones/typhoons, earthquakes, floods, tsunamis and volcanic eruptions. Impoverished populations in informal settlements are generally hit hardest by natural hazards; a single typhoon can kill thousands of residents.

The interconnected character of contemporary national economies and large immigrant diasporas mean that many cities are now networked into a global urban system. Not all cities are beneficiaries—some become the recipients of polluting industries that take advantage of cheap labour and lax environmental regulations and law enforcement (Leichenko and Solecki, 2008). Other cities may have sacrificed local environmental quality to accommodate burgeoning populations and provide employment opportunities. Many Asian cities are among the world's most polluted. Some cities in India and China regularly appear in lists of the most polluted cities on the planet, largely due to poor air quality caused by indoor cooking; burning of coal, wood fires or dung; and a reliance on private automobiles for transport. A few cities, like Beijing, have begun to make inroads into improving environmental quality, by closing down factories, switching to renewable energy and reducing automobile dependence, as well as enforcing stringent emission standards for diesel trucks and cars.

Arguably, some of the biggest challenges that lie ahead for Asian and African cities arise from climate change (Friedman and Sorensen, 2019). Many of the world's poor are spatially concentrated in floodplains and along coastlines (e.g., in Bangladesh) and will bear a disproportionate impact of sea-level rise. Coastal megacities like Shanghai and Jakarta in Asia and Lagos in Africa will be severely impacted by inundation from sea-level rise and flooding—displacing millions of people (Swiss Re Group, 2019). Others, such as Bangkok and Manila, are especially vulnerable to the impact of mega-typhoons that can destroy fragile housing and infrastructure, leaving disease, death and misery in their wake. Informal settlements are particularly prone to damage from severe storms, and their residents are exposed to insect-borne diseases, due to poor sanitation and limited drainage, and mould-related diseases from inhabiting flood-affected structures.

Policy recommendations and ideas for reform

As we have seen in this chapter, the process of urbanisation is associated with numerous environmental impacts. Many of these impacts are distributed unevenly, both spatially and socially. This situation is unjust because the people with the least power in society—those who are most disadvantaged due to ethno-racial characteristics, poverty, gender, (dis)ability, language and education barriers, and the like—tend to bear the highest burden of urban environmental harm. Moreover, such communities are rarely responsible for producing such harm in the first place (see Box 15.3). The causes of urban environmental disparities are deeply structural—reflecting social and economic power relations. The political economy of cities means that urban elites often exert the strongest influence in key matters, such as local politics, land use decisions, infrastructure funding and location, service provision and environmental and planning law formulation and enforcement. In addition, urban growth machines, comprising politicians, urban utility corporations (e.g., electricity, water), local media, land and property developers, professional groups and businesses interests, wield considerable power and are able to strongly influence the economic and environmental trajectories of cities (Logan and Molotch, 2007), usually for their own benefit—a situation termed 'environmental privilege' (Grooms and Boamah, 2018).

When considering remedies, it is important to remember that alternative conceptions of justice may be invoked in different contexts. Broadly speaking, the concept of justice refers to the morally fair and right way of treating others, be they people who are different than us or even other species. Ameliorating the intentional targeting of disadvantaged communities for locally unwanted land uses requires (re)**distributive justice**,[7] overcoming differential law enforcement and biased decision-making calls for **procedural justice**, and a lack of public participation requires efforts to achieve **participatory democracy**.[8] In many cases, justice requires the identification or **recognition**[9] of a problem and affected people in the first place, and the recognition that a fair outcome is not about treating all people the same but may, in fact, necessitate remedial actions that shift power relations, reallocate resources, build capacities and capabilities, develop knowledge and awareness, and create new institutions. We briefly consider some of these in turn.

Since the early 1990s, there have been attempts in the US at state and federal levels to enshrine actions to monitor and undo the impacts of environmental injustices. These have included revising legislation (including city ordinances) regulating waste storage and disposal; the installation or improvement of monitoring systems to track pollutants, 'upgrading safety and emergency programs, and [the enforced] compliance with zoning codes, and emission standards' (Bullard, 1990, p. 66). More recently, the European Union has worked to redress environmental injustices, as has the United Nations. For example, there are now efforts to ban the transnational shipment of hazardous waste from wealthy nations to poorer ones. As many commentators have noted, however, these responses have not been particularly successful. Monitoring pollution, enforcing remediation and compensating affected communities is expensive and necessitates stable government and political commitment. Powerful vested interests can weaken laws and policies. Within government agencies, public servants may balk at taking strong action—for the simple reason that standing up for what is right could mean risking one's job. And monitoring equipment may be scarce in developing countries due to funding constraints and limited opportunities for technical training.

Commentators have suggested a range of solutions to these issues, including:

- the creation of grassroots residents organisations responsible for monitoring the activities of industry and government via citizen science (e.g., so-called bucket brigades);
- providing communities with greater access to information about the type and levels of contamination in the places where they live, work and play;
- capping industries in contaminated neighbourhoods; and
- forcing the siting of new noxious and unwanted land uses away from poor and marginalised neighbourhoods.

Other solutions, promoted by commentators like Bullard (1990), include the provision of affordable healthcare for affected communities, the regulation of both government and corporations to ensure that waste management practices are significantly modified, and the empowerment of marginalised people to participate in decision-making (e.g., multilingual translation services or citizens' juries). Some have also called for government agencies responsible for the monitoring and/or regulation of contaminated sites to employ those most affected by environmental harm—as a way of counteracting complacency, bias and elitism within the public service (Bullard, 1990). At an international scale, efforts are increasing to track the transnational movement of toxic waste, post-consumer electronics and materials destined for recycling. Such solutions fall within the broad categories of institutional reform, community empowerment and corporate social responsibility.

Arguably, steps that focus on the (re)distribution of environmental harm can only ever have limited success; they treat the symptoms, not the cause, and do not redress the structural forces that drive environmental injustice. Oftentimes, clandestine markets, lax law enforcement, corrupt governments, weak (inter)national laws and structural inequalities in global trade mean that if a problem is fixed in one place, it shifts elsewhere. Emerging potential solutions include legislation in Germany that requires companies to take back goods for recycling at the end of their consumer life and experiments in industrial ecology in Scandinavian countries, where the waste output from one industry becomes the raw material input for another (e.g., Kalundborg in Denmark). But such solutions do not address disparities in housing affordability, food deserts, **transit inequality** or disproportionate exposure to natural hazards and climate change impacts; neither do they redress disparities in access to environmental benefits. Solutions must progress beyond limited technical fixes or superficial political reform to encompass broader social, economic and political change. Initiatives to boost civic engagement, foster grassroots organising and strengthen resilience and community cohesion are important steps in this direction (Sampson, 2017). While these remedies sound plausible in theory, they can be significantly harder to achieve in practice.

Conclusion

In this chapter, we have explored different types of environmental inequality that are associated with the growth and operation of cities (e.g., food deserts, air pollution, natural hazards exposure, thermal inequity). We have noted how spatial patterns in the uneven distribution of environmental benefits and harms reflect structural inequalities and social processes with deeply historical roots. We have examined how the socio-demographic characteristics of some populations (e.g., race, class, gender, age, etc.) are associated with disproportionate exposure to environmental harms and circumscribed access to environmental services and benefits. We have considered the processes that result in ethno-racially and class-based injustices (e.g. intentional targeting, property market dynamics, biased decision-making, uneven policy enforcement, lack of recognition). And we noted some of the emerging longer-term trends that are driving urban policy and governance challenges, requiring urgent attention by scholars, practitioners, residents and activists alike (e.g., climate change adaptation).

Importantly, this chapter is only an entry point into these issues, and there is much left unsaid. For instance, we have only briefly touched upon the task of critically assessing urbanisation as a process that distributes environmental benefits and harms among urban populations in very specific ways (via land use planning, siting decisions, service provision and infrastructure [non]delivery, etc.). The very large body of work on environmental justice issues in cities in the developed and developing world points to the need for scholars to work alongside activists to develop meaningful long-term solutions to these problems. More than three decades since the beginning of the environmental justice movement, there is still much to be undertaken in the fight for fair and liveable cities.

Follow-up questions

- What are the different ways that environmental injustices can be expressed in cities given their current development trajectories (e.g. growing, stable or shrinking cities)?
- How does planetary urbanisation pose unique challenges for environmental justice scholars and activists?

- In the age of the Anthropocene, is it useful to talk of environmental justice and multi-species justice separately, or should we find a new approach that brings the two perspectives together?
- Could a global compact like the Montreal Protocol provide the legal, governance and policy framework necessary to address urban trans-national environmental injustices? Explain why or why not.
- What steps could be taken to counter the drivers of environmental inequality in land use planning? Could these steps be retrospective?

Notes

1 As noted earlier, cities are characterised by social stratification. Lancaster Jones (1974) identifies two primary sources for social stratification: achieved and ascribed. Achieved stratification refers to differences in access to resources, wealth and power based on individual achievements (e.g. performance in schooling or employment), whereas ascribed stratification refers to entrenched systematic discrimination or prejudice based on specific 'axes of difference' which shapes people's opportunities and life chances. The two are often interrelated.
2 The idea of intersectionality cautions us to understand that axes of difference are often imbricated. For example, to be a black, disabled, lesbian woman is not the same as if these things were experienced separately (Smith, 1999), and intersectional inequalities warrant closer scrutiny.
3 On environmental racism, see Chapter 17 of this volume.
4 On energy (in)justice, see Chapter 13 of this volume.
5 On food and environmental justice, see Chapter 14 of this volume.
6 On toxic pollution and environmental justice, see Chapter 10 of this volume.
7 On distribution, see Chapter 3 of this volume.
8 On participation, see Chapter 4 of this volume.
9 On recognition, see Chapter 5 of this volume.

References for further reading

Agyeman, J., Bullard, R.D. and Evans, B. (eds.) 2003. *Just Sustainabilities: Development in an Unequal World*. Cambridge, Massachusetts, MIT Press.
Byrne, J. 2013. Environmental justice. In *Oxford Bibliographies Online, Geography*, Retrieved from www.oxfordbibliographies.com/view/document/obo-9780199874002/obo-9780199874002-0008.xml, accessed 23 September 2019.
Cutter, S.L. 2012. *Hazards Vulnerability and Environmental Justice*. New York, Routledge.
Gleeson, B. and Low, N. 2002. *Justice, Society and Nature: An Exploration of Political Ecology*. London, Routledge.
Tsing, A.L., Bubandt, N., Gan, E. and Swanson, H.A. (eds.) 2017. *Arts of Living on a Damaged Planet: Ghosts and Monsters of the Anthropocene*. Minneapolis, University of Minnesota Press.

References used in this chapter

Been, V. 1994. Locally undesirable land uses in minority neighbourhoods: Disproportionate siting or market dynamics? *Yale Law Journal*, pp. 1383–1422.
Bolin, B., Barreto, J.D., Hegmon, M., Meierotto, L. and York, A. 2013. Double exposure in the sunbelt: The sociospatial distribution of vulnerability in Phoenix, Arizona. In C.G. Boone and M. Fragkias (eds.), *Urbanization and Sustainability: Linking Urban Ecology, Environmental Justice and Global Environmental Change* (pp. 159–178). Dordrecht, Springer.

Bullard, R. 1990. *Dumping in Dixie*. Boulder, CO, Westview Press.

Byrne, J., Ambrey, C., Portanger, C., Lo, A., Matthews, T., Baker, D. and Davison, A. 2016. 'Could urban greening mitigate suburban thermal inequity?: The role of residents' dispositions and household practices. *Environmental Research Letters*, 11(9), p. 095014, (electronic).

Byrne, J. and Houston, D. 2020. Urban ecology. In A. Kobayashi (ed.), *International Encyclopedia of Human Geography* (2nd ed., pp. 47–58). Oxford, Elsevier.

Callewaert, J. 2002. The importance of local history for understanding and addressing environmental injustice. *Local Environment*, 7(3), pp. 257–267.

Currie, M.A. and Sorensen, J. 2019. Repackaged 'urban renewal': Issues of spatial equity and environmental justice in new construction, suburban neighbourhoods, and urban Islands of infill. *Journal of Urban Affairs*, 41(4), pp. 464–485.

Derickson, K.D. 2018. Urban geography III: Anthropocene urbanism. *Progress in Human Geography*, 42(3), pp. 425–435.

Fragkias, M. and Boone, C.G. 2013. Towards a new framework for urbanization and sustainability. In C.G. Boone and M. Fragkias (eds.), *Urbanization and Sustainability: Linking Urban Ecology, Environmental Justice and Global Environmental Change* (pp. 1–10). Dordrecht, Springer.

Friedmann, J. and Sorensen, A. 2019. City unbound: Emerging mega-conurbations in Asia. *International Planning Studies*, 24(1), pp. 1–12.

Friesen, J., Taubenböck, H., Wurm, M. and Pelz, P.F. 2019. Size distributions of slums across the globe using different data and classification methods. *European Journal of Remote Sensing*, 52(2), pp. 99–111.

Goodman, R. 1971. *After the Planners*. New York, Simon and Schuster.

Grooms, W. and Frimpong Boamah, E. 2018. Toward a political urban planning: Learning from growth machine and advocacy planning to 'plannitize' urban politics. *Planning Theory*, 17(2), pp. 213–233.

Hall, P. 1996. *Cities of Tomorrow: An Intellectual History of Urban Planning and Design in the Twentieth Century* (3rd ed.). Oxford, Blackwell.

Harvey, D. 1973. *Social Justice and the City*. London, Edward Arnold.

Haughton, G. 1999. Environmental justice and the sustainable city. *Journal of Planning Education and Research*, 18, pp. 233–243.

Koehler, D.H. and Wrightson, M.T. 1987. Inequality in the delivery of urban services: A reconsideration of the Chicago parks. *The Journal of Politics*, 49(1), pp. 80–99.

Lancaster Jones, F. 1974. Social stratification in Australia: An overview of a research program. *Information*, 13(1), pp. 99–118.

Lederbogen, F., Kirsch, P., Haddad, L., Streit, F., Tost, H., Schuch, P., Wüst, S., Pruessner, J.C., Rietschel, M., Deuschle, M. and Meyer-Lindenberg, A. 2011. City living and urban upbringing affect neural social stress processing in humans. *Nature*, 474(7352), pp. 498–501.

Leichenko, R.M. and Solecki, W.D. 2008. Consumption, inequity, and environmental justice: The making of new metropolitan landscapes in developing countries. *Society and Natural Resources*, 21(7), pp. 611–624.

Lim, H., Kim, J., Potter, C. and Bae, W. 2013. Urban regeneration and gentrification: Land use impacts of the Cheonggye stream restoration project on the Seoul's central business district. *Habitat International*, 39, pp. 192–200.

Logan, J.R. and Molotch, H.L. 2007. *Urban Fortunes: The Political Economy of Place*. Berkeley, University of California Press.

Malin, S.A. and Ryder, S.S. 2018. Developing deeply intersectional environmental justice scholarship. *Environmental Sociology*, 4(1), pp. 1–7.

Martinez-Fernandez, C., Audirac, I., Fol, S. and Cunningham-Sabot, E. 2012. Shrinking cities: Urban challenges of globalization. *International Journal of Urban and Regional Research*, 36(2), pp. 213–225.

Pastor, M., Sadd, J. and Hipp, J. 2001. Which came first? Toxic facilities, minority move-in, and environmental justice. *Journal of Urban Affairs*, 23(1), pp. 1–21.

Pulido, L., Sidawi, S. and Vos, R.O. 1996. An archaeology of environmental racism in Los Angeles. *Urban Geography*, 17(5), pp. 419–439.

Ritchie, H. and Roser, M. 2019. Urbanization. *OurWorldInData.org*, Retrieved from https://ourworldindata.org/urbanization, accessed 16 September 2019.

Sampson, R.J. 2017. Urban sustainability in an age of enduring inequalities: Advancing theory and ecometrics for the 21st-century city. *Proceedings of the National Academy of Sciences*, 114(34), pp. 8957–8962.

Sharma-Wallace, L. 2016. Toward an environmental justice of the rural-urban interface. *Geoforum*, 77, pp. 174–177.

Sjoberg, G. 1960. *The Preindustrial City: Past and Present*. New York, The Free Press.

Slater, T. 2013. Your life chances affect where you live: A critique of the 'cottage industry' of neighbourhood effects research. *International Journal of Urban and Regional Research*, 37(2), pp. 367–387.

Smith, D. 1999. *Writing the Social: Critique, Theory and Investigations*. Toronto, University of Toronto Press.

Stanley, B.W., Dennehy, T.J., Smith, M.E., Stark, B.L., York, A.M., Cowgill, G.L., Novic, J. and Ek, J. 2016. Service access in premodern cities: An exploratory comparison of spatial equity. *Journal of Urban History*, 42(1), pp. 121–144.

Swiss Re Group 2019. Risky cities: A deep dive on exposed metropolises. Retrieved from www.swissre.com/our-business/public-sector-solutions/thought-leadership/risky-cities.html, accessed 19th October 2019.

Turner, V.K. and Kaplan, D.H. 2019. Geographic perspectives on urban sustainability: Past, current, and future research trajectories. *Urban Geography*, 40(3), pp. 267–278.

United Nations Department of Economic and Social Affairs. 2018. World urbanization prospects 2018. Retrieved from www.un.org/development/desa/publications/2018-revision-of-world-urbanization-prospects.html, accessed 30 September 2019.

Wolch, J.R., Byrne, J. and Newell, J.P. 2014. Urban green space, public health, and environmental justice: The challenge of making cities 'just green enough'. *Landscape and Urban Planning*, 125, pp. 234–244.

Young, A.F. 2013. Urbanization, environmental justice and social-environmental vulnerability in Brazil. In C.G. Boone and M. Fragkias (eds.), *Urbanization and Sustainability: Linking Urban Ecology, Environmental Justice and Global Environmental Change* (pp. 95–116). Dordrecht, Springer.

16 Water justice

Blatant grabbing practices, subtle recognition politics and the struggles for fair water worlds

Rutgerd Boelens

Learning outcomes

- Explore the multiple ways in which water injustices are expressed.
- Learn that water conflicts are as much about quantities and qualities of water as about meanings, values, truths and knowledges.
- Consider the consequences of presenting water problems as merely technical and managerial issues and 'natural harms'.
- Think about the intimate connection between people, space, identity and struggles for water.
- Develop familiarity with the specificities of water justice research.

Introduction[1]

Today, the governance of water resources primarily concerns the combined socio-political, biophysical and technological issues of how to conserve, make available, allocate and organize water among competing uses and users. Water's decreasing availability and quality, and its unequal distribution, lead to growing pressures on society and nature, threatening future availability and intensifying conflicts. At the same time, the water pollution problem is also growing, as is vulnerability to flood risks, partly triggered by climate change. All these factors raise new questions about differential access to socio-environmental health, protection and security.

For this reason, issues of sustainability and **ecological integrity** in water governance cannot be dealt with in isolation from questions of fairness, solidarity and justice. Instead, the focus must be on the pressing societal problems of how rights and access to water and water-related decision-making are distributed along lines of class, gender, caste and ethnicity in the Global South and North. In many countries, contemporary water policies and legislative measures have tended to aggravate historically rooted inequalities rather than solving them; small-holder irrigator communities, Indigenous territories, or local drinking water committees, often with context-based water practices, are constantly overruled by bureaucratic water administrations, market-driven water policies, desk-invented legislation and top-down project intervention practices.

In this context, Eduardo Galeano's words from *News about the Nobodies* give a fitting background to the issue of water justice:

> Up till recently, poverty was the fruit of injustice. But times have changed greatly: now, poverty is the just punishment that inefficiency deserves, or simply a way of expressing the natural order of things. The world has never been so unfair in dividing up the resources, but the system that governs the world—now discretely called 'the market economy'—takes a daily dip in the bath of impunity.
>
> (Galeano, 1996, p. 1)

As for justice in general, many of today's water dispossessions and unfair accumulation practices find their legitimization in discourses of efficiency and arguments of rationality. And, commonly, those blamed for water governance problems are not the water-grabbing elites and industries, but rather those who suffer from large-scale water injustices—the presumed 'inefficient nobodies'. At the same time, water scarcity tends to be presented as a natural phenomenon rather than as something stemming from deeply political distributive choices. Social norms, policy agreements and scientific standards in water governance naturalize and normalize injustices and inequities, with water policies often endorsing rather than questioning the concentration of water rights in the hands of a few private powerful actors (e.g., Boelens et al., 2018; Perreault, 2014; Venot and Clement, 2013). For instance, the activities of agro-export companies, extractive industries and large-scale hydropower developments become ever more difficult to control as their operations transcend traditional jurisdictions of national water authorities (Menga and Swyngedouw, 2018; Yacoub et al., 2015; Zwarteveen, 2015). In contemporary water debates and governance frameworks, legalistic prescriptions, technocratic water expert formulas and/or neoliberal rationalities and concepts that presume that water is and should be a commodity, preferably a private right, to be managed by the market laws of supply and demand are presented as objective, neutral or even 'natural'. They are now so dominant that they have come to be accepted as normal and inevitable. This makes it difficult to recognize them and identify their true nature: of deeply ideological views and normative ideas that give a strongly biased representation of water realities, their problems and their solutions (Zwarteveen and Boelens, 2014).

Therefore, in line with Galeano's observations, there is a need to reflect on how water policy discourses, models and associated knowledge commonly justify far-reaching redistributions of water which often entail a profoundly unequal distribution of costs and benefits for different groups. Water (in)justices are as much about quantities and qualities of water and modes of access and distribution, as they are about meanings, discourses, truths and knowledges. They also include matters of authority and legitimacy, which extend into questions of culture, territory and identity (Roth et al., 2005, 2015; Zwarteveen et al., 2005).

Notions of justice, fairness or equity in water control and governance are commonly hidden in specific rules, categories or cultural codes of conduct that tend to remain unquestioned and outside of critical scrutiny because they appeal to what is considered 'normal', 'true' or 'natural'. Exposing and examining these '**truth regimes**' is central for questioning water injustices for the following reasons. As influential discourses, they interconnect power and knowledge—and the more power they represent, the more 'truthful' become the water knowledge claims they manifest and, consequently, the more other forms of water knowledge (for example, of peasant and Indigenous communities, women or poor neighborhoods) are subjugated or obliterated (Foucault, 1975). From this analysis, alternative ways of governing water and ordering societies may emerge based on important ideas from protest movements and actions 'from below' which suggest that alternative modes of understanding and dealing with water are not only possible, but necessary.

Identifying and describing particular water realities or water allocations as either just or unjust cannot be done from the stance of an independent outsider, but, rather, always implies engagements that need to be made as explicit as possible (Baviskar, 2007; Haraway, 1991). An examination of the multiple layers of water injustice, ranging from brutal, 'visible' water-grabbing and pollution practices to the subtle politics of misrecognition and exclusion, as well as the covert techniques of equalization and subjugation, requires making our own claims to truth about equitable water governance and management manifest and explicit, in particular in interactions with those who experience water injustices in their everyday lives. This chapter seeks a water justice perspective that is grounded and relational, and engages with the realities, questions, needs and opportunities of those who have less voice and fewer rights in terms of water.

Complementary modes and interactive practices of 'water injustice'

In general terms, new concerns of water governance have triggered a new mix of scientific fields and intervention professionals populating the water policy-making, management and project development arena, be it with a strong presence from the domains of economics and natural resource governance. Environmental concerns have become increasingly prominent, and many water policies and projects have been converted (substantially or discursively) into climate change policy agendas. At the same time, water development and management are no longer the exclusive realm of the state, with knowledge and decision-making powers only concentrated in powerful public agencies and their associated engineers. Partly because of neoliberal reforms, private companies and civil society organizations have now become important actors in water policy arenas (Bakker, 2010; Molle et al., 2009; Swyngedouw, 2015), and the scales of water governance and notions of territorial sovereignty have changed profoundly. Water allocation and policy-making increasingly reside with global policy institutes and international companies, and water management accountability is ever more linked to anonymous transnational companies and global market forces, rather than to governments or local water-user collectives (Vos and Hinojosa, 2016; Zwarteveen, 2015).

Therefore, although competition for and conflicts about water are increasing, the question of who is responsible for the governance, allocation and control of water is becoming ever less clear: multiple new private-sector actors, invisible market mechanisms and global de-territorialized policy institutes now have a far stronger influence on local, national and transnational rule-making. This has a deep impact on aggravating already existing water injustice patterns. Manifestations of these developments range from brutal water grabs to much more subtle forms of misrecognition and exclusion, the effects of which may nevertheless be as dramatic or even more so.

The resulting water control conflicts and contradictions make clear that distributive, cultural and representational water injustices are closely entwined (see also Fraser, 2000; Schlosberg, 2004). Respectively, they deal with the (unequal) division of burdens and benefits of distributing water resources and services (see also Chapter 3 of this volume); (dis)respecting the socio-cultural diversity of water norms, rules and knowledge systems and the ways in which water-based environments are used, managed and controlled (see also Chapter 5 of this volume); and who is included or excluded from (co)decide-making; and on what basis of legitimacy, authority or power (see also Chapter 4 of this volume). The three issues are intimately connected to struggles for socio-ecological integrity ('sustainability') against injustice, in terms of socio-environmental transformations: people and water, humans and nature, co-constitute and depend on each other, and so do present-day and future generations. **Water justice**, therefore, conceptually and politically, can be situated in the field of the political ecology of water, which may be defined as: "the politics and power relationships that shape human knowledge of and intervention in the water world, leading to forms of governing nature and people, at once and at different scales, to produce particular hydrosocial order" (Boelens, 2015a, p. 9).

Water distributive inequities

A violent, spectacular form of growing water injustice is **water-grabbing**: the phenomenon involving wealthy investors and transnational companies buying up land in countries in the Global South to produce food, flowers or green fuels for export, as discussed in Chapter 14. This land is worth little if not accompanied with access to water. In most cases, therefore, land grabs are in fact water grabs, a process that works to displace and literally dry out existing users and producers (Veldwisch et al., 2018). This 'hydro-colonialism' goes beyond classic North—South opposition, as the research NGO GRAIN explains. In recent years, companies from Asia have bought more than 10 million hectares in the Nile basin to grow export crops that need water far beyond the entire water availability of the basin: a "hydrological suicide" (GRAIN, 2013, p. 25).

Box 16.1 Agro-export or food security? Unequal water allocation and access patterns

In general, national water policies, in particular in the Global South (but similarly in Northern hemisphere regions on occasion), are elite- and business-driven: they tend to respond to the (often short-term, extractive) commercial interests of market players looking for economic gains for their companies, rather than catering for the needs of national and local food and water security. So, in law, policy and project intervention practice, water is allocated to where 'its marginal returns are highest', to promote commercial export crops that earn money, but also replacing food crops and so endangering people's food security. A case in point is Ecuador: Gaybor (2011) explains how water allocation policies in the country work to accumulate water in the hands of the already wealthy. Official figures show how, nationally, capitalist farmers who constitute the private sector—with, for instance, large banana plantation companies, flower and other luxury crop farms, and wealthy animal husbandry 'hacienda' estates—represent only 1% of farms, but consume not less than 67% of the total available water flow. Peasant-Indigenous families working with community irrigation systems—whose produce is for home consumption and the local and national food security markets—represent 86% of water users, but only own 22% of irrigated land. What is even worse, they have access to a mere 13% of the total allocated water flow. In some provinces, water allocation inequality is outright appalling. In Imbabura Province in northern Ecuador, for example, the landlords account for 91% of the total flow (Gaybor, 2011, p. 200). According to Gaybor, actual water distribution is even more burdensome than these official figures show, as more than half of the water used by large-scale agribusiness companies is not registered but illegally tapped.

On a worldwide scale, from India to South Africa to Israel, majorities of subsistence farmers face dispossession of their water in favor of commercial farming sectors that produce high-value export products (Veldwisch et al., 2018; Woodhouse, 2012; Zeitoun et al., 2009). Poor families are gradually deprived of the water supplies that they used to access because of declining water levels. In this race to the bottom, to the deepest layers of groundwater, only those who can afford the cost of powerful pumping stations can access this resource. Such encroachments are often actively encouraged by national policies, supporting the 'more productive use of water' and investments in infrastructure by private companies. For instance, in Peru's Ica desert valley, with its fertile soils and strategic location near the capital Lima, rainfall is close to zero. Groundwater is the main resource for thousands of small farmers, and the valley has always been the 'food basket' for the megacity and its surroundings.[2] For the past decade, however, the aquifer has been dramatically over-extracted with a water table (groundwater level) dropping nearly 1 meter per year (Cárdenas, 2012). With significant government support and public funds, new agro-export companies have purchased most of the valley's land to produce high-consumptive export crops such as asparagus, grapes and vegetables. Small and medium-sized farmers are marginalized, and their wells have dried up. They cannot compete with the large owners' powerful water-pumping technologies. The agro-exporters, constituting 0.1% of the users, have amassed 36% of the water, while the smallholders, making up 71% of users, have access to only 9% (Cárdenas, 2012; Damonte and Boelens, 2019).

Encroachment of water territories by extractive industries is another illustrative example. Around the world, in the global North and South, mining companies appropriate upstream water sources, thereby diverting and polluting the downstream flows on which smallholders and Indigenous communities often depend. By buying up water rights and through their powerful presence in the area, mining companies take over de facto water control (e.g. Budds, 2010; Sosa and Zwarteveen, 2011; Stoltenborg and Boelens, 2016). Similarly, in many places, extractive hydrocarbon industries are increasingly dominating water control issues. As Bebbington et al. show, in Ecuador's Amazon region, approximately half of the total area is allotted in concessions to oil companies. In neighboring Peru, it is even worse, with nearly three-quarters of the Amazon region having been allotted to or subject to leasing by hydrocarbon transnationals (Bebbington et al., 2010, pp. 309–311). More generally, extractive industries place increasing pressure on water, requiring it for mining, hydrocarbon, agro-export and hydropower development and exploitation practices. These often systematically combine in capitalist 'Water-Food-Energy Nexus' projects, entailing large-scale hydraulic interventions and territorial transformations—thereby polluting or drying out downstream regions (Allouche et al., 2015).

Box 16.2 Rural to urban water transfers: drinking water for the poor?

Another visible (and often violent) water expropriation practice is when, in many other places in the world, the thirst of cities and industries is quenched at the expense of rural smallholders. Giving cities and industries higher allocational priorities is not necessarily inequitable per se, although it does often result from effective political–economic power lobbies (Hommes et al., 2019). Duarte-Abadía and Boelens (2019) show how Andalusian rural users suffer water scarcity because luxurious tourist residences and golf courses in Málaga get priority. Large groups of subsistence farmers have lost their water rights. In principle, in nearly all countries, drinking water has priority over water for agriculture or for other water use sectors, and this is important to foster human rights and access to water. In practice, however—as Ioris (2016) explains for Lima; Goldman and Narayan (2019) for Bangalore; and Hommes et al. (2019) for cases around the world—these rural–urban water transfers, which are supposed to enhance 'water for all' and therefore quite frequently cite the 'Millennium Development Goals' and 'Human Right to Water' discourses, are often particularly beneficial to the elite city neighborhoods of the Global South. Radonic (2017) illustrates how in northwestern Mexico, the human right-to-water argument was mobilized to undermine Indigenous opposition to large-scale urban-biased infrastructure development, denying Indigenous rural water rights and fostering unequal water rights and access. Once inside the cities, access to water differs hugely between poor and rich urban neighborhoods, and between citizens and industries (e.g., Bakker, 2010; Damonte and Boelens, 2019). Many illustrative cases, such as the heavy metal poisoning of drinking water for the poorest neighborhoods in Flint, Michigan, make clear that the lack of water quantity and quality for particular population groups goes hand in hand with fundamental issues of ethnic discrimination, lack of economic power and political representation, as well as unscrupulous capitalist power-plays (Mohai, 2018; Perreault et al., 2018

The politics of recognition and participation: misrecognizing water normative diversity and controlling decision-making

As mentioned previously, water injustice combines issues of distribution with those of cultural recognition and political representation, in often complex and sometimes paradoxical ways (see Chapters 3–5 of this volume). Cultural, ethnic or gender discrimination cause not just exclusions and suffering, but also often constitute the implicit or explicit foundations for privileging the allocation of water rights to some over others. Exclusion from water-related decision-making may be quite open, based on caste, gender or ethnicity, but often it is less obvious—hidden, for instance, in the membership criteria of water-user organizations, or in behavioral norms in water-system operation and maintenance, or when accessing and using water (e.g., Bhushan Udas and Zwarteveen, 2005; Roth et al., 2005, 2015).

In on-the-ground water control, the existence of legal and institutional pluralism is a given. It produces a myriad of manifestations in everyday realities: **water rights**, rules, principles and authorities—of different origin and legitimization—co-exist and interact in the same water control territory. Local water institutions constitute a dynamic mixture of rules, principles and organizational forms from diverse normative sources. In many cases, they combine local, national and global rules, and mix Indigenous, colonial and current norms (see also Chapters 7 and 20 of this volume). Water-user collectives 'reconstruct' these norms in territory-grounded local law. Systems that at first glance seem to be irrational and disorderly, in practice are in fact a form of organized complexity. These '*living* water rights' often defend non-commodity water institutions as the backbone of the local community, while taking a strategic approach to the market. Despite internal injustices and struggles, they too seek collective control.

In terms of **recognition**, in most countries, this diversity of context-based, 'intangible' water rights poses a tremendous problem for water bureaucrats, planners, and international companies. The existence of diverse water-user authorities, territorial autonomies and community rules that forbid water transfer to outside companies make state domination or free market operation very difficult, as to achieve this, the authorities need a level playing field. In practice, this situation triggers profound conflicts over legitimacy and political order. The question of which mechanisms of acquiring water rights and which allocation principles and management rules are to be considered legitimate is often an intrinsic part of struggles over water. Acceptance (rather than denial or rejection) of the existence of these plural normative/legal orders constitutes a pragmatic starting point for thinking about what water justice is or can be.

But existing forms of accessing, distributing and regulating water use (including Indigenous practices) often do not fit with newly proposed legal and policy arrangements, and recognition and participation then become tacit strategies to *make them fit*. Recognition politics may involve major pitfalls. By definition, recognition (and institutional participation) imply defining and creating institutions of access, property and control that represent particular forms of hierarchy between the 'recognizers' and the 'recognized', between the 'authorizers' and the 'authorized'.[3] Moreover, simply adding recognition and participation to distribution, without examining their precise characteristics, and without scrutinizing the normalizing powers that sustain them, may be detrimental to supporting initiatives for greater water justice.

Box 16.3 Integrated water resources management

A clear and influential example of an a-critical, a-contextual and a-historical 'adding on' of these three key notions can be found in the globalizing policy discourse of integrated water resources management (IWRM). Currently, IWRM is possibly the most significant water policy approach, mainstreamed in most countries of the world and placing great emphasis on recognition and participation of multiple stakeholders. In practice, however, IWRM is frequently used to legitimize 'participatory decisions' made by water experts or technocratic agencies which have crucial implications for distribution. Most IWRM strategies remain narrowly modernist and intervention-focused (Allen, 2006; Molle et al., 2009). They adhere to a market-based capitalist development model, construction of large-scale infrastructure and the formalization/uniformization ('recognition') of diverse water rights, to allow water to become transferable across uses and users. Following global templates, ironically, they 'reframe' the participation-recognition-distribution triangle to align it with the three basic ingredients of neoliberal typologies, popular in the policy-makers' world: namely, decentralized decision-making (multi-stakeholder participation), private property rights (recognition of users' rights) and markets (for equitable and efficient redistribution) (Boelens and Zwarteveen, 2005).

Equalization

The ideas of the prominent policy thinker and World Bank consultant Hernando de Soto serve to illustrate this subtle power game of entwining participation, recognition and redistribution. De Soto advocates the participation of the poor by recognizing their extralegal property rights, suggesting that this will lead to redistribution and overall welfare. Extralegal property rights (such as locally defined water rights) are known only to insiders and cannot be exchanged. Therefore, he argues, "these must be woven into a single system from which general principles of law can be drawn" (De Soto, 2000, p. 162). Helping developing countries to build formal property systems that embrace all their people, he states, is the new, civilizing mission of advanced nations and academics. His argument is that extralegal property owners wish to be recognized and participate, thus gaining freedom from their insecure, collective property relations. In short, they would like to join the win-win game of popular capitalism. His strategy of participation, integration and recognition of local rights and cultures appeals to common-sense notions of justice and equality. The hidden fundament, however, is one of the active destructions of **otherness**. De Soto posits that "private property is the most important institution of social and political integration, the vehicle for leading the population masses to respect for law and order, and preservation of the status quo" (2000, p. 196): the purpose of investigating local rights pluralism is, therefore, to include such rights in the uniform, formal property system that sustains the modern private property market economy.

In former days, the power strategies applied were based on force, inequality and the exclusion of the common people. Modern **equality** ideologies have radically changed the face of power: now 'others' need to be subtly seduced, included and made equal. Indeed, in modern water policies, everybody is potentially equal and *should be* equal. Evidence from around the world regarding water allocation and administration make clear, however, that this ideology of 'equality of all' is used not to abolish the enormously unequal distribution of water property or stop water-grabbing. In reality, making water users equal means oppressing their deviation from the formal rules, norms and rights. Despite the fact that the reference model of 'being equal' is, in practice, usually based on the class, gender, cultural standards and water interests of a powerful minority, the myth of a neutral legal justice framework is strong. Meanwhile, modern water policies are subtly imposing 'equalization' and 'commensuration' (Boelens, 2015b).

Alignment

At the same time, this imperative of equality makes it easy to measure to what extent 'others' deviate from the model (Foucault, 1975). Contemporary calls for inclusion and participatory stakeholder processes therefore need to be treated with caution. Indeed, 'making them normal'—according to outside standards and models—is at the heart of most modern water policies. Thus, neoliberal water policy discourses, for example, not only assume the existence of universally valid values and rules, but also actively establish them (Achterhuis et al., 2010). The production of water knowledge, disciplines and truths—and the ways these inform the shaping of particular water systems, rules and technologies—increasingly concentrates on the effort to align local people, mindsets, identities and resources with the interests and water worlds as imagined by national and global water-power hierarchies.

Across the world, modernist water development projects deploy new forms of governmentality (Foucault, 1991): as Foucault argued, they induce 'governors-mentality' and 'government-rationality', different subtle and less subtle 'arts of government'. In fact, through introducing new water technologies and institutions, they entail a re-patterning of water space and territory: reshaping rules and authority, redirecting labor and production, inducing new norms and values, and rearranging people in new, externally driven water governance environments. Many of the designs that underlie these water development projects, far beyond just installing new hydraulic technology, introduce new management hierarchies, commoditized (or privatized) water resources or services, and new legal frameworks, hostile to the survival of existing water-user collectives. It is plain to see that the norms embedded in these new designs—in irrigation and drinking water systems, watershed projects, **payment-for-environmental services** programs, etc.—though subtly framed in terms of progress and inclusion, fundamentally externalize users' knowledge, agency and authority. These designs are therefore not just 'technical' but simultaneously hydraulic, legal, political and organizational. New hydraulic power grids, commonly linked to nation-state authority, markets and companies, de-pattern and re-pattern local water control systems to establish so-called rational frames of 'water order' (Hommes et al., 2016; Duarte-Abadía et al., 2015). So, a fundamental question is: how is socio-natural order produced (and contested) via the control over water resources, infrastructure, investments, knowledge, truth and, ultimately, water users and authorities? Different groups imagine and realize their wished-for 'territory' differently; they compete and struggle over the definition and composition of its boundaries, elements and meanings to shape their territorial puzzle. Interconnected by their water flows, hydraulic technologies and water governance, these territorial components and definitions are simultaneously social, technological and natural, a **hydro-social territory**[4]: the territorial elements, boundaries, interlinkages and meanings are always, and necessarily, disputed. A fundamental question, therefore, is how governmentality projects try to re-pattern diverse water worlds and align humans, nature and thought within dominant hydro-social territories (Hommes et al., 2016). The next question is: how do people challenge and re-moralize these techno-political water systems to make their *own* water societies? And how do opposing and overlapping configurations shape 'territorial pluralism'? (Hoogesteger et al., 2016).

In contemporary water policy worlds, participatory and equalizing governance techniques to include, control and contain water-user collectives often do not replace the brutal force of former top-down imposition but, in many instances, they have the effect of complementing it. Most water expert institutes maintain the conviction that it is morally necessary to fight irrational water use and make out that all people speak the same water language. Consequently, nowadays, 'making water use and rights rational' has become a missionary process of substituting relationships of community, local property, knowledge and ethics.

In practice, the non-functionality of most mainstream water policy models and their support for large-scale water injustices does not seem to weaken their force. It is often assumed that water policies' non-adaptation to local reality is a proof of their incapacity. However, usually, their aim is not to adapt to local contexts, but to transform and control them: it is the water users' world that must be adapted, and local rights are often seen as systems that are beyond justice and control.

While most water justice studies focus on large-scale water-grabbing, there is also an urgent need to focus also, and in particular, on the manifold, invisible, everyday, bottom-up forms of encroachment upon local water societies. In many such cases, water dramas do not occur intentionally, and injustices are not committed on purpose. On the contrary, the intentions and interventions are generally highly moral, rational, development-oriented and, as current policy-language goes, 'pro-poor'.

A relational and engaged understanding of water justice: entwining layers, scales and actors

Water, as a fundamental life-enabling and life-threatening force, may generate strong collaboration and intense conflicts. Its deeply contested nature—in terms of meaning, values and ways to allocate and govern diverse waters—mean that water control conflicts are everywhere. These disputes occur over how water is to be used, distributed, managed, treated or talked about. While some of these conflicts may be open and visible, mostly they occur in subtler, less directly discernible ways. Both open water and subsurface water disputations evidence the fact that such contestations are not just about accessing and allocating water quality and quantity. Analytically, four interrelated echelons may be distinguished ('echelons of rights analysis' [ERA]: Boelens, 2015b; Boelens and Zwarteveen, 2005; Zwarteveen et al., 2005).

At a basic level, there is the dispute concerning access to and use of water-related resources: namely, which users and use-sectors have access to water, hydraulic infrastructure and the material and financial means to use and manage water resources. At the next level, there is contestation over the contents of rules and rights: the formulation and substance of water rights, management rules and laws that determine water distribution and allocation. At the third echelon, we see the struggle over the authority and **legitimacy** to make and enforce those water rights and rules: that is, who has the decision-making power about questions of water use, allocation and governance. And fourth, there is the conflict among *discourses*: the power-knowledge regimes that articulate water problems and solutions and make fixed linkages and standard logical relations between concepts, actors and objects, defining their identity, substance, position and hierarchies. These four echelons are intrinsically related; conflict and outcomes at one echelon define the contents and contestations at the next echelon. The struggle over discourses, the fourth echelon, is about inducing a coherent regime of representation that strategically links the previous echelons together and makes their contents and linkages appear natural, as the moral or scientific 'order of things'. Together, contestations over these four echelons range from defending/opposing current distributive inequalities and undemocratic forms of representation to conserving/challenging the very politics of truth itself, including the identities that are imposed upon marginalized water cultures and user groups by state and market-based governmentalities.

Attention to the cultural embeddedness, plurality and complexity of water rights requires a shift of focus, away from exclusive attention to formal structures and regulations towards an interest in how and by whom water rights and governance forms are produced, reproduced and transformed in particular ecological and cultural settings. It also deals with how people experience water laws (formal and informal) in the context of their own local society and how they use them as a crucial resource in their day-to-day aspirations and struggles (Roth et al., 2005, 2015). In their efforts, these water cultures continually re-invent rules and identities and traditions. Water-user collectives and federations know that their existence depends on defending their water rights and rule-making spaces, and so they continue to create non-conformity and complexities, while at the same time trying to achieve representation and bring about changes in the policy institutes, intervention projects and state institutional network.

Most water-user communities integrate with national and international policies, markets and partnerships, embedding the local in the global and the global in the local. Conflicts over water governmentality involve community–state contradictions, and conflicts among local smallholders and new water lords, as well as with the transnational extractive industries that operate across spatial scales. In many regions, grassroots organizations build multi-actor federations to contest the neoliberalization of water, the negative effects of dams, water pollution, the separation of water rights and decision-making powers from local livelihoods, and policies and actions that attack rights pluralism, polycentrism and the integrity of their territories (e.g. Duarte-Abadía et al., 2019; Hoogesteger and Verzijl, 2015; Romano, 2017). Such networks also show that state, scientific and policy-making communities are not monolithic. Many state employees, professionals and scientists struggle *from within*, forming alliances with water-user groups to capture cross-scale opportunities. As Benford and Snow (2000) observe, social movements also need to frame their demands in ways that align with the values and ideas of national political parties and/or the general public.

Fundamentally, struggles over water are contests over resources and legitimacy, the right to exist as water control communities, and the ability to define the nature of water problems and solutions. By connecting materiality with cultural–political struggles, they demand both the right to be equal *and* the right to be different: affected water-user communities tend to combine their struggle against highly unequal resource distribution with their demands for greater autonomy and sharing in water authority. In other words, the intimate connections between people, water, space and identity fuses their struggles for material access to and control of water use systems (**distributive justice**) and the ecological defense of their neighborhoods and territories (socio-ecological integrity) with their battle over the right to culturally define and politically organize these socio-natural systems (cultural and representational justice). Therefore, to understand grounded water justice, there is a need to move from universalist, descriptive theories that prescribe what water justice should be to focus on understanding how people on the ground experience and define water justice, in actual practice. Understanding the embeddedness of particular ideals of justice—the way these get constituted through social practices—requires a grounded, comparative and historical approach.

In the formal water policy and governance world, liberal, socialist, or neoliberal models of equality have generally tended to reflect the dominant water society's elitist, capitalist or scientific-expert mirror—ignoring peasant, Indigenous and women's interests and views. Beyond abstract de-humanized models, but also beyond localized romanticism, it is urgent to systematically explore the sources of water injustice, local views on fairness and the impacts of formal laws and justice policies on human beings and ecosystems. Indeed, understanding water justice calls for a contextual, grounded, relational approach (Joy et al., 2014; Perreault, 2014; Roth et al., 2005, 2018; Zwarteveen and Boelens, 2014).

Appeals for greater water justice call for identifying, comprehending and acting upon both overt water-grabbing and those forms of suffering that are concealed by the modernist water science–policy nexus. This requires combining grassroots, academic, activist and policy action. Water justice theories, as well as water justice movements, have much to gain through the critical integration of heterogeneity, bringing together a plurality of contexts, experiences, views, tools and strategies. A critical pluralism approach to water justice, therefore, abstains from universalizing counter-discourses, claims and concepts, but is not relativist, calling for critical articulation and engagement across contexts and differences (cf. Schlosberg, 2004). Accordingly, we may understand water justice as: "the interactive societal and academic endeavor to critically explore water knowledge production, allocation and governance and to combine struggles against water-based forms of material dispossession, cultural discrimination, political exclusion and ecological destruction, as rooted in particular contexts" (Boelens, 2015a: 34).

Water societies based on greater social justice cannot be engineered by scientists or policy-makers; they result from interweaving cross-cultural water knowledge and cross-societal pressures from the bottom up. Research and action, therefore, should engage diverse water actors to discover multiple water-truths and worldviews and co-create transdisciplinary knowledge about understanding, transforming and distributing natural resources. The aim is to explore connections among the diverse and disparate ways of viewing and struggling for water justice. In short, water justice involves critical engagement with water movements, dispossessed water societies and, step by step, on the ground, the interactive formulation and construction of alternative hydro-social orders.

Follow-up questions

- How is water scarcity constructed and by whom?
- How are water dispossessions and unfair decision-making practices legitimized in discourses of efficiency and arguments of rationality?
- How can water justice studies, beyond the focus on large-scale water-grabbing, scrutinize the manifold invisible, bottom-up forms of encroachment upon local water societies; in particular, those as deployed by neoliberal and state-centred recognition politics, alignment policies and participation projects?
- How can the power dimensions of truth claims in water science and development be unravelled?
- How to link diverse (natural/social science, policy, vernacular, Indigenous, subaltern) bodies of knowledge, including those of activists, women and smallholder water users, and engage them on equal terms?

Notes

1 This chapter is largely based on Boelens et al. (2018); Roth et al. (2005); Zwarteveen and Boelens (2014).
2 On food, agriculture and environmental justice, see Chapter 14 of this volume.
3 On justice as recognition, see Chapter 5 of this volume.
4 Hydrosocial territory can be defined as: "the contested imaginary and socio-environmental materialization of a spatially bound multi-scalar network in which humans, water flows, ecological relations, hydraulic infrastructure, financial means, legal-administrative arrangements and cultural institutions and practices are interactively defined, aligned and mobilized through epistemological belief systems, political hierarchies and naturalizing discourses" (Boelens et al., 2016:2).

References for further reading

Boelens, R. 2014. Cultural politics and the hydrosocial cycle: Water, power and identity in the Andean highlands. *Geoforum* 57, 234–247.
Damonte, G.H. 2019. The constitution of hydrosocial power: Agribusiness and water scarcity in Ica, Peru. *Ecology and Society* 24(2), 21. doi:10.5751/ES-10873-240221
Dupuits, E. 2019. Water community networks and the appropriation of neoliberal practices: Social technology, depoliticization, and resistance. *Ecology and Society* 24(2), 20. https://doi.org/10.5751/ES-10857-240220
Jackson, S. 2018. Water and indigenous rights: Mechanisms and pathways of recognition, representation, and redistribution. *WIREs Water* 5, e1314. https://doi.org/10.1002/wat2.1314

Seemann, M. 2016. Inclusive recognition politics and the struggle over hydrosocial territories in two Bolivian highland communities. *Water International* 41(1), 157–172.

Sultana, F. and Loftus, A. (eds.) 2019. *Water Politics. Governance, Justice and the Right to Water*. London: Routledge.

Swyngedouw, E. and Boelens, R. 2018. '. . . And Not a Single Injustice Remains': Hydro-territorial colonization and techno-political transformations in Spain. In R. Boelens, T. Perreault and J. Vos (eds.), *Water Justice* (pp. 115–133). Cambridge: Cambridge University Press.

Wilson, N. 2019. 'Seeing Water Like a State?': Indigenous water governance through Yukon first nation self-government agreements. *Geoforum* 104, 101–113.

Ženko, M. and Menga, F. 2019. Linking water scarcity to mental health: Hydro–Social interruptions in the lake Urmia Basin, Iran. *Water* 11, 1092. doi:10.3390/w11051092

References used in this chapter

Achterhuis, H., Boelens, R. and Zwarteveen, M. 2010. Water property relations and modern policy regimes: Neoliberal utopia and the disempowerment of collective action. In R. Boelens, D. Getches and A. Guevara (eds.), *Out of the Mainstream. Water Rights, Politics and Identity* (pp. 27–55). London and New York: Earthscan.

Allen, T. 2006. IWRM: The new sanctioned discourse? In P. Mollinga, A. Dixit and K. Athukorala (eds.), *Integrated Water Resources Management: Global Theory, Emerging Practice and Local Needs* (pp. 38–63). New Delhi: Sage.

Allouche, J., Middleton, C. and Gyawali, D. 2015. Technical veil, hidden politics: Interrogating the power linkages behind the Nexus. *Water Alternatives* 8, 610–626.

Bakker, K. 2010. *Privatizing Water. Governance Failure and the World's Urban Water Crisis*. Ithaca, NY: Cornell University Press.

Baviskar, A. 2007. *Waterscapes: The Cultural Politics of a Natural Resource*. Delhi: Permanent Black.

Bebbington, A., Humphreys, D. and Bury, J. 2010. Federating and defending: Water, territory and extraction in the Andes. In R. Boelens, D. Getches and A. Guevara (eds.), *Out of the Mainstream: Water Rights, Politics and Identity* (pp. 307–327). London and Washington, DC: Earthscan.

Benford, R. D. and Snow, D. A. 2000. Framing processes and social movements: An overview and assessment. *Annual Review of Sociology* 26(1), 611–639.

Bhushan Udas, P. and Zwarteveen, M. 2005. Prescribing gender equity? The case of the Tukucha Nala irrigation system, central Nepal. In D. Roth, R. Boelens and M. Zwarteveen (eds.), *Liquid Relations*. New Brunswick, NJ: Rutgers University Press.

Boelens, R. 2015a. *Water Justice in Latin America: The Politics of Difference, Equality, and Indifference*. Amsterdam: CEDLA and University of Amsterdam.

Boelens, R. 2015b. *Water, Power and Identity. The Cultural Politics of Water in the Andes*. London and Washington, DC: Routledge/Earthscan.

Boelens, R., Hoogesteger, J., Swyngedouw, E., Vos, J. and Wester, P. 2016. Hydrosocial territories: A political ecology perspective. *Water International* 41(1), 1–14.

Boelens, R., Perreault, T. and Vos, J. (eds.) 2018. *Water Justice*. Cambridge: Cambridge University Press.

Boelens, R. and Zwarteveen, M. 2005. Prices and politics in Andean water reforms. *Development and Change* 36(4), 735–758.

Budds, J. 2010. Water rights, mining and indigenous groups in Chile's Atacama. In R. Boelens, D. Getches and A. Guevara (eds.), *Out of the Mainstream. Water Rights, Politics and Identity* (pp. 197–211). London and Washington, DC: Earthscan.

Cárdenas, A. 2012. *La carrera hacia el fondo. Acumulación de agua subterránea por empresas agroexportadoras en Ica, Perú*. Justicia Hídrica. www.justiciahidrica.org.

Damonte, G. and Boelens, R. 2019. Hydrosocial territories, agro-export and water scarcity: Capitalist territorial transformations and water governance in Peru's coastal valleys. *Water International* 44(2), 205–222.

De Soto, H. 2000. *The Mystery of Capital. Why Capitalism Triumphs in the West and Fails Everywhere Else*. New York: Basic Books.

Duarte-Abadía, B. and Boelens, R. 2019. Colonizing rural waters. The politics of hydro-territorial transformation in the Guadalhorce Valley, Málaga, Spain. *Water International* 44(2), 147–167.

Duarte-Abadía, B., Boelens, R. and Du Pre, L. 2019. Mobilizing water actors and bodies of knowledge. The multi-scalar movement against the Río Grande Dam in Málaga, Spain. *Water* 11, 410. doi:10.3390/w11030410

Duarte-Abadía, B., Boelens, R. and Roa-Avendaño, T. 2015. Hydropower, encroachment and the repatterning of hydrosocial territory: The case of Hidrosogamoso in Colombia. *Human Organization* 74(3), 243–254.

Foucault, M. 1975. *Discipline and Punish. The Birth of the Prison*. New York: Vintage Books.

Foucault, M. 1991. Governmentality. In G. Burchell, C. Gordon and P. Miller (eds.), *The Foucault Effect: Studies in Governmentality* (pp. 87–104). Chicago: University of Chicago Press.

Fraser, N. 2000. Rethinking recognition. *New Left Review*, May–June, 107–20.

Galeano, E. 1996. Noticias de los Nadies. *El País*, January 27.

Gaybor, A. 2011. Acumulación en el campo y despojo del agua en el Ecuador. In R. Boelens, L. Cremers and M. Zwarteveen (eds.), *Justicia Hídrica: Acumulación, Conflicto y Acción Social* (pp. 195–208). Lima: IEP.

Goldman, M. and Narayan, D. 2019. Urban transformation through the analytic of water crisis: A study of Bangalore's hydrosocial regimes. *Water International* 44(2), 94–113.

GRAIN. 2013. Secando el continente africano: detrás de la acumulación de tierras está la acumulación del agua. In A. Arroyo and R. Boelens (eds.), *Aguas Robadas*. Quito: Abyayala.

Haraway, D. 1991. *Simians, Cyborgs and Women: The Reinvention of Nature*. New York: Routledge.

Hommes, L., Boelens, R., Harris, L.M. and Veldwisch, G.J. 2019. Rural—urban water struggles: Urbanizing hydrosocial territories and evolving connections, discourses and identities. *Water International* 44(2), 81–93.

Hommes, L., Boelens, R. and Maat, H. 2016. Contested hydro-social territories and disputed water governance: Struggles and competing claims over the Ilisu Dam development in southeastern Turkey. *Geoforum* 71, 9–20.

Hoogesteger, J., Boelens, R. and Baud, M. 2016. Territorial pluralism: Water users' multi-scalar struggles against state ordering in Ecuador's highlands. *Water International* 41(1), 91–106.

Hoogesteger, J. and Verzijl, A. 2015. Grassroots scalar politics: Insights from peasant water struggles in the Ecuadorian and Peruvian Andes. *Geoforum* 62, 13–23.

Ioris. 2016. Water scarcity and the exclusionary city: The struggle for water justice in Lima, Peru. *Water International* 41(1), 125–139.

Joy, K. J., Kulkarni, S., Roth, D. and Zwarteveen, M., 2014. Re-politicizing water governance: Exploring water reallocations in terms of justice. *Local Environment* 19(9), 954–973.

Menga, F. and Swyngedouw, E. 2018. *Water, Technology and the Nation-State*. London and New York: Routledge, Earthscan.

Mohai, P. 2018. Environmental justice and the flint water crisis. *Michigan Sociological Review* 32, 1–41.

Molle, F., Mollinga, P. and Wester, F. 2009. Hydraulic bureaucracies and the hydraulic mission: Flows of water, flows of power. *Water Alternatives* 3(2), 328–349.

Perreault, T. 2014. What kind of governance for what kind of equity? Towards a theorization of justice in water governance. *Water International* 39(2), 233–245.

Perreault, T., Boelens, R. and Vos, J. 2018. Introduction: Re-politicizing water allocation. In R. Boelens, T. Perreault and J. Vos (eds.), *Water Justice* (pp. 34–42). Cambridge: Cambridge University Press.

Radonic, L. 2017. Through the aqueduct and the courts: An analysis of the human right to water and indigenous water rights in Northwestern Mexico. *Geoforum* 84, 151–159.

Romano, S.T. 2017. Building capacities for sustainable water governance at the grassroots: 'Organic Empowerment' and its policy implications in Nicaragua. *Society & Natural Resources* 30(4), 471–487.

Roth, D., Boelens, R. and Zwarteveen, M. (eds.) 2005. *Liquid Relations. Contested Water Rights and Legal Complexity*. New Brunswick, NJ: Rutgers University Press.

Roth, D., Boelens, R. and Zwarteveen, M. 2015. Property, legal pluralism, and water rights: The critical analysis of water governance and the politics of recognizing "local" rights. *Journal Legal Pluralism and Unofficial Law* 47(3), 456–475.

Roth, D., Zwarteveen, M., Joy, K. J. and Kulkarni, S. 2018. Water governance as a question of justice: Politics, rights, and representation. In R. Boelens, T. Perreault and J. Vos (eds.), *Water Justice* (pp. 43–58). Cambridge, MA: Cambridge University Press.

Schlosberg, D. 2004. Reconceiving environmental justice: Global movements and political theories. *Environmental Politics* 13(3), 517–540.

Sosa, M. and Zwarteveen, M. 2011. Acumulación a través del despojo: El caso de la gran minería en Cajamarca. In R. Boelens, L. Cremers and M. Zwarteveen (eds.), *Justicia Hídrica: Acumulación, Conflicto y Acción Social* (pp. 381–392). Lima: IEP.

Stoltenborg, D. and Boelens, R. 2016. Disputes over land and water rights in gold mining: The case of Cerro de San Pedro, Mexico. *Water International* 41(3), 447–467.

Swyngedouw, E. 2015. *Liquid Power. Contested Hydro-Modernities in 20th Century Spain*. Cambridge, MA: MIT Press.

Veldwisch, G.J., Franco, J. and Mehta, L. 2018. Water grabbing: Practices of contestation and appropriation of water resources in the context of expanding global capital. In R. Boelens, T. Perreault and J. Vos (eds.), *Water Justice* (pp. 59–70). Cambridge: Cambridge University Press.

Venot, J.P. and Clement, F. 2013. Justice in development? An analysis of water interventions in the rural South. *Natural Resources Forum* 37, 19–30.

Vos, J. and Hinojosa, L. 2016. Virtual water trade and the contestation of hydrosocial territories. *Water International* 41(1), 37–53.

Woodhouse, P. 2012. New investment, old challenges. Land deals and the water constraint in African agriculture. *Journal Peasant Studies* 39(3–4), 777–794.

Yacoub, C., Duarte, B. and Boelens, R. 2015. *Agua y Ecología Política. El extractivismo en la agroexportación, la minería y las hidroeléctricas en Latino América*. Quito: Abya-Yala.

Zeitoun, M., Messerschmid, C. and Attili, S. 2009. Asymmetric abstraction and allocation: The Israeli—Palestinian water pumping record. *Ground Water* 47(1), 146–160.

Zwarteveen, M. 2015. *Regulating Water, Ordering Society: Practices and Politics of Water Governance*. Inaugural lecture. Amsterdam: University of Amsterdam.

Zwarteveen, M. and Boelens, R. 2014. Defining, researching and struggling for water justice: some conceptual building blocks for research and action. *Water International* 39(2):143–158.

Zwarteveen, M., Roth, D. and Boelens, R. 2005. Water rights and legal pluralism: Beyond analysis and recognition. In D. Roth, R. Boelens and M. Zwarteveen (eds.), *Liquid Relations* (pp. 254–268). New Brunswick, NJ: Rutgers University Press.

Part III
Actors and subjects of environmental justice

17 Racial minorities in the United States

Race, migration, and reimagining environmental justice

Lisa Sun-Hee Park and Stevie Ruiz

Learning outcomes

- Identify the ways in which race and racism shape environmental inequality.
- Understand the types of institutional disparities that result in environmental injustice.
- Explore the ways in which immigrants have participated in environmental justice movements.

Introduction

Scholars and activists have, for over four decades, presented evidence that people of colour, as well as poor, working-class, and Indigenous communities face greater threats from pollution and industrial hazards than other groups. Environmental threats include municipal and hazardous waste incinerators, garbage dumps, coal-fired power plants, polluting manufacturing facilities, toxic schools, occupationally hazardous workplaces, substandard housing, uneven impacts of climate change, and the absence of healthy food sources. Marginalized communities tend to confront a disproportionate volume of these threats, what researchers and advocates have labelled environmental injustice and **environmental racism**.

As discussed in Chapter 2 of this volume, Afton, North Carolina is a classic example. In 1978, Governor Jim Hunt was struggling to clean up transformer fluid that had been illegally dumped along North Carolina's highways for years by the Ward Transformer Company. This fluid contained polychlorinated biphenyls (PCBs), a toxic chemical compound banned by Congress the following year. Governor Hunt's solution was to move this hazardous waste to Afton, a predominantly low-income African American town in Warren County (Bullard, 2000).

In the eyes of African American residents, the decision to move this poison into their neighbourhood was an explicit example of environmental racism legislated by the governor's office. In response, 500 African American residents and activists joined forces to block the transportation of the initial shipment of PCB-contaminated soil (Bullard, 2000). The police assaulted, beat, and tackled these environmental justice activists. Explicit forms of state-sponsored violence were used against those who spoke against the governor's authority. The National Guard, Sheriff's Department, and Afton Police were called in to quash civil disobedience. In addition to physically blocking the trucks carrying PCB, the protestors called national newspapers to make these events public, networked with organizations across North Carolina (and the Southeastern corridor), educated themselves about the legislative process, and pushed for greater regulation and oversight by the Environmental Protection Agency (EPA).

The assault on activists demonstrated three things: 1) dumping PCB in the backyard of marginalized racial minority neighbourhoods is an act of violence by the state and is legal; 2) those who challenge state-sanctioned violence will be punished further, and in this case, it was in the form of physical assaults by police; and 3) the environmental clean-up of one community resulted in the contamination of another.

The confrontation of PCB contamination in Afton, North Carolina, is one among many stories that comprise the history of environmental justice. Afton is marked as the beginning of the environmental justice movement in the United States (see Chapter 2 of this volume), but the activists never intended to make history. Members of this working-class community of colour came together for their own survival, like so many others within the environmental justice movement. The list of environmental justice struggles is dauntingly long and diverse, given that working-class communities of colour are more likely to be impacted by extractive industrial operations, such as mining, large dams (see Chapter 16 of this volume), and timber harvesting, as well as "natural disasters" like flooding, earthquakes, and hurricanes. These patterns occur at local, regional, national, and global scales, and the damage to public health, cultures, economies, and ecosystems from such activities is well documented. For example, immigrants and people of colour in California's Silicon Valley live in communities with disproportionately high concentrations of toxic Superfund sites and water contamination and work in jobs that expose them to disproportionately high volumes of hazardous chemicals (Pellow and Park, 2002). In the city of Chicago, African Americans and Latinos live in neighbourhoods with disproportionately high numbers of garbage dumps and other environmental hazards (Pellow, 2002), and we see this pattern holding true for Asian-Americans, Native Americans, and working-class whites nationally (Bullard et al., 2007).

The field of environmental justice studies has emerged as a means to consider the historical and contemporary drivers of environmental inequalities and their many manifestations and as a vehicle to address this problem through research, action, and policy. As illustrated in this volume, environmental justice studies span the fields of history, sociology, anthropology, geography, law, communication, economics, literature, ethnic studies, public health, architecture, medicine, and many more. Activists and policy-makers have also produced a great deal of research on environmental justice issues and have drawn on the work of scholars to pass laws and introduce state and corporate policies that would confront some of the most glaring aspects of environmental injustice in the United States and globally. Scholars have also demonstrated how communities have responded to such ecological violence creatively, through protest, art, science, and sustainable development projects (see Chapter 9 of this volume). Such work underscores how environmental injustices shape the politics of race, class, and indigeneity (see Chapters 7 and 20 of this volume); citizenship, gender, and sexuality (see Chapter 18 of this volume); and culture (see Chapter 5 of this volume).

Nativist environmentalism

There are distinctive ways in which racial minorities experience racism and colonialism in the United States. Environmental racism, for example, impacted African Americans through zoning laws that allowed for the dumping of toxic hazards in communities where they lived. Asian immigrants historically were denied naturalization and citizenship rights under Alien Land Laws which impacted their ability to acquire land in the American West. Mexican-Americans and Mexican immigrants were disenfranchised of land rights under the Treaty of Guadalupe Hidalgo, when half of Mexico's northern territory was consumed by the United States. Native Americans were stripped of land and sovereignty through settler colonialism. In each community, race and citizenship were powerful institutions directed at depriving racial minorities of environmental equity.

This chapter expands our understanding of environmental justice movements by including Japanese-American internment, Mexican-Americans' eviction from Chavez Ravine in Los Angeles, and population control against Latino immigrants in Aspen as significant examples. In each case study, we analyze how native-born and immigrant racialized minorities were impacted by federal policies, urban-planning, and population control movements that were viewed as environmental solutions for one group but contributed to environmental injustice of another. At the intersections between race, citizenship, and immigration, we show how race and nativism operate, while appearing colour-blind and simultaneously producing devastating consequences for immigrant and native-born racialized minorities.

Drawing on the work of scholars studying race and immigration in the United States, nativism is part of a system of discourses and actions that seek to promote the interests of mainstream (native-born white middle-class) peoples in opposition to other populations on the grounds of their foreignness. **Foreignness** need not be strictly defined as non-native-born, since people of colour born in the United States have been, throughout our history, defined as foreign in cultural terms. And while some forms of nativism may not be overtly racist, the justification for inclusion or exclusion of certain groups almost always comes down to race. Indeed, race cannot be separated from nativism because the meaning of legal and full social citizenship (i.e., belonging) in the United States has always been racialized.

Extending this concept, **nativist environmentalism** is a political movement that seeks racial exclusivity in places deemed to have special ecological and racial or cultural significance (see Park and Pellow, 2011). Nativist environmentalism is a form of racism rooted in a sense of entitlement to places imbued with particular socio-ecological importance. In other words, while traditional nativists defend *their* nation's borders because they believe they are the truly rightful inhabitants, nativist environmentalists do the same when it concerns the confluence of environmental and cultural entitlements. It is environmentalism with a racial inflection and nativism with an ecological inflection. Nativist environmentalism is the ideological force at the nexus of the nativist and environmental movements, a force that threatens to damage both our social fabric and our planet.

A number of scholars have described this kind of language as a core part of the "new racism," a racism that no longer relies on outdated and abhorrent biological notions of superior versus inferior peoples. Instead, the new racism is based on the idea that there are insurmountable and incompatible *cultural* distinctions between peoples (Balibar, 1991). As sociologist Howard Winant (2001) notes: "the reinterpretation of racialized differences as matters of culture and nationality, rather than as fundamental human attributes somehow linked to phenotype, turns out to justify exclusionary politics and policy far better than traditional white supremacist arguments can do." The new racism is a critical component of the "postracial" (or "colour-blind") approach to race that has been sweeping the nation for some time—since civil rights legislation formally outlawed public acts of racism. Postracial perspectives on racial inequality deny the existence of race, and therefore are inherently blind to racism. People and institutions approaching race in this way can, at the same time, employ the new racism to argue that certain groups are unassimilable based on other characteristics (language, religion, nationality, etc.).

> **Box 17.1 Re-envisioning environmentalism**
>
> Examples abound of the links between efforts to "save" the earth and efforts to control certain groups of people. Both of these practices intersect with discomforting frequency. Environmental and nativist movements in the United States have been historically racist, classist, and patriarchal, and this is evident in a number of initiatives that have been rooted in biological, natural, and social-scientific ideas of how the world should be. For example, the creation of many US national parks was made possible through the explicit removal or containment of Native American tribes. This is what feminist scholar Betsy Hartmann (2004) calls coercive conservation, the violent expulsion of local people from what become wilderness preserves. The environmental group Conservation International (CI) is notorious for such practices. Hartmann reports that CI works with the World Wildlife Fund and USAID, both of which are infamous for their focus on population control in the Global South (Hartmann, 2004).

Case study 1: Japanese-American internment

Under Executive Order 9066, Japanese immigrants and citizens of Japanese descent were evacuated from their homes in 1942. After the attacks on US naval bases in the occupied islands of Hawaii (Pearl Harbour), white hysteria grew across the mainland about the possibility of a second attack. President Franklin D. Roosevelt issued his executive order to evacuate and resettle 120,000 civilians of Japanese descent living in the so-called "exclusion zone," comprising Oregon, California, western Washington, and southern Arizona, into internment camps. In the process of their evacuation, members of the Japanese-American community were stripped of their rights to their possessions, including property rights, gardens/nurseries, land, and their homes (Chiang, 2010). Families were imprisoned in ten different camps in "dust-choked" landscapes throughout the United States as a means to isolate and humiliate this community (Limerick, 1992). Under these difficult conditions, concerns about their environment were central to what Asian-Americans historians refer to as "camp life."

At Camp Minidoka, for example, Japanese Americans used their gardening skills to improve their environmental conditions. Prior to WWII, it is estimated that 43% of Nikkei (Japanese emigrants and their descendants born outside Japan) were involved in agriculture by 1934. One-third were employed as full-time landscapers/gardeners in Southern California (Tamura, 2012). Most of the internment camps were located in isolated deserts in the western region of the United States, a stark contrast to the coastal landscapes where these internees had lived. Despite arid conditions, Japanese Americans transformed their landscape at Camp Minidoka to include victory gardens.

During WWII, the federal government encouraged US families to plant "victory gardens" to grow their own vegetables to offset agricultural shortages (Chiang, 2010, p. 243). Ironically, this shortage was due in part to the incarceration of Japanese-Americans who were the primary producers of so many West Coast crops, including tomatoes, snap beans, sugar beets, and strawberries. The victory garden campaign characterized gardening as a patriotic duty for those on the home front, so that there would be enough food to feed the military. For some Japanese-Americans forcibly removed from their farms as enemies against the state, victory gardens were an effort to assert their status as law-abiding citizens. It also served as a testament to the resilient nature of the community to take back whatever autonomy it could within the confines of its new, hostile landscape. Victory gardens afforded their tenders a place where people could gather to discuss politics, family conflicts, and model new forms of adaptation. Over time, all ten internment camps developed agricultural operations to improve the notoriously terrible diet served in the giant mess halls.[1]

Case study 2: evictions in Chavez Ravine

In 1957, 7,500 Mexican and Mexican-American residents were evicted from the 300-acre community of Chavez Ravine. Located east of downtown Los Angeles, Chavez Ravine was prime real estate for economic development and, in 1951, city planners saw its potential. However, there was one obstacle: its Mexican residents. By 1952, most houses had been destroyed, set ablaze by the local fire department and police officers who used the neighbourhood as target practice during routine trainings in the "pit" of the city. By 1957, the area was handed over to Walter O'Malley, owner of the Brooklyn Dodgers.

In the final days of forced eviction, police found themselves battling against unexpected opponents: grandmothers and mothers. Mexican-American and Mexican immigrants did not leave without giving the fight of their lives. Unincorporated from the formal City of Los Angeles, Chavez Ravine possessed a special meaning for Mexican-American families. It was one of the few residential neighbourhoods in Los Angeles adjacent to Elysian Park and the surrounding embankments of the Los Angeles River. When the city threatened to forcibly remove them, families networked across the kitchen table, on each other's doorsteps, across windowsills, and in alleyways, fighting collectively against the violence of the city.

Abrana Arechiga, for example, refused to leave her home. With her children, she waited for police to arrive and confronted the officers and yelled her protest in Spanish. A few doors down, sheriff's deputies carried Aurora Vargas from her Chavez Ravine home. Families upon families were evicted simultaneously as the entire community was set on fire. Networked and organized, individuals used their physical bodies by lying down on the ground, holding onto the land they had cultivated and stewarded, and where they had raised their children. When the Angustain family was removed, their daughters were dragged kicking and screaming, holding onto the rails down each flight. When police reached 73-year-old Mrs. Alice Martin and her 85-year-old best friend Mrs. Ruth Rayford, the women reached for their broomsticks (Laslett, 2015). In the encounter between the Sheriff's Department and Chavez Ravine residents, the community's **disobedience** mirrored the same behaviour that was witnessed in Warren County; the residents planted their bodies in the land that was rooted in a historical legacy of structural inequality.

The eviction of Mexican and Mexican-American families is an example of environmental injustice, and nativist environmentalism specifically, because Chicana/o communities have been forcibly removed from their ancestral homelands as successive of waves of Spanish, Mexican, and US colonists came to California. Each settler society used development as a rationale to strip the Gabrielino and Mexican residents of land rights (Hernandez, 2017, p. 24). In this way, settler colonialism structured the environmental inequities that were apparent the day of the Chavez Ravine eviction. The legacy of colonialism persisted for generations as Los Angeles developed into a modern cosmopolitan space, included the building of Dodger Stadium in the mid-20th century. In the case of Chavez Ravine, Chicana/o residents were denied decision-making power to determine governance over natural resources, such as access to a uniform system of potable drinking water, as well as public sanitation. Indeed, Chavez Ravine was excluded from urban-planning development in Los Angeles upon its designation as an unincorporated zone. In this way, environmental inequity is structured by both historical and geographical processes that are determined by the residual colonialism that continues to impact the ways in which city planners view certain racialized populations to be an obstacle to a city's future.[2]

Eviction from Chavez Ravine set a legal and cultural precedent in city planning that continues to haunt Mexican-Americans' relationship to the urban landscape of Los Angeles till this day. The types of nativist policies that drove the expulsion of Mexican-American residents from the city core provided new opportunities to introduce white capital and investment into the downtown area. As Gaye Theresa Johnson so eloquently acknowledges, the eviction of Mexican-Americans brought in new forms of spatial entitlements where white investors continued to evict and expel working-class communities throughout Los Angeles under eminent domain (Johnson, 2013, p. 60). Evictions carry racial overtones in restricting where Mexican-Americans can live, play, and work, by historically denying them access to parks, recreation facilities, and swimming pools, while at the same time, limiting their residential options to the dirtiest air quality, next to freeways that continue to have harmful consequences upon the environmental health of the community.

Case study 3: population control in Aspen, Colorado

In 2000, the city council of Aspen, Colorado, a town that sits more than 600 miles from the US–Mexican border, unanimously passed an anti-immigration resolution (Park and Pellow, 2011). The resolution called on the federal government to implement greater restrictions on immigration in order to preserve the economic, cultural, and ecological integrity of the nation and this premier city. Aspen is an exclusive resort town with an international reputation for high-end service and a stunning landscape of pristine mountains, all configured to welcome wealthy skiers in the winter and wealthy hikers in the summer. Ninety-five percent of the residents are white and, like many tourist-dependent towns and cities in the United States, Aspen depends upon cheap immigrant labour—a majority of whom are people of colour—to fuel its local service economy.

And, like many states, Colorado has struggled with how to maintain its labour force for **natural resource extraction** and tourism, while protecting the ecosystems threatened by those industries. Aspen responded with nativist environmentalism. In June 1999, an organization calling itself the Valley Alliance for Social and Environmental Responsibility was formed in an effort to restrict immigration on the basis of ecological sustainability. It argued that immigrants were the leading threats to the environment: "Population is the No.1 factor in the encroachment of environmental degradation—immigration is the No. 1 factor contributing to population growth." (Stiny, 1999) This effort had strong support from local, state, and national entities including former Colorado Governor Dick Lamm, a Democrat and part-time resident of Aspen. Lamm became a populist environmental hero in Colorado in the 1970s when he stopped the Olympic Games from coming to the state on the grounds that it would produce an enormous negative ecological footprint. He was also a board member of FAIR (the Federation for American Immigration Reform, a leading anti-immigration organization) and has written extensively on immigration and population growth as ecological threats. He was one of the nativist candidates running for election to the Sierra Club Board of Directors during its internal crisis over immigration in 2004.

The Valley Alliance for Social and Environmental Responsibility also counts outspoken nativist Republican Congressman Tom Tancredo among its supporters. Tancredo is virulently anti-immigrant, Islamophobic, and pro-war. At the 2010 Tea Party Convention, he called for a return to literacy tests for voters, in order to prevent the election of candidates like Barack Obama, whose ascendancy he blamed on the "cult of multiculturalism." He has argued that if the United States bombed holy Muslim sites in the Middle East, such a practice would deter Islamic extremists from attacking domestic US targets.³ He has repeatedly earned the respect of many racists across the country for his uncompromising stance on these issues. The Alliance is one of many nativist groups in the state of Colorado and in the Rocky Mountain region that is well networked, a group that includes the Colorado Alliance for Immigration Reform (CAIR), a state affiliate of FAIR. The Alliance has friends and supporters in Congress, at FAIR, NumbersUSA, and other nationally active and highly visible nativist groups.

While this initiative in Aspen, Colorado, is just one initiative in a small town, it is part of a much longer history of efforts aimed at protecting the earth's finite resources for the benefit of a powerful minority. For instance, in the early 20th century, many European-American elites turned their sights on newcomers and identified immigrants as a primary cause of ecological problems in the cities. Specifically, immigrants were blamed for the rise in urban pollution, when, in fact, these populations were associated with polluted spaces because they were forced to live in smoke-choked neighbourhoods and work in occupationally hazardous factories and sweatshops (Gottlieb, 2005). More recently, since the 2010 census revealed increasing number of states turning "majority minority," predictions that the United States will one day constitute a white minority has generated growing anxiety. Reports of these demographic shifts have produced fear among native-born whites, concerned that it is not just "their jobs" but now "their *country*" that is in danger of being taken over by foreigners. This anxiety is directed toward immigrants, as whites and other citizens try to restrict access to education, health and social services, decent jobs, housing, and a broad range of environmental amenities. These are longstanding and continuing nativist concerns, and environmentalism has been successfully used by nativist proponents to convey a level of legitimacy by those who seek to soften their racist inclinations.

Conclusion

To understand how environmental justice impacts racial minorities, we must acknowledge the long history of struggles related to access to land, dispossession, forced removal, and internment. Otherwise, we will continue to narrowly misconstrue each environmental challenge as a singular event that affects everyone equally. The history of environmental justice shows us that, just as environmental catastrophes are human-made, so is social inequality, and the two are deeply intertwined. To tackle one without addressing the other will only function to perpetuate the environmental problem as we fortify the social inequalities that reinforce racial divisions. In addition, the historic efforts of racial minority communities fighting political disenfranchisement, denial of citizenship, dispossession, and internment may hold important insights into how to organize a broad-based resistance against state-sponsored violence and take seriously the ways in which mainstream environmental solutions may cause further environmental vulnerability for racial minorities.

The cases outlined in this chapter illustrate the importance of understanding nativist environmentalism; for instance, to recognize how racism has been historically embedded within environmental movements and how we must expand our definition of what constitutes the environment and the methods of its protection. In this regard, environmentalism could have been a transformative force that embraces justice for all, given our shared global ecosystem. Instead, we have constructed political, economic, and social borders to protect only certain people's ecosystems and human communities. These borders are not only artificial; they are also the *source* of environmental devastation. Cleaning one neighbourhood by polluting, threatening, or removing another is antithetical to not only environmental logic, but also fundamentally unjust.

> **Follow-up questions**
>
> - How does race and racism impact environmental inequalities at an institutional level?
> - What types of environmental justice advocacy have immigrants engaged in?
> - In what ways does nativist environmentalism shape environmental activism?

Notes

1 On environmental justice issues related to food and agriculture, see also Chapter 14 of this volume.
2 On urban environmental justice struggles, see also Chapter 15.
3 Associated Press, "Tancredo: If They Nuke Us, Bomb Mecca," Fox News, July 18, 2005, www.foxnews.com/story/tancredo-if-they-nuke-us-bomb-mecca.

References for further reading

Finney, C. (2014) *Black Faces, White Spaces: Reimagining the Relationship of African Americans to the Great Outdoors*. Chapel Hill, University of North Carolina Press.
LaDuke, W. (2013) *The Militarization of Indian Country*. East Lansing, Makwa Enewed.
Pellow, D.N. (2018) *What Is Critical Environmental Justice?* Cambridge, Polity Press.
Pulido, L. (1996) *Environmentalism and Economic Justice: Two Chicano Struggles in the Southwest*. Tucson, University of Arizona Press.
Taylor, D.E. (2016) *The Rise of the American Conservation Movement: Power, Privilege, and Environmental Protection*. Durham, Duke University Press.

References used in this chapter

Balibar, E. (1991) Is there a neo-racism? In E. Balibar & I. Wallerstein (eds.), *Race, Nation, Class: Ambiguous Identities*. New York, Verso, pp. 17–28.
Bullard, R. (2000) *Dumping in Dixie: Race, Class and Environmental Quality*. Boulder, Westview Press.
Bullard, R., Mohai, P., Saha, R. & Wright, B. (2007) Toxic wastes and race at twenty: Why race still matters after all these years. *Environmental Law* 38(2).
Chiang, C. (2010) Imprisoned nature: Toward an environmental history of world war II Japanese American incarceration. *Environmental History* 15(2), 236–267.
Gottlieb, R. (2005) *Forcing the Spring: The Transformation of the American Environmental Movement*. Washington, DC, Island Press.
Hartmann, B. (2004) Conserving Racism: The Greening of Hate at Home and Abroad. *Different Takes: A Publication of the Population and Development Program at Hampshire College* 27, 1–4.

Hernandez, K.L. (2017) *City of Inmates: Conquest, Rebellion, and the Rise of Human Caging in Los Angeles, 1771–1965*. Chapel Hill: The University of North Carolina Press.

Johnson, G.T. (2013) *Spaces of Conflict, Sounds of Solidarity: Music, Race, and Spatial Entitlement in Los Angeles*. Berkeley, University of California Press.

Laslett, J.H.M. (2015) *Shameful Victory: The Los Angeles Dodgers, the Red Scare, and the Hidden History of Chavez Ravine*. Tucson, University of Arizona Press.

Limerick, P.N. (1992) Disorientation and reorientation: The American landscape discovered from the west. *Journal of American History* 79, 1021–1049.

Park, L.S. & Pellow, D.N. (2011) *The Slums of Aspen: Immigrants vs. the Environment in America's Eden*. New York, New York University Press.

Pellow, D.N. (2002) *Garbage Wars: The Struggle for Environmental Justice in Chicago*. Boston, MIT Press.

Pellow, D.N. & Park, L.S. (2002) *The Silicon Valley of Dreams: Environmental Injustice, Immigrant Workers, and the High-Tech Global Economy*. New York, New York University Press.

Stiny, A. (1999). Former governor Lamm heads growth conference. *Aspen Daily News*, 7 October.

Tamura, A. (2012) Gardens in camp. *Densho Encyclopedia*, 25 June, 22, 19 UTC. Available from: https://encyclopedia.densho.org/Gardens%20in%20camp/ [Accessed 20 June 2019].

Winant, H. (2001) *The World is a Ghetto: Race and Democracy Since World War II*. New York, Basic Books, p. 35.

18 Gender matters in environmental justice

Sherilyn MacGregor

Learning outcomes

- Gain an understanding of gender as a concept and analytical lens.
- Learn how gender injustice shapes environmental injustice.
- Become aware of the role that women have played in activist struggles for environmental justice.
- Develop familiarity with ecofeminist environmental justice.

Introduction

Ecofeminist scholars have played a central role in informing contemporary thinking about both justice and human-environment relations. Women activists have led grassroots movements for environmental justice since the concept was coined in the late 20th century. The inclusion of a chapter focusing on women in this volume does not mean that women as academics and activists, gender as a social construct and ecofeminism as a political theory are not tightly woven throughout the rest of the chapters. Why, then, is it necessary to have a chapter on women and gender when there is no chapter dedicated to the role of men? Perhaps a debate about this question will lead to productive scrutiny on some of the problematic assumptions and blind spots that have traditionally been apparent in the academic field of environmental justice.

This chapter is written from an ecofeminist perspective[1] and contains a set of key themes and examples that together give an overview of how and why gender should be central to the study of environmental justice. The purpose is twofold: to redress the tendency of environmental justice scholars to background gender, and to provide critical tools for unsettling dominant assumptions about the connections between women, gender and environmental justice. These tools are important, not only for thinking like a scholar, but also as prevention against blindly sustaining the relations and ideas that contributed to the crisis of environmental injustice in the first place.

Gender inequality and environmental injustice

The best place to start the discussion is with a review of how questions of environmental justice are shaped by gender. Understanding gender inequalities is important for a full analysis of environmental injustice. At the same time, it stands to reason that there are significant connections between the politics of gender justice and the search for more sustainable societies.

Gender is defined as the social construction of differences between femininity and masculinity, which tend to be associated with women and men, respectively.[2] Although gender differences are sometimes celebrated, they also cause inequalities and power imbalances. In most societies, women have lower social status than men, and from this imbalance flow inequitable workloads; fewer opportunities, resources and assets; and discrimination, abuse and violence in women's everyday lives. A lot of social science research on gender is concerned with analyzing the problems women experience by virtue of being female and socialized to be feminine. But women are never only just women: they are also members of racialized and class-defined groups, belonging to placed-based communities who hold a kaleidoscope of concerns and experiences. They are workers and elders and criminals; some (not many) are political leaders and chief executive officers. When a spotlight is shone on women in this chapter, it is the job of the reader to mentally resist a one-dimensional image. At the same time, the temptation to equate gender with women is also in need of resistance. Men are also gendered beings, even though the operation of masculinities in social and political life is regularly overlooked in the environmental social sciences. The role that hegemonic masculinities play in creating and sustaining all manner of environmental injustices needs much greater critical scrutiny than it has received to date (Hultman, 2017; Connell, 2017).

It has become common to think about environmental justice as trivalent, incorporating distributive, procedural and recognition aspects (Walker, 2012). **Gender inequality** is a major driver of all three intersecting axes of injustice. Examples and evidence can be given to illustrate its relevance in the environmental sphere. But it is also necessary to consider additional ways that gender is connected to environmental injustice that are not easily captured by this standard approach. Feminists contend that the concept of justice itself has developed in response to gendered concerns of the (male) theorists who have written it into existence, which means that they have tended to overlook issues of androcentrism, embodiment and care (Fraser, 2013).

In environmental justice scholarship, **distributive justice** is often taken to mean the distribution of environmental goods, such as clean water and green space, and bads, such as toxic pollution and exhaustion of common resources (see Chapter 3 of this volume). It is a central claim that poor environmental quality is directly linked to socio-economic inequality; poor people tend to live in poor environments (Holifield et al., 2018). Gender plays a role in determining distribution of goods and bads, primarily due to economic inequalities. Current statistics indicate that women in every country have less financial wealth than men, and therefore are less able to use their economic resources to make themselves resilient to threats from environmental change (UN Women, 2019a). It is well documented that women, especially those whose inequality is compounded by poverty, racism and bodily fragility, are most likely to be killed or hurt by hurricanes, droughts and extreme heat (Neumayer and Plümper, 2007). Women whose relative poverty restricts their choices, mobility and adaptation capabilities suffer prolonged hardship in the aftermath of natural disasters.

Lack of participation and representation in political structures is a sign that collective procedures are not functioning fairly. **Procedural justice** is another way of saying that decisions are arrived at fairly. The question that environmental justice scholars ask is: who gets to participate in making decisions about policies and plans that shape living environments? (see Chapter 4 of this volume). Asking this question from a gender perspective leads to considerations of how sexism shapes politics and leadership within society. Women from all social groups are under-represented in most democratic systems and marginalized completely in authoritarian regimes (UN Women, 2019b). At the level of national and local state politics, men make up the majority of those making decisions that affect environmental quality, from land use planning and property development to transport systems and air quality standards. It is often reasoned that, when an unrepresentative group is empowered to use its own expertise, experience and values to make decisions for the general public, the outcomes do not serve the common good. Research on transport routes and housing design shows that they are often based on the life patterns of employed men and nuclear Western families, and therefore leave a wide range of needs unmet (Jarvis et al., 2009). The adult male body is used as the default in medical education and research, which results in a lack of knowledge about how drugs and chemicals affect the bodies of women, children and elder people (Westervelt, 2015). So, lack of participation and representation can be deadly. The key point here is that abstract ideals of equal opportunity and scientific objectivity mask the existence of deep, structural injustices in the design, maintenance and governance of human societies.

Why is it that the needs of the vast majority of people are deemed irrelevant? The answer usually points to evidence and arguments about the lack of respect and value given to particular groups in society, and the need to think about **recognition** as being fundamental to environmental justice (see Chapter 5 of this volume). In the environmental justice field, misrecognition due to race and class has been central to explanations of why poor people of colour live in degraded and toxified environments that would be intolerable for the white middle classes. For evidence of how these axes of misrecognition intersect with gender, environmental justice scholars point to the high rate of police violence towards and mass incarceration of men of colour in the US (Pellow, 2016; see also Chapter 22 of this volume). The claim that African American men are treated as having lesser value than white people has inspired the Black Lives Matter movement in the US. As several environmental justice scholars have argued, this misrecognition also permeates the mainstream environmental movement, which seems to have greater concerns about recycling and wildlife conversation than it does for the environmental conditions that structure Black lives (Pellow, 2016; Nixon, 2011).

Theorizing how the problem of misrecognition relates to gender injustice is a central project of feminist scholarship. Nancy Fraser (2013) offers a definition that is useful for identifying the specificities of **gender justice** as they pertain to women, the aspects that cannot be mixed easily with race and class to the point of dissolution. This definition stems from her critique of, and related desire to move beyond, the polarized choice between equality (treating likes alike, or women like men) and difference (treating unlikes unalike, treating women differently insofar as they are different from men). She treats justice as a more complex concept than can be captured in simple words like 'equality'. It is, for her, best conceptualized as being comprised of a set of distinct normative principles (see Box 18.1) that must be achieved at once in order for gender justice to be realized: 'Failure to satisfy any one of them means failure to realise the full meaning of gender justice' (Fraser, 2013, p. 116).

Box 18.1 Principles of gender justice

The anti-poverty principle: first and foremost, prevent poverty; at the very least, relieve suffering caused by inability to meet basic needs. Particularly relevant to gender justice: 60 per cent of the world's poor are women (UNEP, 2016).

The anti-exploitation principle: prevent the exploitation of vulnerable people; mitigate exploitable dependence because when people are dependent on others for money, they can easily be exploited.

Income equality: equalize the distribution of real per capita income; substantially reduce the gap between men's and women's incomes (the average gender wage gap is currently 23 per cent, according to the International Labour Organization [www.ilo.org]).

Leisure-time equality: equalize the distribution of work so that women do not continue 'to suffer disproportionately from "time poverty"' (Fraser, 2013, p. 119). ('In virtually every country, men are able to fit in valuable extra minutes of leisure each day while women spend more time doing unpaid housework.' [Ferrant et al., 2014])

Equality of respect: ensure that men and women are equally free from being trivialized, objectified and deprecated; requires equality of recognition of the value of all people's personhood, contributions and paid and unpaid work.

The anti-marginalization principle: promote equal participation of men and women in all areas of society—in employment, civil society, politics, etc. This entails the provision of the necessary conditions for participation (childcare, accessibility, 'family-friendly' schedules and spaces, etc.), as well as the 'dismantling of masculinist work cultures' (Fraser, 2013, p. 120) and environments that are hostile to women.

The anti-androcentrism principle: stop treating men and men's life patterns as the norm to which women must assimilate. 'Restructure androcentric institutions so as to welcome human beings who can give birth and who often care for relatives and friends, treating them not as exceptions but as ideal-typical participants.' (Fraser 2013, p. 121)

Source: based on and adapted from Fraser (2013)

In the field of environmental justice, more needs to be done to consider the specificities of **environmental sexism** (Gaard, 2018). To do so, it is useful to apply Fraser's principles when thinking of environmental problems, as well as the solutions that are advocated by green visionaries. In what ways does misrecognition of women and all things feminized shape the phenomenon in question? How do androcentrism and misogyny function in environmentalism? As two examples, consider the objectification of women's bodies in anti-fur campaigns or plans for a new green deal that prioritize the creation of jobs in male-dominated sectors (such as engineering and construction) and leave the unfair distribution of feminized caring and welfare services untouched. These examples can legitimately be criticized as failing on gender justice.

Before moving to a discussion of women's roles as activists in the next section, it is important to mention that, in addition to making theoretical and empirical links between gender inequality and environmental injustice, links also exist between greater gender equality and environmental quality. In *The Spirit Level*, Wilkinson and Pickett (2009) present a convincing case for why reducing social inequality will result in greater environmental sustainability.

There is a measurable link between levels of gender equality and environmental protection. The lower the status of women in a country, the worse that country's performance in tackling environmental problems, such as excessive carbon emissions. Ergas and York (2012) use data from 160 countries and statistical regression models to assess the effects of women's political status on per capital CO_2 emissions (using key control variables). Previous research has linked women's political status to environmental treaty ratification (Norgaard and York, 2005) and to political designation of protected land area within nations (Nugent and Shandra, 2009). On the basis of these findings, they argue: 'improving women's status around the world may be an important part of efforts to curtail greenhouse gas emissions and prevent dramatic climate changes from undermining the long-term prospects of societies' (Ergas and York, 2012, p. 974).

Gender balance has been linked with good local environmental decision-making, too. It seems that, when women and men work together, and blend their respective experiences and insights, they are more likely to arrive at plans and policies that improve sustainability. For example, Buckingham et al.'s (2005) research on waste-management systems in Europe found that the local governments with the highest recycling rates had a higher than average percentage of women employees. Their staff also tended to include fewer engineers and more decision-makers from diverse professional backgrounds, such as education and social work. This evidence does not prove causation, but suggests that there may be a link between gender balance, environmental justice and pro-environmental decision-making (Buckingham, 2010).

Action for environmental justice: why gender matters

Gender plays a role in the extent to which environmental issues are problematized, as well as how people respond to them in public and private spheres. Much of the scholarship on political agenda-setting highlights the role of elites; viewed through a gendered environmental justice lens, the fact that the majority of elite people who hold the power to create and react to environmental injustices are privileged white men is problematic.

Research has shown that there are significant differences in the way men and women in general perceive and respond to environmental problems (Kennedy and Dzialo, 2015; Strapko et al., 2016). Aggregated data suggest that men tend to be less risk-averse, less likely to be concerned about environmental harms and less likely to engage in pro-environmental actions in daily life than women (Hunter et al., 2004; McCright and Xiao, 2014). For example, there is strong opinion poll data from the US to suggest that men, and white conservative men especially, are less concerned about and even more likely to deny the existence of **anthropogenic climate change** than women and other demographic groups (McCright and Dunlap, 2011). Sociologists have theorized that a tendency to be disengaged with environmental challenges stems from the desire to protect a masculine identity and the social privilege it affords. One study by social psychologists has analyzed survey data that suggests that hegemonic masculinity may prevent some men from engaging in 'green' activities and identifying with environmentalism because both are culturally associated with socially devalued hegemonic femininity (Brough et al., 2016). In short, there are men who do not want to appear 'unmanly' by embracing environmentalism.

These patterns, where taking action to remedy environmental harms is gendered, point towards a discussion about the role of women as the leaders and workers who pursue action in the name of achieving environmental justice. That women turn to activism stands to reason because women's position on the receiving end of gender injustice makes them particularly interested in addressing the causes and effects of environmental injustice. And since formal procedures discriminate against women, the battles take place outside the system, through local organizing and protest. There is a substantial body of literature that brings gender into environmental justice by focusing on women in a way that celebrates their grassroots activism (Krauss, 1993; Rainey and Johnson, 2009; see MacGregor, 2006 for a discussion). Examples of women organizing to fight against environmental harms and injustices can be found in countries the world over. Although difficult to capture comprehensively here, it is possible to categorize the types of struggles as focusing on three concerns: bodily health and well-being; land and livelihood; and global environmental justice. What follows is a brief discussion and illustration of these types, followed by some caveats against the temptation to romanticize women's environmental justice activism.

Protecting human health

The American biologist Rachel Carson is typically credited with making the links between chemicals in the ecosystem and threats to all lifeforms that sparked the modern environmental movement. It is often claimed that it was no coincidence that she was a woman scientist (Seager, 2017). Women have been at the frontlines not only in exposing the risks of scientific control of nature, but also of activist struggles against threats to human health from toxic pollution. They have taken the lead not because they feel an innate sensitivity to the natural world but because, as members of families and communities, women shoulder a disproportion amount of responsibility for taking care of the well-being of others.

The concentration of toxic contamination in areas populated by working-class and racialized people is perhaps the most important example of environmental injustice.[3] Feminist research has been successful in showing how chemical pollution affects bodies, particularly bodies that are vulnerable due to age, disability and normal conditions such as pregnancy (Scott, 2015). Toxic pollution also occurs in the private sphere of homes, as well as in public spaces and workplaces. As those who take primary responsibility for the everyday care of bodies (human and non-human), and who tend to spend more time 'at home', women are in the position to notice, question and react to problems with their immediate environment and effects on the living bodies for whom they care. In many cases, women have claimed to be involved in protest out of concern for their loved ones, such as children and elderly parents, who are the most physically vulnerable to the effects of toxic pollution (MacGregor, 2006).

As discussed in Chapter 10 in this volume, one of the first, and arguably most iconic, struggles for environmental justice happened at **Love Canal**, New York, in the early 1980s (Blum, 2008). After a small group of working-class women residents of a new housing development noticed a growing number of illnesses and miscarriages in the neighbourhood, their research led them to discover that their houses were built on a chemical dump. It took Lois Gibbs and her fellow campaigners many years to get the Hooker Chemical Company and local authorities to recognize and pay compensation for the health impacts. Along the way, the women were famously dismissed as 'hysterical housewives' by male politicians who refused to accept their evidence as valid (Seager, 1996). Their efforts to gather proof about the harmful impacts of pollution have since been deemed a form of 'citizen science' or popular epidemiology upon which many grassroots struggles for environmental justice depend.

Industrial pollution can make not only the ground under our feet, but also the air we breathe and the food and water we ingest, a threat to health and **well-being** As those most responsible for providing everyday sustenance and taking care of children, women are typically more in tune with the quality of the local environment than men. And when women are socially and culturally positioned as being responsible for these necessities, they emerge as leaders of activist movements. For example, Japanese women started a consumer movement against dangerously high levels of mercury in food (Insook, 2011). Indigenous women in Latin and North America demand action and compensation from mining and oil companies whose extractive operations have contaminated their communities' water (Jenkins, 2014; Privott, 2019). A further example is Sze's (2007) research on campaigns for better air quality in New York, where women from racialized communities lead anti-asthma organizations. Sze's work is particularly useful in showing how gender, race and class intersect in grassroots, neighbourhood-level environmental justice activism (see also Sze, 2017).

Defending land and livelihoods

The power of corporations to grab and enclose land and then expel people from their settlements in order to extract resources through mining, drilling, logging and intensive farming is a matter of serious injustice, particularly in colonized, resource-rich countries. A second kind of environmental justice struggle in which women have emerged as leaders is the defence of land and **livelihoods**. Here again, the reason is not necessarily that women feel closer to nature, but rather that they depend acutely on land and place for the daily survival of their households. Whereas male members of subsistence communities often travel away from home for paid, seasonal work, women are rooted in place and responsible for growing food and maintaining community well-being over time (Babugura, 2017). When their ability to provide this continuity is threatened by the activities of multinational corporations, states and militaries/militia, women have organized to fight back.

The Chipko movement in the Uttarakhand region of India is often held up as the iconic example of women's struggle to defend land from exploitation, specifically deforestation. While there are debates over how it has been portrayed, most agree that the Chipko movement has been important in highlighting the gender dimensions of environmental politics in India (Brown, 2014; Shiva, 1989). The Kenyan Greenbelt Movement is another such example, in this case founded by a woman (Nobel Laurate Wangari Maathai), of a women's movement that combines environmental conservation and women's empowerment (Muthuki, 2014; Nixon, 2011) (Box 18.2). The Federation of Ogoni Women has a long history of protest and occupation against the environmental and economic crimes of companies such as Shell and Chevron in the Niger Delta (Ikelegbe, 2005). To show that these types of women-led environmental justice protests are not exclusive to the Global South, we can consider the fact that anti-fracking struggles in the Global North (in Australia and the UK, for example) are being led by older women who present fracking as a threat to land, livelihoods and climate, particularly to the rights of younger generations to a tolerable future (Larri and Newlands, 2017; Pidd, 2015).

Box 18.2   The Green Belt Movement

Founded in Kenya in the late 1970s, the Green Belt Movement is an important example of an environmental justice struggle shaped by the intersections of gender, poverty, colonialism and resource depletion.

The movement was started by Dr Wangari Maathai, who was a scientist, feminist and environmentalist and in 2004 became the first African woman to be award the Nobel Peace Prize. What began with planting a few trees to commemorate the achievements of women environmental activists eventually became a grassroots organization of women engaged in community-based tree reforestation and poverty reduction in Kenya and other African countries. The movement now operates in 30 countries and has planted over 50 million trees.

The Green Belt Movement was initiated by women and responds to women's needs for skills, independent income and political representation. The planting of trees is a form of non-violent civil disobedience that slowly reclaims common land while challenging the sexist association of forests and forestry with men. It can be understood as an 'intersectional' environmental justice movement because it has never been single-issue (only about conservation or restoration), but rather it has always tried to advance the interlinked causes of environmental, women's and human rights while simultaneously addressing the 'impact of accumulative resource mismanagement on biodiversity, soil quality, food security, and the life prospects of rural women and their families' (Nixon, 2011, p. 139).

While tempting to celebrate this example, it is also important to note the significant level of criticism that Maathai endured, mainly by male politicians who dismissed her as unpatriotic, not a proper (i.e., submissive) African woman and a 'madwoman' (even though she was the first woman in East and Central Africa to earn a PhD). Over the course of her activist career she faced eviction, unemployment and an acrimonious divorce, and was imprisoned several times.

Sources: Maathai (2007); Schell (2013); Muthuki (2014); Nixon (2011)

Demanding global climate justice

A third kind of struggle has taken shape at the global level, where women have framed the disproportionate effects of climate change and state population and immigration policies on women's lives as matters of environmental justice. Women's organizations that work transnationally, such as the Women's Environment and Development Organization, have been active in lobbying governments and institutions to take gender inequality seriously as a driver of climate change, as well as exposing the uneven gendered impacts of climate-related disasters. Women's climate activism is a good place to look for examples, not only of the evidence presented to support justice claims-making, but also examples of gendered leadership. It is also increasingly possible to find lobbying efforts within official United Nations (UN) circles, such as the work of the Women's and Gender Constituency (WGC) of the UN Framework Convention on Climate Change (UNFCCC). In 2017 the WGC successfully pressured global climate diplomats to agree a Gender Action Plan, which prioritizes gender equality and human rights in the just transition to decarbonized societies (UNFCCC, 2017). The Troika+ was set up to foster a greater role for women and to boost the importance of a social justice agenda in high-level climate negotiations. A joint initiative of the Mary Robinson Foundation and the government of Mexico, this group includes more than 55 high-profile women leaders, including women ministers, deputy ministers and senior women leaders from various international organizations. Another key goal is to use women's involvement to counteract the adversarial, macho style of deliberation that holds sway in a male-dominated climate governance arena. Many women leaders believe that other ways of working would be more effective in the search for climate compromises. Christiana Figueres, former executive secretary of the UNFCCC, has stated that her style of leadership played a huge role in brokering the Paris Agreement, a style informed by her experience as a woman (keynote lecture, London 2018; on climate justice, see also Chapter 12 of this volume).

In addition to climate change, women activists make significant efforts to demand reproductive justice as part of the environmental agenda (Bhatia et al., 2019). Di Chiro (2008) explains the intersection of environmental justice and women's **reproductive rights** (see also Chapter 24 of this volume) through the example of Asian and Pacific Islander immigrant women's activism in the US. For the activists she studied, there is a clear, expressed link between reproductive freedom and a range of other issues including 'immigrant rights, workers' rights, queer rights, environmental justice, educational justice, bringing an end to violence against women, and the empowerment of youth' (Shen, quoted in Di Chiro, 2008, p. 288). More recently, feminist environmental justice scholars have noted the risk that global efforts to reduce carbon emissions and mitigate climate change will involve neo-Malthusian policies on so-called over-population. As Bhatia et al. (2019) have argued, in the context of the increasingly visible effects of unsustainable development and the rise of far-right populism across the globe, there is renewed need for feminist activism to expose and resist populationist solutions that reaffirm racialized, misogynist, neocolonial and neoliberal relations.

Against romanticizing women's environmental justice activism

So far, this section has given some selected examples (they are by no means the only ones) that capture some of the history but also provide an indication of what environmental justice activists who are women have accomplished and the challenges they have faced along the way. But it is important to note also some key caveats that should prevent the temptation to unduly romanticize women's activism in struggles for environmental justice.

The first caveat is that, while women may be over-represented in grassroots environmental groups, men are more likely to hold leadership and paid roles in formal environmental organizations (Buckingham, 2017). As Gaard observes, 'at the level of community activism, environmental justice has been powered by people of color, focusing on race and class, with grassroots women doing much of the activism and prominent male leaders serving as spokespersons and theorists' (2018, p. 4). White men take up the most space as leaders and agenda-setters. For example, in her comprehensive review of the 'state of diversity' in US environmental organizations, Taylor (2014) finds that, although some gains have been made by white women, men are still much more likely than women to hold positions of power. We might add to this the question of whether it is fair that women should shoulder a heavier load of the frontline struggle to clean up an ecological mess that they played a smaller role in making than men.

A second caveat is that, in many societies, women engage in political activism at considerable risk to their own safety and security. Global Witness reports that the number of people killed or imprisoned each year for trying to defend their environment is rising at unprecedented rates (Butt et al., 2019). Women are particularly vulnerable to violence due to their gender. The sexual violence against women perpetrated by mostly male military and police forces is well documented. The vulnerability of women to arrest and detention is compounded by the consequences for those who depend on their care, including children and elderly family members. Police often use threat of arrest and removal of children as a way of discouraging women from getting involved in direct action (Monk et al., 2019).

A third caveat is that recognizing women's environmental justice activism should not underplay differences among women or women's involvement in environmentally destructive practices. The discussion in this section has pointed out repeatedly that women tend to be drawn into environmental justice activism because of their social and cultural positions rather than their innate or essential female traits. This means that their activism is context-dependent and differs according to what kind of society and political-economic system they live in and how their lives are shaped by colonialism, patriarchy, **heteronormativity** and so on. The struggles of women in rural areas of the Global South, for instance, are vastly different from those in urban Europe. When women form transnational environmental organizations, there are often productive but also sometimes painful conflicts over aims, strategies and visions.

Not all women environmental activists are feminists, and many are critical of the narrow mobilization of gender found in the liberal feminism of white privileged women. The fact that we can point to and even celebrate many examples of some women's heroic environmental justice activism should never prevent criticism of other women's complicity in environmental destruction around the world. Ecological feminism is a distinct strand of feminism precisely because it does not embrace the liberal goal of giving women the same rights as men to exploit and overconsume the earth's finite resources. As will be explained in the next section, **ecofeminism** offers a radical perspective on environmental justice that attempts simultaneously to correct liberal feminism's ignorance of planetary limits and environmentalism's blindness to patriarchal privilege.

Ecofeminist environmental justice

Ecofeminism has a long history which is often traced back to the provocative political writings of Francoise D'Eubonne, who analyzed capitalist-patriarchy as a system that oppresses women at the same time as destroying the planetary life-support system (D'Eubonne, 1974). She argued that it would not be possible to survive the ecological crisis without fundamental cultural change, which in her 1970s view necessarily involved the move away from male-dominated militaries, nuclear technology and the embrace of unlimited economic growth. Her point, and the claims of other early ecofeminists, was not that it was women's natural goodness that would save the planet, but **patriarchy**'s exploitation of all things feminine that would destroy it.

From the polemical writings of D'Eubonne has evolved a comprehensive ethico-political ecofeminist perspective—drawing on the scholarship of Val Plumwood, Karen J. Warren, Donna Haraway, Bina Agarwal, Vandana Shiva, Greta Gaard and many others—that inspires new generations of environmentally engaged scholars and activists (for an overview, see MacGregor, 2017). However, nearly four decades later, whether due to an unfounded caricature of ecofeminism as simplistic and essentialist, or due to wilful academic ignorance, there remains a blindness to the role of gender and the relevance of ecofeminist theory in the environmental social sciences, including in the environmental justice field. This is unfortunate because ecofeminism offers a fundamental critique of power relations through the analysis of dualisms and the logic that underpins most forms of domination—not just the male domination of women, but also human domination of the more-than-human, white people of racialized people of colour, rich over poor, straight over queer, and so on.

Their critical analyses of philosophical rationalism, neoclassical economics, anthropocentrism and epistemic remoteness leads ecofeminist theorists to understand socio-ecological injustice as a condition that is produced by multiple intersecting forces.

But ecofeminist thinking is not only about critique; there are also core principles and an overall vision that can contribute important dimensions to environmental justice as movement and field of scholarship. At the heart of this vision is the aim to build a sustainable society that does not exploit people and 'earthothers', even those who perform care out of love and/or duty. Fraser's anti-exploitation and equality of leisure-time principles (Box 18.1) effectively capture the spirit of an ecofeminist vision of a caring society that is organized around meeting the needs of embodied beings in a fair and democratic way. Care and justice go hand in hand in this vision, so that, while valuing the work of nurturing and sustaining life, this work is not taken for granted or left for only some people to carry out. Ecofeminist climate justice advocates are increasingly calling for a gender-just, socially sanctioned reduction in paid working hours in order to bring about individual and societal freedom from the need for economic growth: 'Being "time wealthy" is both an aim of, and prerequisite for, a resource-light life' (Alber et al., 2017, p. 77).

An ecofeminist interpretation of environmental justice involves reimagining the concept of sustainability to promote flourishing of and dialogues between all earth's inhabitants. Gaard (2018) insists that justice from an ecofeminist perspective is always intersectional and interspecies in its outlook, meaning that Fraser's people-cantered feminist principles must be rewritten to incorporate an interspecies ethic. Environmental justice has tended to be about justice for marginalized people and, in some cases, it has put women at the centre of concern and action. This focus remains important; justice, after all, is a humanist concept that few ecofeminists would wish to reject. In fact, some have raised concerns that posthumanist environmental scholarship may be providing new reasons to ignore gender and dismiss feminist insights about intersectional differences (Gaard, 2017). At the same time, being consistent with ecofeminist analyses, a vision of just sustainability will require 'transspecies diversity' and 'an inclusive, ecological, economical, and participatory democracy' (Gaard, 2018, p. 23).

Conclusion

Pellow argues that the inclusion of gender, sexuality and other axes of social difference is now part of the second generation of environmental justice scholarship: it should no longer be treated as an optional addition but embedded in the very foundations of the field (Pellow, 2016; see also Chapters 22 and 24). In this chapter, the point has been that while the use of an intersectional lens is called for when examining examples of environmental injustice, it is also important to understand the specific ways that gender operates and how people are unfairly harmed, burdened or advantaged because of their gender identities. The examples given of women's activism, fuelled by their concerns for health, livelihoods and the uneven effects of climate emergency, highlight not only how gender intersects with class, 'race', dis/ability and age to produce diverse local struggles, but also that for many activists being a woman carries specific political meanings as well as embodied risks.

It has been a central aim of this chapter to show how and why gender matters in environmental justice research, action and alternative imaginaries. Gender is not just an empirical category or identity (and not reducible to male/female), but also a discursive social construction that organizes the world. It is a concept that 'structurally organizes virtually every aspect of social life in all cultures' (Peterson and Runyan, 1999, p. 31). The study of gender politics should involve the analysis of power relations between men and women, and the discursive and social constructions of hegemonic masculinities and femininities that shape the way we interpret, debate, articulate and respond to social/natural/technological phenomena like war, economic crisis and climate change. The lens of gender brings into acute focus the processes, norms and power relations through which we can recognize the workings of hegemonic masculinities and hegemonic femininities in all social phenomena. It is this kind of critical questioning that ecofeminist perspectives offer to environmental justice. As others have argued, environmental justice 'may be more effective in collaboration' with ecofeminism than it is currently (Gaard, 2018, p. 4). There is more work to be done to embed gender and feminist insights into the field of environmental justice, so that chapters like this one will not be necessary in future.

Follow-up questions

- How can gender, as a lens through which to see mobilizations of masculinities as well as femininities in all aspects of the socio-environmental nexus, become embedded (as opposed to optional) in environmental justice scholarship?
- If women and people of colour make up the majority of members of grassroots environmental justice struggles, then why is it that white men hold most of the leadership positions?
- What are some examples of how misrecognition on the grounds of gender operates within environmentalism?
- If care—as in caring for people and the planet—is symbolically feminized, and therefore devalued in most societies, then how effective is it likely to be as an overarching frame for a just transition?

Notes

1 In this chapter, ecofeminism is treated as related to but distinct from feminism in that it is always concerned with the intersections of environmentalist and feminist political goals. Much of the theoretical base of ecofeminism comes from feminist theory, such as feminist theories of justice and patriarchy.
2 Gender studies research increasingly addresses gender-queer, three-spirit, non-binary and fluid performances and experiences of gender. Sexualities, on a continuum from asexual to heterosexual to LGBT+, are also important to include in a comprehensive review. However, due to lack of space and lack of research on gender non-binary and LGBT+ interface with environmental justice issues (but see Sandilands and Erickson 2010; Seymour 2017), this chapter looks primarily at gender differences as they align to dominant understandings and performances of the male–female/masculinity and femininity binary.
3 On toxic contamination and environmental justice, see Chapter 10 of this volume.

References for further reading

Agarwal, B. (2016) *Gender Challenges* (vols 1, 2, 3). Oxford: Oxford University Press.
Buckingham, S. and Le Masson, V. (eds.) (2017) *Understanding Climate Change through Gender Relations*. London: Routledge.
Detraz, N. (2016) *Gender and the Environment*. New York: Policy Press.

Gaard, G. (2018) *Critical Ecofeminism*. New York: Lexington Books.
Harcourt, W. and Bauhardt, C. (eds.) (2018) *Feminist Political Ecology and the Economics of Care: In Search of Economic Alternatives*. London: Routledge.
MacGregor, S. (ed.) (2017) *The Routledge Handbook of Gender and Environment*. London: Routledge.
Stein, R. (ed.) (2004) *New Perspectives on Environmental Justice: Gender, Sexuality and Activism*. New Brunswick, NJ: Rutgers University Press.

References used in this chapter

Alber, G., Calhoon, K. and Röhr, U. (2017) Gender and urban climate change policy: tackling cross-cutting issues towards equitable, sustainable cities. In S. Buckingham and V. Le Masson (eds.), *Understanding Climate Change through Gender Relations*, pp. 64–86. London: Routledge.
Babugura, A. (2017) Gender equality, sustainable agricultural development, and food security. In S. MacGregor (ed.), *The Routledge Handbook of Gender and Environment*, pp. 357–371. London: Routledge.
Bhatia, R., Sasser, J.S, Ojeda, D., Hendrixson, A., Nadimpally, S. and Foley, E. (2019) A feminist exploration of 'populationism': engaging contemporary forms of population control. *Gender, Place & Culture*, https://doi.org/10.1080/0966369X.2018.1553859 (accessed 02-04-2019).
Blum, E. (2008) *Love Canal Revisited: Race, Class, and Gender in Environmental Activism*. Lawrence, KA: University of Kansas Press.
Brough, A.R., Wilkie, J., Jingjing, Ma, Isaac, M.S. and David Gal, D. (2016) Is eco-friendly unmanly? The green-feminine stereotype and its effect on sustainable consumption. *Journal of Consumer Research* 43(4): 567–582.
Brown, T. (2014) Chipko legacies: sustaining an ecological ethic in the context of Agrarian change. *Asian Studies Review* 38(4): 639–657.
Buckingham, S. (2010) Call in the women. *Nature* 468:502.
Buckingham, S. (2017) Gender and climate change politics. In S. MacGregor (ed.), *The Routledge Handbook of Gender and Environment*, pp. 384–397. London: Routledge.
Buckingham, S., Reeves, D. and Batchelor, A. (2005) Wasting women: the environmental justice of including women in municipal waste management. *Local Environment* 10(4): 427–444.
Butt, N., Lambrick, F., Menton, M. and Renwick, A. (2019) The supply chain of violence. *Nature Sustainability* 2: 742–747.
Connell, R. (2017) Masculinities in the socioscene. In S. MacGregor and N. Seymour (eds.), *Men and Nature: Hegemonic Masculinities and Environmental Change*, pp. 5–8. Munich: Rachel Carson Center, Ludwig Maximilians University.
D' Eaubonne, F. (1974) *Le Féminisme ou la mort*. Paris: Pierre Horay.
Di Chiro, G. (2008) Living environmentalisms: coalition politics, social reproduction and environmental justice. *Environmental Politics* 17(2): 276–298.
Ergas, C. and York, R. (2012) Women's status and carbon dioxide emissions: a quantitative cross-national analysis. *Social Science Research* 41: 965–976.
Ferrant, G., Pesando, L. and Nowacka, K. (2014) *Unpaid Care Work: The Missing Link in the Analysis of Gender Gaps in Labour Outcomes*. OECD Development Centre, www.oecd.org/dev/development-gender/Unpaid_care_work.pdf (accessed 25-11-19).
Fraser, N. (2013) *The Fortunes of Feminism*. London: Verso.
Gaard, G. (2017) Posthumanism, ecofeminism and inter-species relations. In S. MacGregor (ed.), *The Routledge Handbook of Gender and Environment*, pp. 115–129. London: Routledge.
Gaard, G. (2018) *Critical Ecofeminism*. Lanham: Lexington Books.
Holifield, R., Chakraborty, J. and Walker, G. (eds.) (2018) *The Routledge Handbook of Environmental Justice*. London: Routledge.
Hultman, M. (2017) Exploring industrial, ecomodern and ecological masculinities. In S. MacGregor (ed.). *The Routledge Handbook of Gender and Environment*, pp 239–252. London: Routledge.

Hunter, L.M., Hatch, A. and Johnson, A. (2004) Cross-national gender variation in environmental behaviors. *Social Science Quarterly* 85(3): 677–694.

Ikelegbe, A. (2005) Engendering civil society: oil, women groups and resource conflicts in the Niger Delta region of Nigeria. *Journal of Modern African Studies* 43(2): 241–270.

Insook, Jo (2011) *The Empowerment of the Women in the Kanagawa Seikatsu Club Movement*. PhD thesis, Department of Geography, Royal Holloway University of London.

Jarvis, H., Kanter, P. and Cloake, P. (2009) *Cities and Gender*. London: Routledge.

Jenkins, K. (2014) Unearthing women's anti-mining activism in the Andes: Pachamama and the 'mad old women'. *Antipode* 47(2): 442–460.

Kennedy, E.H. and Dzialo, L. (2015) Locating gender in environmental sociology. *Sociology Compass* 9(10): 920–929.

Krauss, C. (1993) Women and toxic waste protests: Race, class and gender as resources of resistance. *Qualitative Sociology* 16(3): 247–262.

Larri, L. and Newlands, M. (2017) Knitting Nannas and Frackman: a gender analysis of Australian anti-coal seam gas documentaries and implications for environmental adult education. *Journal of Environmental Education* 48(1): 35–45.

Maathai, W. (2007) *Unbowed: A Memoir*. New York: Anchor Books.

MacGregor, S. (2006) *Beyond Mothering Earth: Ecological Citizenship and the Politics of Care*. Vancouver: University of British Columbia Press.

MacGregor, S. (2017) Introduction. In S. MacGregor (ed.), *The Routledge Handbook of Gender and Environment*, pp. 1–24. London: Routledge.

McCright, A. and Dunlap, R. (2011) Cool dudes: the denial of climate change among conservative white males in the United States. *Global Environmental Change* 21(4): 1163–1172.

McCright, A. and Xiao, C. (2014) Gender and environmental concern: insights from recent work and for future research. *Society and Natural Resources* 27(10): 1109–1113.

Monk, H., Gilmore, J. and Jackson, W. (2019) Gendering pacification: policing women at anti-fracking protests. *Feminist Review* 122(1): 64–79.

Muthuki, J.M. (2014) Ecological activism highlighting gendered relations: Wangari Maathai's greenbelt movement in Kenya. In T. Doyle and S. MacGregor (eds.), *Environmentalism Around the World: Shades of Green in Politics and Culture*, pp. 195–226. Santa Barbara CA: Praeger/ABC-Clio.

Neumayer, E. and Plümper, T. (2007) The gendered nature of natural disasters: the impact of catastrophic events on the gender gap in life expectancy, 19812002. *Annals of the Association of American Geographers* 97(3): 551–66.

Nixon, R. (2011) *Slow Violence and the Environmentalism of the Poor*. Boston: Harvard University Press.

Norgaard, K. and York, R. (2005) Gender equality and state environmentalism. *Gender and Society* 19(4): 506–522.

Nugent, C. and Shandra, J.M. (2009) State environmental protection efforts, women's status, and world polity. *Organization & Environment* 22(2): 208–229.

Pellow, D.N. (2016) Toward a critical environmental justice studies: black lives matter as an environmental justice challenge. *Du Bois Review* 13(2): 221–236.

Peterson, V.S. and Runyan, A. S. (1999) *Global Gender Issues* (2nd ed.). Boulder, CO: Westview Press.

Pidd, H. (2015) Anti-fracking Nanas: the government is all out for shale—we're all out to stop it. *The Guardian*, www.theguardian.com/environment/2015/jul/03/nanas-shale-cuadrilla-fracking-lancashire (accessed 05-05-2019).

Privott, M. (2019) An ethos of responsibility and indigenous women water protectors in the #NoDAPL movement. *American Indian Quarterly* 43(1): 74–100.

Rainey, S. and Johnson, G. (2009) Grassroots activism: an exploration of women of color's role in the environmental justice movement. *Race, Gender & Class* 16(3/4): 144–173.

Sandilands, C. and Erickson, B. (eds.) (2010) *Queer Ecologies: Sex, Nature, Politics, Desire*. Bloomington: Indiana University Press.

Schell, E. (2013) Transnational environmental justice rhetorics and the green belt movement: Wangari Muta Maathai's ecological rhetorics and literacies. *JAC* 33(3/4): 585–613.

Scott, D. (ed.) (2015) *Our Chemical Selves: Gender, Toxics and Environmental Health.* Vancouver: University of British Columbia Press.

Seager, J. (1996) 'Hysterical housewives' and other mad women: grassroots environmental organizing in the USA. In D. Rocheleau, E. Wangari and B. Thomas-Slater (eds.), *Toward a Feminist Political Ecology: Global Perspectives from Local Experience.* New York: Routledge.

Seager, J. (2017) Rachel Carson was right – then, and now. In S. MacGregor (ed.), *The Routledge Handbook of Gender and Environment*, pp. 27–42. London: Routledge.

Seymour, N. (2017) Transgender environments, In S. MacGregor (ed.), *The Routledge Handbook of Gender and Environment*, pp. 253–269. London: Routledge.

Shiva, V. (1989) *Staying Alive: Women, Ecology and Development.* London: Zed Books.

Strapko, N., Hempel, L., MacIlroy, K. and Smith, K. (2016) Gender differences in environmental concern: reevaluating gender socialization. *Society & Natural Resources* 29(9): 1015–1031.

Sze, J. (2007) *Noxious New York: The Racial Politics of Urban Health and Environmental Justice.* Cambridge, MA: MIT Press.

Sze, J. (2017) Gender and environmental justice. In S. MacGregor (ed.), *The Routledge Handbook of Gender and Environment*, pp. 159–168. London: Routledge.

Taylor, D. (2014) *The State of Diversity in Environmental Organizations: Mainstream NGOs, Foundations, Government Agencies' Green 2.0.* http://orgs.law.harvard.edu/els/files/2014/02/FullReport_Green2.0_FINALReducedSize.pdf (accessed 15-11-2018).

UNEP (2016) *The Global Gender and Environment Outlook 2016.* Nairobi, Kenya: UN Environment.

UNFCCC (2017) *Gender Action Plan.* https://unfccc.int/topics/gender/workstreams/the-gender-action-plan (accessed 21-11-2019).

UN Women (2019a) *Progress on the Sustainable Development Goals: The gender snapshot 2019.* United Nations Entity for Gender Equality and the Empowerment of Women (UN Women); Department of Economic and Social Affairs (DESA). www.unwomen.org/en/digital-library/publications/2019/09/progress-on-the-sustainable-development-goals-the-gender-snapshot-2019

UN Women (2019b) *Women in Politics 2019.* United Nations Entity for Gender Equality and the Empowerment of Women (UN Women); Inter-Parliamentary Union. www.unwomen.org/en/digital-library/publications/2019/03/women-in-politics-2019-map

Walker, Gordon (2012) *Environmental Justice: Concepts, Evidence and Politics.* London: Routledge.

Westervelt, A. (2015) The medical research gender gap: how excluding women from clinical trials is hurting our health. *The Guardian,* www.theguardian.com/lifeandstyle/2015/apr/30/fda-clinical-trials-gender-gap-epa-nih-institute-of-medicine-cardiovascular-disease (accessed 21-11-2019).

Wilkinson, R. and Pickett, J. (2009) *The Spirit Level: Why Equality is Better for Everyone.* London: Penguin.

19 Labour unions and environmental justice

The trajectory and politics of just transition

Dimitris Stevis

Learning outcomes

- Learn about labour unions and the history of just transitions.
- Explore the variability of just transitions.
- Consider the complementarity between just transitions and environmental justice.
- Consider just transitions as labour unions' socioecological justice strategy.

Introduction

The strategy of just transition was originally developed by the Oil, Chemical and Atomic Workers' Union (OCAW) during the late 1980s as a solution to the adverse impacts of environmental policies on workers and communities dependent on environmentally harmful practices, particularly those involving chemicals. A number of factors make this a compelling fusion of social and environmental justice. First, the union representing this group of workers took the lead in calling for the orderly discontinuation of its sources of employment. Second, it did so because of the inequitable environmental impacts of those activities upon the workers themselves and frontline communities and society at large. Finally, it called for a transition that was deliberate and socially just.

Thirty years later, the inclusion of just transition in the 2015 **Paris Agreement**, followed by the December 2018 Silesia Declaration on Solidarity and Just Transition, has made the concept highly visible and highly contested. Yet, much of the emergent literature on just transitions appears unaware of its origins or views it as a corrective to the study of sociotechnical transitions (Stevis et al., 2020). In fact, when most people think about trade unions and the environment, they assume that unions will automatically choose jobs over the well-being of the environment.

My goal in this chapter is to outline how organic intellectuals within the world of labour have sought to promote **just transition** as a creative solution that fuses social and environmental justice, while at the same time recognizing that just transition is increasingly contested within the world of labour and beyond. In the first part of this chapter, I outline the parameters of just transition and clarify the analytical scheme employed in tracing its trajectory, variability and promise. In the second and main part, I divide the trajectory of just transition into four distinct periods, and in the final part, I close by briefly sketching out broader implications for the study of just transition and environmental justice.

Parameters and analytical scheme

Parameters

The fact that just transition has often been associated with environmental transition is an accident of history. Just transitions are also needed in response to relocations, automation, obsolescence and any other processes of change. Moreover, its increasing association with coal, energy and climate policies should not obscure the need for just transitions across the entire environment. As far as justice goes, just transition is a subcategory of environmental justice as it concerns all dimensions of justice, including the **distribution** of harms and benefits (see Chapter 3), the correction or prevention of these maldistributions, and how we treat nature itself. Moreover, just transitions also deal with transitions and, in that sense, they add an important dimension to justice itself (Newell and Mulvaney, 2013).

Just transitions, as specific policies, are most needed in the absence of a solidaristic political economy. In countries with a significant social safety net, just transition strategies are part of the fabric of society. It is not surprising that the quest for just transition first emerged in a liberal country as it was becoming neoliberal, and that it is now emerging in erstwhile solidaristic societies as they are also becoming neoliberal. This interpretation is supported by the historical record. The 1960s and 1970s were an era of fermentation and experimentation during which what was viewed as 'possible'—both nationally and globally—was much more ambitious than it has been since the 1980s (in relation to environmental justice, see Rector, 2014, 2018; Purdy, 2018). It is not surprising that during that period, the competing superpowers crushed the two most hopeful experiments in socialist democracy—those of Chile and Czechoslovakia. The point here is that the trajectory of just transition should not be viewed as a smooth path, setting off from a low level of ambition and travelling upwards to an increasingly higher one.

A third clarification is that the strategy of just transition was not conceived as a corrective to sociotechnical or other transition approaches. Its inspiration came from organic intellectuals within the labour and environmental justice movements before transition studies took off in the late 1990s, and it has followed its own trajectory within environmental labour studies.[1] The earliest literature on just transitions was by activists and scholars in industrial relations and, to a lesser degree, environmental justice (Young, 1998; Cohen-Rosenthal, 1998; Gould et al., 2004). It is noteworthy that otherwise comprehensive and astute recent reviews of transition literatures do not mention (Loorbach et al., 2017) or analyse just transitions (Temper et al., 2018). While the study of just transition may be fruitfully combined with various transition approaches, particularly just sustainability transitions, it should not be thought of as an academic corrective of any shortcomings that these approaches may have. Rather, just transitions should be viewed as more grounded in social and environmental justice (Stevis and Felli, 2016; Evans and Phelan, 2016).

Voice and choice

In examining various forms of environmental justice, and justice in general, analysts frequently turn to the concepts of recognition, participation and distribution (Schlosberg, 2007; Agyeman et al., 2016). This is a useful heuristic, provided that these dimensions are not reified and associated with particular categories of people; e.g., recognition for gay or indigenous people and distribution for worker and socialist movements (for discussions, see Olson, 2008). In reality, every act of justice or injustice involves historical combinations of (mis)recognition, (non)participation and (mal)distribution. Labour unions, for instance, sought recognition and participation in order to achieve distribution. Moreover, the distribution they often sought was immaterial, such as shorter working hours and better education. The goals of gay people are not necessarily exhausted once recognition is achieved, but, also, involve participation and the right to redistribute power away from those who discriminate against them.

This is more apparent once we recognize that no category of people is homogeneous. Liberal unionists, like liberal feminists or environmentalists, may well prioritize recognition because they already have access to participation and are satisfied with the existing distribution of power. Accordingly, I employ these intersectional dimensions but reorganize them to ask who has **voice** (recognition and participation) and who has **choice** (what issues are on the agenda, including but not limited to conventional distributional issues). The terminology comes from industrial relations, where it is used to examine employee participation and flexible working (Tapia et al., 2015; Barry and Wilkinson, 2016), and more strongly connects procedure and substance while foregrounding a relational approach to justice and power. Here, and in earlier work, I have expanded these terms to apply to the broader political economy (Stevis, 2000).

Voice refers to who is recognized, particularly by those responsible and powerful, and who can participate in deliberations affecting themselves or others, including the most powerful. Mapping who is recognized amongst people and in the natural world forces us to expand our ethical horizons and is a necessary precondition for a more inclusive approach to justice and democracy (Low and Gleeson, 1998; Young, 2006). Recognition may be a precondition, but it does not ensure participation, while participation does not necessarily bring with it equality of voice. Therefore, it is possible to have broad recognition and broad participation—broad voice—but within more or less narrow choice.

Choice requires that we reflect upon participation. On the one hand, challengers of an order may be allowed to participate only if they temper their demands, or particular categories of challengers may be 'represented' by those more comfortable with managing as opposed to changing the existing order. Such arrangements allow narrow choice but can serve to legitimate policy outputs and outcomes by producing a hegemonic order that is supported by enough stakeholders. Broad choice, on the other hand, means that powerful stakeholders come to the table with the clear understanding that their privileged position may, in fact, be taken away from them through redistribution or even the reorganization of the relationship. Additionally, broad choice also means that, in the language of policy-making, the preferences of weaker stakeholders are now considered realistic policy outputs. My focus thus remains within what commentators recognize as the procedural dimension of justice, but stretches that dimension by highlighting the fact that procedural politics are also about the politics of distribution. In my view, moreover, the broadening of voice and choice is the result of political organization and contestation and the formation of compelling counterhegemonic alliances. It is rarely, if ever, the result of more deliberation and, when it is discretionary, it serves as a strategy of co-optation.

Voice, choice and ambition in just transitions

Who proposes just transition, and for whom is it proposed? These questions address **recognition** at two levels.[2] Are the proposals only by and for some workers affected, or are they for most of those affected, including communities and society at large? Are they solely focused on the distribution of harms and benefits amongst humans, or is there also explicit recognition of their impacts on nature? Combined, these questions should aim at the delineation of the spatial and temporal scale of a just transition proposal, as well as its scope: namely, who is actually recognized across space and time (Mulvaney, 2014; Mertins-Kirkwood and Deshpande, 2019). Conflicts over environmental justice at one location are often resolved by displacing their impacts across space, or across time by privileging some stakeholders and marginalizing others. White workers in extraction or manufacturing, for instance, become the working class and the model workers—in the process making invisible women and those larger numbers of men working along the supply chain; or white women who were campaigning for emancipation, but were not accepting of black women, came to represent all women during the late 19th century and much of the 20th century.[3]

The second dimension of voice is that of **participation**.[4] By whom will a just transition proposal be adopted and implemented, and how will this be done? In general terms, public policies in minimally democratic societies allow broader participation than private policies. Citizens can decide on the basis of criteria and values other than profit or the overestimated efficiency of corporate practices, for example, with respect to religion, nationalism, ideology, health or family, and so on.

As noted, choice is not only about the distribution of material harms and benefits, but also about power. Broad choice, in that vein, implies that the range of policy outputs is not predestined by the existing order of things but can envision both their redistribution, along with the possible reorganization of the social relations that produce environmental inequality. Here, we are expanding the realm of procedural justice by asking what policy options are on the agenda and which ones are marginalized. Must it be within the parameters of the existing order of things, or does it envision more or less dramatic changes? Will the just transition policy compensate those left behind by environmental policies, while also providing just opportunities for all in the emerging green economy (Zabin et al., 2016; Piggot et al., 2019), or will it promote a green growth strategy in which just transition is a collateral but not guaranteed benefit for those left behind (Pollin et al., 2008; Jacobs, 2012; Tienhaara, 2014)? And what about the means to be employed? Will they be public or private (Sweeney and Treat, 2018)? And will the measures and standards adopted encourage green growth or an alternative approach to our relations with nature and each other (Lohman, 2009)?

Transformative changes are not necessarily egalitarian or ecological (Hopwood et al., 2005). Those that are egalitarian and ecological have involved particular historical configurations and tensions of voice and choice (Ciplet and Harrison, 2019). Accordingly, more environmental justice is likely if nature is valued and policies are aimed at a more egalitarian distribution of harms and benefits amongst humans *and* between humans and nature—what is termed '**ecological justice**'. The question is then: does a just transition proposal address both social and environmental justice, and how does it achieve this? Does the proposal aim to address only market or state crises, or does it envision a more equitable reorganization of the broader political economy? A number of analysts have sought to separate out different types of just transitions. Sweeney and Treat (2018) differentiate between social dialogue and social power approaches. Paul Hampton (2015) distinguishes between neoliberal, ecological modernization and Marxist/socialist approaches. The Just Transition Research Collaborative (2018) employed similar end points but sought to discriminate between reforms whose aim is to manage and stabilize the existing order—in this case, neoliberalism—and structural reforms, or revolutionary reforms in Luxemburg's language (O'Brien, 2019), and non-reformist reforms in Andre Gorz's language (Gorz, 1962; Bond, 2008), which are part of a 'war of position' towards more profound change. In general, the account that will be presented here highlights the variability of just transitions and cautions against the use of the '**transformation**' label to cover all forms of change and transition.

Labour unions and environmental justice: a prehistory

The conventional history traces environmental justice back to the early 1980s protests over siting hazardous facilities in Warren County, North Carolina, which forced the US to recognize the impacts of environmental racism. These protests, in turn, were part of disputes about hazardous facilities that can be traced further back into the 1970s.[5] During the 1970s, the term environmental justice was used by two other movements. One was a network of progressive religious activists who started using the term 'eco-justice' during the very early 1970s (Hessel, 2007). Eco-justice envisioned justice amongst humans and between humans and nature in a manner that is similar to what we associate today with critical and ecological justice. The term is still used in the US both amongst religious organizations and by organizations and networks that fuse social and ecological justice.[6]

A second line of thought came out of **social environmentalism** (for a historical overview, see Gottlieb, 2005). During the 1950s, social environmentalism was manifested in opposition to the adverse impacts of industrial production and nuclear tests. During the 1960s, it was associated with the growing movement to hold industrial producers accountable and to advance the cause of occupational health and safety.[7] Both were part of the mobilizations that permeated the US (and other parts of the world) at the time. Major promoters included some trade unions, as well as activists in Ralph Nader's network of organizations (Page and Sellers, 1970–71; Rector, 2018). Union concerns about the environment are therefore neither recent nor homogeneous (see Dewey, 2019). What is important here is that in 1976, the United Auto Workers hosted the Working for Economic and Environmental Justice and Jobs Conference (Bryant, 1997; Rector, 2014, 2017, 2018; interview with participant). This event brought together a variety of activists associated with the labour, urban environmental and social justice movements, and clearly sought to combine notions of social and environmental justice and direct them towards structural change within the parameters of a still conventionally liberal US.

What also became apparent during that era was that there were two polar opposite worldviews within the US trade union movement. On one side were those who called for comprehensive reform to address social and ecological challenges, such as the Autoworkers and the OCAW. On the other side were the more conservative and liberal unionists who were sceptical of social and environmental reform (Donahue, 1977). The interactions between unions, environmentalists, and social justice and community activists continued through the 1970s but declined during the 1980s. This turn was the result of conflicts between unions (Logan and Nelkin, 1980), as well as because of changes in the leadership of key unions, such as the Autoworkers and the OCAW.

The emergence of just transition: from the late 1980s to ~2001[8]

In the late 1980s, a shift in the internal politics of the OCAW—as well as the rapid decline in its membership and a successful strike against BASF Corp that drew support from environmentalists (Minchin 2002)—allowed the resurgence of labour environmentalism within the union. In 1989, it called for a single-payer national health care programme for the US, and at its 1991 Convention it "passed [a] resolution calling for 'A New Social, Political, and Economic Agenda' which set goals for the 1990s, including national health care, a Labor

Party alternative, environmental protection, a Superfund for Workers, and international trade unionism" (United Steelworkers Local 608 & 712, 2020). In fact, the Superfund for Workers (OCAW, 1991; Wykle et al., 1991) was the first incarnation of just transition. A key figure in this process was Tony Mazzochi, who was aware of and influenced by social environmentalism (Mazzocchi, 1993; Leopold, 2007; Montrie, 2018). During the 1960s and 1970s, he led efforts by the OCAW and the union movement towards occupational health and safety. In his view, some jobs were too damaging to society and nature, and ought to be discontinued. He also realized that workers and unions would not automatically recognize these connections, and that union environmentalism would require extensive work within their ranks.

The term 'just transition' was first used in 1995, when Les Leopold and Canadian unionist Brian Kohler (Leopold, 1995), both associated with the OCAW network, employed it during a presentation to the International Joint Commission on Great Lakes Water Quality.[9] The venue was not accidental as the vast bulk of US and Canadian manufacturing was taking place around these bodies of water. In Leopold's words (p. 83):

> We propose that a special fund be established, a just-transition fund which we've called in the past a Superfund for Workers. Essentially this fund will provide the following:
>
> 1 Full wages and benefits until the worker retires or until he or she finds a comparable job;
> 2 Up to four years of tuition stipends to attend vocational schools or colleges plus full income while in school;
> 3 Post-educational stipends or subsidies if no jobs at comparable wages are available after graduation;
> 4 Relocation assistance.

The fact that a network within and around the OCAW had a much broader vision is evident in the role of the union in the formation of the Just Transition Alliance (JTA) in 1997 and in the Blue/Green Working Group set up by the American Federation of Labor and Congress of Industrial Organizations (AFL-CIO) in 1996/1997. The JTA brought together environmental and social justice organizations that represented the most vulnerable and marginalized populations in the US. In addition to its participation in innovative training initiatives, the JTA was also involved in a number of specific local campaigns that sought to bring together workers and frontline communities to enhance unionization and raise awareness about their exposure to anti-union and environmentally unjust practices (Harvey, 2018; Slatin, 2009; Public Health and Labor Institutes, 2000; Young, 1998; interviews). While largely anthropocentric, this collaboration did not see nature as merely a route towards more social equality. While primarily motivated by the maldistribution of environmental harms, it did not seek to solve them by simply displacing the problem across space or time, but instead sought to address the key causes of those harms, including by downsizing the production of toxins.

For a short period of time during the late 1990s, it seemed as if the US environmental and labour movements would, in fact, adopt a common environmental agenda in which just transition would be a major component. In 1997, despite the opposition of the United Mineworkers and the building and construction unions, the AFL-CIO established a Blue/Green Working Group coordinated by Jane Perkins, a unionist and the first woman executive director of Friends of the Earth, US. The Group brought together unions and environmentalists over six years and sought to produce a comprehensive synthesis in which just transition played a central role (Barrett et al., 2002). However, it turned out that some of the participants were more interested in preventing just transition policies, rather than advancing them. Their opposition, along with the election of an anti-environmentalist president and the nationalism that followed the events of September 11, 2001 ensured that just transition fell off the agenda of the US union and environmental movements. While the collaboration between unions and environmentalists did not end completely, it did move in the direction of green growth and away from emphasizing justice.

In terms of the analytical scheme employed here, the OCAW and the Blue/Green Working Group sought to expand the voice of workers and environmental justice advocates and broaden choice by placing equitable transition, through large-scale public policy, on the agenda. The strategic goal was to promote structural reform and stem the tide of neoliberalism in a manner that included both social and environmental justice. Despite discussions between OCAW, and then the Blue/Green Working Group, with the administration of President Bill Clinton, just transition did not become an official policy even after the institutionalization of environmental justice within the Environmental Protection Agency. Nor did any elements of US business break with its strongly anti-union and anti-environment attitudes—even symbolically, as the World Business Council for Sustainable Development had done in 1992, paving the way for business environmentalism. As a result, the voices of unions and environmental justice advocates were constrained by the unwillingness of the most powerful social forces to engage with them. Mazzocchi and his supporters were very much aware of that reality and sought to respond by forming a Labor Party, albeit without success.

The globalization of just transition: ~2001–2013

Before its demise in the US, the concept of just transition had already found its way into the efforts of **global union organizations**[10] which were seeking to shape a labour union policy with respect to sustainable development and climate change (Gereluk and Royer, 2001; Silverman, 2006; Rosemberg, 2020; Hampton, 2015). The results of these efforts became apparent in 2009 when the International Trade Union Confederation (ITUC) explicitly connected just transition to climate policy and included it in its broader set of demands in preparation for the Copenhagen climate negotiations (ITUC, 2009). Overall, the role of global union organizations—coordinated by Anabella Rosemberg of the ITUC, in collaboration with a network of national unions—was highly significant (Rosemberg, 2010).

A central force was the Spanish Comisiones Obreras (CCOO) which, under the leadership of Joaquin Nieto, had adopted a proactive approach to the environment and climate as far back as the late 1980s. Environmentalism was institutionalized within the union with the formation, in 1996, of El Instituto Sindical de Trabajo, Ambiente y Salud (ISTAS) (Gil, 2012). ISTAS served as a research and education arm that would allow the union to integrate environmental and health and safety priorities into its practices and collective agreements. In 2004, CCOOs/ISTAs formed Sustainlabour, a labour environmentalist NGO (Martin Murillo, 2012) which, over its 12-year existence, played a critical role in promoting labour environmentalism at a global level. One important step was the 2006 Trade Union Assembly on Labour and the Environment (Rosemberg, 2020). The meeting brought together unions from around the world, as well as global union organizations. Sustainlabour and the ITUC continued their collaboration, producing training events, including a second Trade Union Assembly in 2012, and material that increasingly integrated the narrative of just transition (Sustainlabour and UNEP, 2008).

Another important lead player in the diffusion of just transition was the British Trades Union Congress (TUC), which had engaged with sustainable development and the environment since the late 1980s.[11] Under the guidance of Philip Pearson, the TUC (2008) published 'A Green and Fair Future: For a Just Transition to a Low Carbon Economy', the result of systematic research and union deliberation (Hampton, 2015). Just transition debates also emerged in Australia, which, as an extractive economy, faces significant challenges related to climate, as well as pollution (much of its coal is lignite), and during the early years of the millennium, unions were active in exploring just transition solutions to these problems (Evans, 2007; Australian Manufacturing Workers' Union, 2008; Snell and Fairbrother, 2012). While just transition began largely as a strategy promoted by unions in the Global North, it did have some diffusion into the Global South, one prominent example being a systematic training programme undertaken by the National Union of Metalworkers of South Africa (Cock, 2011; personal communications).

From 2001–2007, union proposals were offered as alternatives to neoliberalism and as necessary steps towards a green transition that was effective and just. But the Great Recession, from late 2007 until the early 2010s, led many to argue that the global financial crisis provided an opportunity for ambitious programmes that would address, primarily, the economy and the environment (Pollin et al., 2008; Renner et al., 2008; TienHaara, 2014). National and global unions were active participants in this turn, some pushing for a just transition and others sidestepping the issue (Räthzel and Uzzell, 2012; Felli, 2014; Stevis and Felli, 2015). The AFL-CIO, for instance, set up an Energy Working Group which avoided the term.

During this third period, we again see the tension between strategies that sought structural transitions that fused justice and green industrial policy, such as those proposed by the Comisiones Obreras and the TUC, and strategies that centred around green industrial policies, such as those emanating from a number of US unions, intergovernmental organizations and various proponents of green new deals. The second group aimed at the managerial reform of a capitalism in crisis with justice reduced to a collateral outcome of green growth.

At this time, the voice of unions with respect to just transition expanded marginally because of the positive responses of some **intergovernmental organizations** (IGOs), particularly the United Nations Environment Programme and the International Labour Organization (ILO), and some participation by the International Organization of Employers.[12] However, neither any of the major IGOs or employer organizations, nor any states, engaged in a dialogue over just transition, while most green economy proposals tended towards green capitalism. Civil society was also largely uninvolved. Nevertheless, this was a still a period during which unions, primarily those with a deeper commitment to social and environmental justice, were seeking to raise their voices so as to influence choice, increasingly with respect to climate policy. While the initiatives by the supportive IGOs did open a small window to broader voice and choice, it was largely symbolic, especially with respect to social and ecological justice.

The proliferation of just transition: ~2013–present

The narrative of just transition has now proliferated across a range of social forces, including more **labour unions**, environmentalists, social justice organizations, advocacy organizations, IGOs, corporate alliances and think-tanks, and climate negotiations (Stevis et al., 2020; Mousu, 2020; International Labour Organization, 2015; Harvey, 2018; World Bank,

2018; International Labour Office, 2018; Jenkins, 2019). But there are serious limitations to this proliferation because the narrative is still less prominent or even completely absent in many countries—and also many union movements—particularly in the Global South (Just Transition Research Collaborative, 2018; Satgar, 2018; Hirsch et al., 2017; Morena et al., 2020), while many in the Global North, including unions, are opposed to just transition or do not consider it relevant for them. There are also cases of unions which at one time were leaders in the movement but have now backtracked significantly or completely. And in the US, where just transition has re-emerged, it has not done so at the same pace or in the same form as before. Finally, there is a more pernicious strategy that is being utilized by capital and its allied unions; namely, the use of just transition to stall climate or other green policies.

Global union organizations have continued to promote just transition (Rosemberg, 2020; Stevis et al., 2020), and in 2016, the ITUC formed the Just Transition Centre which works closely with two global business networks (The B Team and We Mean Business) and places great emphasis on collaborative industrial relations (Just Transition Centre, 2017; Just Transition Centre and The B Team, 2018). The active role of corporate associations has the effect of expanding the recognition of workers and unions by the social actors most directly in contact with them. But this shift has come at the cost of closing Sustainlabour, and thus weakening efforts at formulating a more autonomous labour voice on the environment. A number of national unions have also continued to engage with just transition—not always smoothly—including unions in some European countries, Canada, Argentina, South Africa, Australia, New Zealand and elsewhere. In some prominent cases, like Germany, it is not clear that what is promoted as a just green transition is actually so (Reitzenstein et al., 2020). Overall, however, just transition still remains predominantly a Global North policy agenda amongst labour unions, with pockets of engagement from the Global South (Satgar, 2018; Hirsch, 2017).

Significantly, community and environmental justice organizations and networks, as well as philanthropic organizations, particularly in the US, have been prominent in promoting just transition. The case of the resurgence of just transition in the US is interesting because unions are still quite sceptical—despite the efforts of networks such as the Labor Network for Sustainability (2016), the People's Climate Movement and the Climate Justice Alliance (2017)—about placing just transition on the agenda of the union movement. The People's Climate Movement, for example, does include some unions, while others have played a key role at sub-federal levels—including a visionary if failed policy proposition in Washington state and a successfully adopted policy in Colorado. The promotion of a Green New Deal (US Congress, 2018) in the US provides a potential common platform for progressive unions and activists, even though conservative unions have come out against it. That just transition remains a sensitive issue amongst US unions is evident by the public positions of the Blue/Green Alliance—the result of the efforts of the late 1990s to bring unions and environmentalists together. While questions of equity are important for the Blue/Green Alliance (2019), it has largely avoided the narrative of just transition because many unions, including some of its members, are opposed to it.

As the narrative of just transition proliferates, one would expect to see more proposals that fuse compensatory provisions for those left behind with provisions for just green economies. Such proposals continue to emanate from various elements of civil society and some unions, but the most prominent proposals and policies addressing climate focus on coal (e.g., World Bank, 2018) and are largely compensatory rather than forward-looking. This is not to suggest that workers in coal will not be immediately and deeply affected, but just transitions are also necessary with respect to other fuels, increasingly used in the Global North. Equally worrisome is the association of just transition only with climate and not with the total environment. Unjust transitions in mining, forestry, agriculture or other sectors are also likely to breed divisions amongst workers and unions, and opposition to the strategy of just transition.

During this final period, then, it would seem that there has been a broadening of voice and choice because other social forces have engaged the just transition narrative. However, this is true to a more limited degree than appearances would suggest. While a wider variety of stakeholders have joined the debate, this has not translated into a commensurate broadening of voice and choice, leaning instead in the direction of corporate social responsibility and top-down policies. While there do exist organizations and networks that advocate structural and transformative change, none of the proposals and policies involving governments and IGOs, as well as most global and national unions, advance a reconfiguration of voice and choice beyond managerial reform. In fact, the proliferation of voice and the marginal openings in choice could well mislead us into overestimating the ambition of many actual policies and prominent proposals. Those social forces that do want to advance a more profound just transition—and there are many of them (see JTRC, 2018; Barca, 2016)—are now facing not only the rejection of just transition by various unions and other social forces but, also, its selective appropriation and the weakening of its ambition (for an overview, see Morena et al., 2020; Piggot et al., 2019).

Closing comments

I have examined the trajectory of just transition in terms of voice and choice and their combination, or their ambition. Over the last 30 years, voice—in terms of recognition and participation—has expanded in two ways: first, in terms of the varieties of stakeholders that have adopted or proposed just transitions; and second, in terms of the willingness of states and capitals to engage advocates, albeit within clearly delineated terms. Thus, the broadening of voice has not been accompanied by a commensurate broadening of choice. A key assumption underlying the analytical scheme employed here is that our analysis must be grounded historically. A transformational policy based on profound, if divergent, expansions of voice and choice may well be socioecological, technological, or authoritarian. Our world is currently facing competing paths—also evident in past transformative periods, as a close reading of Polanyi's *Great Transformation* (whose 75th anniversary we have just celebrated) reveals (Hultgren, 2019; Brand et al., 2019). All paths involve significant contestation and conflict as power migrates from some social forces to others. An egalitarian, ecological transformation calls for the kinds of voice and choice that reduce inequality and marginalization within the parameters of nature. In these terms, it is not clear to me whether the most prominent just proposals emerging from unions—as well as states, capital, philanthropists and most social movements—are more transformative than the proposals of the 1990s and early 2000s, when unions and activists with democratic socialist and socialist ideologies were spearheading the effort. Such proposals still exist, if only at the margins of a growing mainstream. Equally important is the fact that some current proposals that seem transformative—because of the varieties of voices and choices associated with them—may actually be rather modest in those terms and in their ambition.

Within those parameters, I would like to close by repeating the key animating points behind this chapter. First, environmental justice analysts and advocates will be well served to engage with labour environmentalism. Despite the crises that labour has been facing—many of its own making—it remains one of the largest movements and, in any event, it will never be possible to achieve just green transitions without the active engagement of unions and a reconsideration of the nature and organization of work. Labour environmentalists are already facing significant opposition within the labour movement. The uncritical adoption of the 'jobs vs environment' myth, therefore, obscures and marginalizes some of the most promising proponents of socioecological justice. Second, and related, just transition has been the product of labour environmentalists and their allies over the past 30 years. It originated in the world of practice and was nurtured by elements of the labour movement. To dismiss their interpretations and efforts is to colonize that world and, in the process, misrecognize it. Finally, at its best, just transition integrates the most compelling and creative combination of social and ecological justice and one that is grounded in political realities and contestations. Not only has it widened the political viewpoints of workers and unions, but it has also begun to expand the horizons of other social forces by placing social and ecological justice at the centre of analysis and practice—exactly where it should be.

Follow-up questions

- What is just transition, and what does it add to environmental justice?
- How has the universe of actors proposing just transition changed over the last several decades?
- Has this broadening of voices been associated with the broadening of choices and, thus, more ambitious just transition proposals?
- Is it desirable to seek a common definition or operationalization of just transition?
- What are the relative merits of evaluating just transition proposals in terms of voice and choice, rather than recognition, participation and distribution (or procedure and substance)?
- Does the analytical scheme help us differentiate different types of just transitions and, thus, environmental justice?

Notes

1 While there was an engagement between labour environmentalism and sociotechnical systems (Cohen-Rosenthal 1997), the sociotechnical systems approach, which was developed in response to labour issues in the UK after WWII, has not been acknowledged by sociotechnical transitions analysts.
2 On recognition, see also Chapter 5 of this volume.
3 On environmental racism, see also Chapter 17 of this volume.
4 On participation and procedural justice, see also Chapter 4 of this volume.
5 On the history of environmental justice, see also Chapter 2 of this volume.
6 On eco-centric approaches of justice, see also Chapter 21 of this volume.
7 Occupational health and safety have been central to labour politics around the world since at least the 19th century. See Gottlieb (2005) and Asher (2014).
8 For a longer history of just transition, on which this part is based, see Stevis et al., (2020).
9 There was significant collaboration between OCAW and Canadian unions. On just transition in Canada during the late 1990s, see Canadian Labour Congress 2000; Bennett 2007.
10 Global union organizations refers to a small number of global organizations that bring together national unions.

11 According to scholars and activists, the 1970s Lucas experiment, and the vision behind it, can be considered one of the pioneering examples of labour environmentalism and just transition (Räthzel et al., 2010).
12 The International Organization of Employers represents employers at the ILO, but its power over business is even more limited that than that of the ITUC over its affiliates.

References for further reading

International Labour Organization. 2015. *Guidelines for a Just Transition Towards Environmentally Sustainable Economies and Societies for All*. ILO. www.ilo.org/wcmsp5/groups/public/—ed_emp/—emp_ent/documents/publication/wcms_432859.pdf (last accessed November 14, 2019).

International Trade Union Confederation. 2009. *A Just Transition: A Fair Pathway to Protect the Climate*. Brussels: ITUC. www.ituc-csi.org/IMG/pdf/01-Depliant-Transition5.pdf

Just Transition Research Collaborative. 2018. *Mapping Just Transition(s) to a Low-Carbon World*. www.unrisd.org/80256B3C005BCCF9/(httpPublications)/9B3F4F10301092C7C12583530035C2A5?OpenDocument

Mazzocchi, Tony. 1993. An Answer to the Jobs-Environment Conflict? www.greenleft.org.au/content/answer-jobs-environment-conflict

Morena, Edouard, Dunja Krause and Dimitris Stevis (eds.). 2020. *Just Transitions: Social Justice in the Shift Towards a Low-Carbon World*. Pluto Press.

References used in this chapter

Agyeman, Julian, David Schlosberg, Luke Craven and Caitlin Matthews. 2016. Trends and Directions in Environmental Justice: From Inequity to Everyday Life, Community and Just Sustainabilities. *Annual Review of Environment and Resources* 41: 421–420.

Asher, Robert. 2014. Organized Labor and the Origins of the Occupational Health and Safety Act. *New Solutions* 24(3): 279–301.

Australian Manufacturing Workers' Union. 2008. *Making Our Future: Just Transitions for Climate Change Mitigation*. Granville, Australia: AMWU National Office.

Barca, Stefania. 2016. Labor in the Age of Climate Change: Any Just Transition to a Green Economy Must Take Place in Labor's Terms—Not Capital's. *Jacobin*, March 18. www.jacobinmag.com/2016/03/climate-labor-just-transition-green-jobs/ (last accessed April 2019).

Barrett, James, J. Andrew Hoerner, Steve Bernow and Bill Dougherty. 2002. *Clean Energy and Jobs: A Comprehensive Approach to Climate Change and Energy Policy*. Washington, DC: Economic Policy Institute and the Center for a Sustainable Economy.

Barry, Michael and Adrian Wilkinson. 2016. Pro-Social or Pro-Management? A Critique of the Conception of Employee Voice as a Pro-Social Behaviour within Organizational Behaviour. *British Journal of Industrial Relations* 54(2): 261–284.

Bennett, David. 2007. Labour and the Environment at the Canadian Labour Congress—The Story of the Convergence. *Just Labour: A Canadian Journal of Work and Society* 10: 1–7.

Blue/Green Alliance. 2019. *Solidarity for Climate Action*. www.bluegreenalliance.org/wp-content/uploads/2019/07/Solidarity-for-Climate-Action-vFINAL.pdf (last accessed November 14, 2019).

Bond, Patrick. 2008. Reformist Reforms, Non-Reformist Reforms and Global Justice: Activist, NGO and Intellectual Challenges in the World Social Forum. In Judith Blau and Marina Karides (eds.), *The World and US Social Forums: A Better World Is Possible and Necessary*. Leiden: Brill, pp. 155–172.

Brand, Ulrich, Christoph Gorg and Markus Wissen. 2019. Overcoming Neoliberal Globalization: Social-ecological Transformation from a Polanyian Perspective and Beyond. *Globalizations*. https://www.Tandfonline.Com/Doi/Full/10.1080/14747731.2019.1644708

Bryant, Bunyan. 1997. The Role of the SNRE in the Environmental Justice Movement. *University of Michigan School of Natural Resources and Environment*, February 1. http://umich.edu/~snre49unyan2/history.html (last accessed April 2019).

Canadian Labour Congress. 2000. *Just Transition for Workers During Environmental Change*. Ottawa: Canadian Labour Congress.

Ciplet, David and Jill Harrison. 2019. Transition Tensions: Mapping Conflicts in Movements for a Just and Sustainable Transition. *Environmental Politics*, published online, March 31. https://doi.org/10.1080/09644016.2019.1595883.

Climate Justice Alliance. 2017. *Our Power Communities: Just Transition Strategies in Place*. https://climatejusticealliance.org/workgroup/our-power/ (last accessed May 2019).

Cock, Jacklyn. 2011. Contesting a 'Just Transition' to a Low Carbon Economy. *Global Labour Column*, 76. University of Witwatersrand. http://column.global-labour-university.org/2011/01/contesting-just-transition-to-low.html (last accessed May 2019).

Cohen-Rosenthal, Edward. 1997. Sociotechnical Systems and Unions: Nicety or Necessity. *Human Relations* 50(5): 585–604.

Cohen-Rosenthal, Edward. 1998. Labor and Climate Change: Dilemmas and Solution. *New Solutions* 8(3): 343–363.

Dewey, Scott. 2019. Working-Class Environmentalism in America. *Oxford Research Encyclopedias: American History*. https://oxfordre.com/americanhistory/abstract/10.1093/acrefore/9780199329175.001.0001/acrefore-9780199329175-e-690?rskey=JRyUtz&result=2 (last accessed November 14, 2019).

Donahue, Thomas. 1977. Environmental and Economic Justice. *EPA Journal* 3: 6–7, 26.

Evans, Geoff. 2007. A Just Transition from Coal to Renewable Energy in the Hunter Valley of New South Wales, Australia. *International Journal of Environment, Workplace and Employment* 3(3–4): 175–194.

Evans, Geoff and Liam Phelan. 2016. Transition to a Post-Carbon Society: Linking Environmental Justice and Just Transition Discourses. *Energy Policy* 99: 329–339.

Felli, Romain. 2014. An Alternative Socio-Ecological Strategy? International Trade Unions' Engagement with Climate Change. *Review of International Political Economy* 21(2): 372–398.

Gereluk, Winston and Lucien Royer. 2001. *Sustainable Development of the Global Economy: A Trade Union Perspective*. Geneva: International Labour Office.

Gil, Begoña María-Tomé Gil. 2012. Moving Towards Eco-unionism: Reflecting the Spanish Experience. In Noras Räthzel and David Uzzell (eds.), *Trade Unions in the Green Economy: Working for the Environment*. London: Routledge, pp. 64–77.

Gorz, Andre. 1962. *Strategy for Labor: A Radical Proposal*. Boston, MA: Beacon.

Gottlieb, Robert. 2005. *Forcing the Spring: The Transformation of the American Environmental Movement* (2nd ed.). Island Press.

Gould, Kenneth A., Tammy L. Lewis and J. Timmons Roberts. 2004. Blue-Green Coalitions: Constraints and Possibilities in the Post 9-11 Political Environment. *Journal of World-Systems Research* 10(1): 91–116.

Hampton, Paul. 2015. *Workers and Trade Unions for Climate Solidarity: Tackling Climate Change in a Neoliberal World*. New York, NY: Routledge.

Harvey, Samantha. 2018. Leave No Worker Behind: Will the Just Transition Movement Survive Mainstream Adoption? *Earth Island Journal*, Summer 2018. www.earthisland.org/journal/index.php/magazine/entry/leave_no_worker_behind/ (last accessed May 2019).

Hessel, Dieter. 2007. *Eco-Justice Ethics*. www.bluegreenalliance.org/wp-content/uploads/2019/07/Solidarity-for-Climate-Action-vFINAL.pdf (last accessed November 14, 2019).

Hirsch, Thomas, Manuela Matthess and Joachim Fünfgelt (eds.). 2017. *Guiding Principles & Lessons Learnt for a Just Energy Transition in the Global South*. Friedrich Ebert Stiftung. https://library.fes.de/pdf-files/iez/13955.pdf (last accessed May 2019).

Hopwood, Bill, Mary Mellor and Geoff O'Brien. 2005. Sustainable Development: Mapping Different Approaches. *Sustainable Development* 13(1): 38–52.

Hultgren, John. 2019. Polanyi Returns to Bennington: The Role of Liberal Arts Colleges at a Moment of 'Great Transformation'. *Europe Now*. www.europenowjournal.org/2019/03/04/polanyi-returns-to-bennington-the-role-of-liberal-arts-colleges-at-a-moment-of-great-transformation/ (last accessed November 14, 2019)

International Labour Office. 2018. *Just Transition Towards Environmentally Sustainable Economies and Societies for All—ILO ACTRAV Policy Brief*. Written by Béla Galgóczi, Senior Researcher at the European Trade Union Institute (ETUI). www.ilo.org/wcmsp5/groups/public/—ed_dialogue/—actrav/documents/publication/wcms_647648.pdf

International Labour Organization. 2015. *Guidelines for a Just Transition Towards Environmentally Sustainable Economies and Societies for All*. ILO. www.ilo.org/wcmsp5/groups/public/—ed_emp/—emp_ent/documents/publication/wcms_432859.pdf (last accessed November 14, 2019).

International Trade Union Confederation. 2009. *A Just Transition: A Fair Pathway to Protect the Climate*. Brussels: ITUC. www.ituc-csi.org/IMG/pdf/01-Depliant-Transition5.pdf (last accessed May 2019).

Jacobs, Michael. 2012. *Green Growth: Economic Theory and Political Discourse*. Center for Climate Change Economic and Policy Working Paper 108 and Grantham Research Institute on Climate Change and the Environment Working Paper 92. www.lse.ac.uk/GranthamInstitute/wp-content/uploads/2012/10/WP92-green-growth-economic-theory-political-discourse.pdf (last accessed May 2019).

Jenkins, Kirsten. 2019. *Implementing Just Transition after COP24*. Climate Strategies. https://climatestrategies.org/wp-content/uploads/2019/01/Implementing-Just-Transition-after-COP24_FINAL.pdf (last accessed May 2019).

Just Transition Centre. 2017. *Just Transition. A Report for the OECD*. Brussels: Just Transition Centre. www.oecd.org/environment/cc/g20-climate/collapsecontents/Just-Transition-Centre-report-just-transition.pdf (last accessed May 2019).

Just Transition Centre and The B Team. 2018. *Just Transition: A Business Guide*. www.ituc-csi.org/IMG/pdf/just_transition_-_a_business_guide.pdf (last accessed April 2019).

Just Transition Research Collaborative. 2018. *Mapping Just Transition(s) to a Low-Carbon World*. Geneva: Rosa-Luxemburg-Stiftung, University of London Institute in Paris and United Nations Research Institute for Social Development. www.unrisd.org/jtrc-report2018 (last accessed April 2019).

Labor Network for Sustainability & Strategic Practice: Grassroots Policy Project. 2016. *'Just Transition'—Just What Is It? An Analysis of Language, Strategies, and Projects*. www.labor4sustainability.org/wp-content/uploads/2016/07/JustTransitionReport-FINAL.pdf (last accessed May 2019).

Leopold, Les. 1995. *Statement at the International Joint Commission's 1995 Biennial Meeting on Great Lakes Water Quality 'Our Lakes, Our Health, Our Future'*. September 22–25, Duluth, Minnesota, pp. 80–84. https://legacyfiles.ijc.org/publications/C46.pdf (last accessed April 2019).

Leopold, Les. 2007. *The Man Who Hated Work but Loved Labor: The Life and Times of Tony Mazzocchi*. White River Junction: Chelsea Green Publishing Company.

Logan, Rebecca and Dorothy Nelkin. 1980. Labor and Nuclear Power. *Environment* 22(2): 6–13, 34.

Lohman, Larry. 2009. Toward a Different Debate in Environmental Accounting: The Cases of Carbon and Cost-Benefit. *Accounting, Organizations and Society* 34(3–4): 499–534.

Loorbach, Derk, Niki Frantzeskaki and Flor Avelino. 2017. Sustainability Transitions Research: Transforming Science and Practice for Societal Change. *Annual Review of Environment and Resources* 42: 599–626.

Low, Nicholas and Brendan Gleeson. 1998. *Justice, Society and Nature: An Exploration in Political Ecology*. New York: Routledge.

Martin Murillo, Laura. 2012. From Sustainable Development to a Green and Fair Economy: Making the Environment a Trade Union Issue. In Nora Räthzel and David Uzzell (eds.), *Trade Unions in the Green Economy: Working for the Environment*. London: Routledge, pp. 29–40.

Mazzocchi, Tony. 1993. An Answer to the Work-Environment Conflict? *Green Left Weekly*, 114, September 8. www.greenleft.org.au/content/answer-jobs-environment-conflict (last accessed November 2019).

Mertins-Kirkwood, Hadrian and Zaee Deshpande. 2019. *Who is Included in a Just Transition? Considering Social Equity in Canada's Shift to a Zero-carbon Economy*. Canadian Center for Policy Alternatives. www.policyalternatives.ca/publications/reports/who-is-included-just-transition (last accessed November 14, 2019).

Minchin, Timothy. 2002. *Forging a Common Bond: Labor and Environmental Activism During the BASF Lockout.* Gainesville, Florida: University of California Press.

Montrie, Chad. 2018. *The Myth of Silent Spring: Rethinking the Origins of American Environmentalism.* Berkeley, California: University of California Press.

Morena, Edouard, Dunja Krause and Dimitris Stevis (eds.). 2020. *Just Transitions: Social Justice in the Shift Towards a Low-Carbon World.* London: Pluto Press.

Mousu, Nils. 2020. Business in Just Transition: The Never-Ending Story of Corporate Sustainability. In Edouard Morena, Dunja Krause and Dimitris Stevis (eds.), *Just Transitions: Social Justice in the Shift Towards a Low-Carbon World.* London: Pluto Press.

Mulvaney, Dustin. 2014. Are Green Jobs Just Jobs? Cadmium Narratives in the Lifecycle of Photovoltaics. *Geoforum* 54: 178–186.

Newell, Peter and Dustin Mulvaney. 2013. The Political Economy of the 'Just Transition'. *The Geographical Journal* 179(2): 132–140.

O'Brien, Robert. 2019. Revisiting Rosa Luxemburg's Internationalism. *Journal of International Political Theory.* https://doi.org/10.1177/1755088219833416

OCAW. 1991. *Understanding the Conflict between Jobs and the Environment. A Preliminary Discussion of the Superfund for Workers Concept.* Denver: OCAW.

Olson, Kevin (ed.). 2008. *Adding Insult to Injury: Nancy Fraser Debates Her Critics.* London, UK: Verso Press.

Page, Joseph and Gary Sellers. 1970–71. Occupational Safety and Health: Environmental Justice for the Forgotten American. *Kentucky Law Journal* 59(1): 114–144. https://uknowledge.uky.edu/cgi/viewcontent.cgi?article=2664&context=klj

Piggot, Georgia, M. Boyland, A. Down and A.R. Torre. 2019. *Realizing a Just and Equitable Transition from Fossil Fuels.* Stockholm Environment Institute (SEI). www.sei.org/publications/just-and-equitable-transition-fossil-fuels/

Pollin, Robert, Heidi Garrett-Peltier, James Heintz and Helen Scharber. 2008. *Green Recovery: A Program to Create Jobs and Start Building a Low-Carbon Economy.* Amherst, MA: University of Massachusetts-Amherst, Department of Economics and Political Economy Research Institute (PERI).

Public Health and Labor Institutes. 2000. *A Just Transition for Jobs and the Environment. Training Manual.* New York: The Public Health and Labor Institutes.

Purdy, Jeremiah. 2018. The Long Environmental Justice Movement. *Ecology Law Quarterly* 44(4): 809–864.

Räthzel, Nora and David Uzzell (eds.). 2012. *Trade Unions in the Green Economy. Working for the Environment.* Abingdon, NY: Routledge.

Räthzel, Nora, David Uzzell and David Elliott. 2010. The Lucas Aerospace experience: Can Unions become Environmental Innovators? *Soundings* 46: 76–87.

Rector, Josiah. 2014. Environmental Justice at Work: The UAW, the War on Cancer, and the Right to Equal Protection from Toxic Hazards in Postwar America. *The Journal of American History* 101(2): 480–502.

Rector, Josiah. 2017. Labor and the Environmental Justice Movement: Why Their Shared History Matters Today. *Process: A Blog for American History.* www.processhistory.org/rector-environmental-justice/

Rector, Josiah. 2018. The Spirit of Black Lake: Full Employment, Civil Rights, and the Forgotten Early History of Environmental Justice. *Modern American History* 1(1): 45–66.

Reitzenstein, Alexander, Sabrina Schulz and Felix Heilmann. 2020. The Story of Coal on Germany: A Model for Just Transition in Europe? In Edouard Morena, Dunja Krause and Dimitris Stevis (eds.), *Just Transitions: Social Justice in the Shift Towards a Low-Carbon World.* London: Pluto Press, pp. 151–171.

Renner, Michael, Sean Sweeney and Jill Kubit. 2008. *Green Jobs: Towards Decent Work in a Sustainable, Low-Carbon World.* Nairobi: UNEP/ILO/IOE/ITUC. www.ilo.org/wcmsp5/groups/public/@dgreports/@dcomm/documents/publication/wcms_098504.pdf (last accessed April 2019).

Rosemberg, Anabella. 2010. Building a Just Transition: The Linkages between Climate Change and Employment. *International Journal of Labour Research* 2(2): 125–161.

Rosemberg, Anabella. 2020. No Jobs on a Dead Planet': The International Trade Union Movement and Just Transition. In Edouard Morena, Dunja Krause and Dimitris Stevis (eds.), *Just Transitions: Social Justice in the Shift Towards a Low-Carbon World*. London: Pluto Press, pp. 32–55.

Satgar, Vishwas (ed.). 2018. Trade Union Reponses to Climate Change and Just Transition. *South African Labour Bulletin* 42(3).

Schlosberg, David. 2007. *Defining Environmental Justice: Theories, Movements, and Nature*. Oxford: Oxford University Press.

Silverman, Victor. 2006. 'Green Unions in a Grey World': Labor Environmentalism and International Institutions. *Organization & Environment* 19(2): 191–213.

Slatin, C. 2009. *Environmental Unions: Labor and the Superfund*. Amityville, NY: Baywood Publishing Company.

Snell, Darryn and Peter Fairbrother. 2012. Just Transition and Labour Environmentalism in Australia. In Noe Räthzel and David Uzzell (eds.), *Trade Unions in the Green Economy: Working for the Environment*. London: Routledge, pp. 146–161.

Stevis, Dimitris. 2000. Whose Ecological Justice? *Strategies* 13(1): 63–76.

Stevis, Dimitris and Romain Felli. 2015. Global Labour Unions and Just Transitions to a Green Economy. *International Environmental Agreements: Politics, Law and Economics* 15(1): 29–43.

Stevis, Dimitris and Romain Felli. 2016. Green Transitions, Just Transitions? Broadening and Deepening Justice. *Kurswechsel* 3: 35–45.

Stevis, Dimitris, Edouard Morena and Dunja Krause. 2020. Introduction: The Genealogy and Contemporary Politics of Just Transitions. In Edouard Morena, Dunja Krause and Dimitris Stevis (eds.), *Just Transitions: Social Justice in the Shift Towards a Low-Carbon World*. London: Pluto Press.

Sustainlabour and United National Environmental Programme. 2008. *Climate Change, its Consequences on Employment and Trade Union Action. A Training Manual for Workers and Trade Unions*. Nairobi: UNEP.

Sweeney, Sean and John Treat. 2018. *Trade Unions and Just Transition. The Search for a Transformative Politics*. TUED Working Paper No. 11. New York: Trade Unions for Energy Democracy, Rosa Luxemburg Stiftung—New York Office and Murphy Institute. http://unionsforenergydemocracy.org/inside-the-just-transition-debate-new-tued-working-paper-examines-union-approaches/ (accessed last November 2019).

Tapia, Maite, Christian Ibsen and Thomas Kochan. 2015. Mapping the Frontier of Theory in Industrial Relations: The Contested Role of Worker Representation. *Socioeconomic Review* 13(1): 157–184.

Temper, Leah, Mariana Walter, Iokiñe Rodriguez, Ashish Kothari and Ethemcan Turhan. 2018. A Perspective on Radical Transformations to Sustainability: Resistances, Movements and Alternatives. *Sustainability Science* 13: 747–764.

Tienhaara, Kyla. 2014. Varieties of Green Capitalism: Economy and Environment in the Wake of the Global Financial Crisis. *Environmental Politics* 23(2): 187–204.

Trades Union Congress. 2008. *A Green and Fair Future: For a Just Transition to a Low Carbon Economy*. Touchstone Pamphlet No. 3. London: Trades Union Congress. www.tuc.org.uk/sites/default/files/documents/greenfuture.pdf (last accessed May 2019).

US Congress. House Resolution 109. 2018. *Recognizing the Duty of the Federal Government to Create a Green New Deal*. 116th Congress. www.congress.gov/bill/116th-congress/house-resolution/109 (last accessed May 2019).

USW Steelworkers Local 608 & 712. 2020. *Who We Are?* https://usw-608.com/who-we-are (last accessed March 30, 2020).

World Bank. 2018. *Managing Coal Mine Closure: Achieving a Just Transition for All*. www.worldbank.org/en/topic/extractiveindustries/publication/managing-coal-mine-closure (last accessed April 2019).

Wykle, Lucinda, Ward Morehouse and David Dembo. 1991. *Worker Empowerment in a Changing Economy: Jobs, Military Production and the Environment*. New York: The Apex Press.

Young, Jim. 1998. Just Transition: A New Approach to Jobs vs Environment. *Working USA* July/August: 42–48.

Young, Iris Marion. 2006. Responsibility and Global Justice: A Social Connection Model. *Social Philosophy & Policy* 23(1): 102–130.
Zabin, Carol, Abigail Martin, Rachel Morello-Frosch, Manuel Pastor and Jim Sadd. 2016. *Advancing Equity in California Climate Policy: A New Social Contract for Low-Carbon Transition.* http://laborcenter.berkeley.edu/pdf/2016/Advancing-Equity-Executive-Summary.pdf

20 Indigenous environmental justice
Anti-colonial action through kinship

Kyle Whyte

Learning outcomes

- Learn about Indigenous peoples, ancient traditions of kinship that are relevant to understanding environmental justice.
- Understand that Indigenous environmental justice struggles are anti-colonial.
- Explore kinship as one way of understanding solutions to environmental injustice.

Introduction

Nearly four hundred million Indigenous peoples live on 22% of the world's land surface. These lands are tied to about 80% of the planet's biodiversity (Sobrevila, 2008, p. xii). The term **Indigenous peoples** refers to a particular social situation in which some populations of the world live. Societies define themselves specifically as Indigenous peoples to express the fact that they exercise their own political and cultural self-determination on homelands of their own (International Labour Organization, 1989, Manuel, 1974, Sanders, 1977, Tauli-Corpuz and Mander, 2006). Self-determination refers to a society's capacity to pursue freely its own plans in ways that support the aspirations and needs of its members (United Nations General Assembly, 2007).

Yet, at some point in recent history, one or more other societies colonized the lands of Indigenous peoples. These colonizing societies now claim—often under the guise of a nation-state like Canada or Australia—to be the dominant governance authority. The colonization process, whether several hundred years ago or more recently, typically involves warfare, forced assimilation, land dispossession, asset seizure, financial deprivation, workforce discrimination, sexual and gendered violence, and food and health insecurities. Today, Indigenous peoples work tirelessly to overcome the impacts of colonialism and to affirm their self-determination (Anaya, 2004, Sanders, 1977, Tauli-Corpuz and Mander, 2006).

Consider my own background as an example. I am Potawatomi, which is an Anishinaabe/Neshnabé group, and a member of the Citizen Potawatomi Nation. The Citizen Potawatomi are an Indigenous people living within the borders of what is currently called the United States. The US is a nation-state which colonized us and numerous other Indigenous peoples in order to acquire the land base it presently claims to exercise sovereignty over. Yet Potawatomi peoples continue, despite the dominance of the US Today, one way the Citizen

Potawatomi Nation expresses self-determination is through having its own government. Our self-governance started generations prior to the year the US was established in 1776. In fact, the US signed treaties with our ancestors. Treaties are political instruments reserved for legal agreements between self-governing societies. So, as a nation, we have governed ourselves over the course of generations. Our contemporary self-governance is apparent through our managing health and environmental programs, holding political elections, and investing in economic growth opportunities, among other governance activities.

We have our own culture, which stems from our homelands in the Great Lakes region. Our culture includes linguistic, artistic, ceremonial, religious, and philosophical traditions. It is a culture that is distinct from the other cultures of populations in the US, though, like all cultures, ours is internally diverse, having multiple variations and embracing persons with wide-ranging beliefs and lifestyles. Part of our culture stems from traditions tied to maintaining woodland ecosystems with meadow-like openings for plants and animals to flourish, planting corn, and protecting freshwater and wetlands for fish, wild rice, and medicinal plants. We have a website today that shows examples of our culture and self-governance for persons who are unfamiliar with Indigenous peoples: www.potawatomi.org.

US colonialism poses great harms to our self-governance and culture. Our original homelands in the Great Lakes region, including the woodlands, openings, and wetlands, were largely destroyed to make way for US and Canadian agricultural, recreational, and industrial zones. US settlers created boarding schools to strip our children of their culture. In the 19th century, the US forced the people who now are members of the Citizen Potawatomi Nation to relocate over a thousand miles away to lands in the Plains region that had little ecological resemblance to the Great Lakes.

Today I (personally) live in what is currently called Michigan, which is a location in our original homelands. Yet, the nation I am part of as an enrolled member is in Oklahoma, with the headquarters of the Citizen Potawatomi Nation being the city of Shawnee, Oklahoma. Potawatomi people still live in the Great Lakes region, with four different Potawatomi nations in what is currently called Michigan who exercise self-government, and even more nations and communities in what are currently called Ontario and Wisconsin.

Given this context in which I work as a Potawatomi relative, advocate, and scholar, I want to introduce the readers of this chapter to a particular dialogue that Indigenous peoples have with each other about environmental justice. As someone involved, both locally and globally, with Indigenous environmental justice advocacy, I hope to convey some of how we Indigenous persons dialogue together regarding environmental justice, seek solidarity with one another, and share knowledge. I hope to refer both to traditions and knowledge stemming from Potawatomi society, but also connect that with the traditions and knowledge systems of others. One such Indigenous tradition of environmental justice I wish to convey in this chapter is **kinship**.

Kinship is a large topic, and I so will focus here on just one aspect of it: reciprocity. In general, kinship refers to qualities of the relationships we have with others—whether others are humans, plants, animals, fishes, insects, rocks, waterways, or forests. Indigenous peoples with kinship traditions are unique cultures that have their own distinct ways of naming and relating to beings and entities in the world. Holders of these traditions talk quite a bit with each other about some of the different ways they have come to understand kinship (for example, see TallBear, 2019, Todd, 2017, Napoleon, 2013, Borrows, 1997).

Indigenous traditions focus on the details of how our bonds with others really work. Kin relations are like ideal family or friendship bonds, and are composed of types of relationships and qualities of relationships. In a good friendship, the friends may be mutually responsible, for example, to support each other's well-being. Mutual responsibility is a type of relationship—that is, a general category of a relationship. But what makes this relationship truly a kinship relationship is if the mutual responsibility has certain inherent qualities.

Qualities are dimensions of relationships like trust, consensuality, transparency, reciprocity, and accountability. For example, just because two people declare themselves to be mutually responsible for each other does not mean they are genuine friends. Without qualities like trust or accountability in the bond, it will not be possible for either of the persons to really take responsibility for each other's well-being. While I certainly have a responsibility for the well-being of anyone I interact with, what makes a friend a friend is that our bond together is laden with the aforementioned qualities, like trust and accountability.

These qualities take time to develop and nurture. But when they are inherent in a relationship, that relationship is critically important to us. We depend on friends and family for our well-being psychologically, culturally, socially, existentially, economically, and politically. In times of crisis, it is our friends and family who can act quickly on our behalf and provide the support we need. For we know, having spent years building our relationships with them, that they are trustworthy, respect our consent, are transparent with us, will genuinely try to do what's best for us, and behave accountably.

Reciprocity is an important kinship quality. A bond has the quality of reciprocity when each relative (or friend) believes the others to be in a long-term gift-receiving and gift-giving relationship. In other words, each relative respects the gifts that they can give to others in the kinship group to support their well-being. With that respect comes the expectation that they will be intrinsically motivated to give gifts of their own (Kimmerer, 2013). To be in a reciprocal bond with one another is to take to heart that one is committed to honoring gift givers and devoted to taking actions that support the well-being of the givers (i.e., giving back). Reciprocity is a kinship quality and, whenever it is present in a relationship, the relatives or friends in the relationship are in a better position to live interdependently.

As I will show in this very introductory chapter, this idea of kinship and reciprocity is one facet of how some Indigenous peoples understand their connections to non-humans and the environment. In this chapter, reciprocity is often invoked using the language of gift giving or interdependency. Though my examples here concern humans, I could have easily discussed the exact same quality of reciprocity with respect to bonds between humans and fish, or between birds and mountains.

Ecosystems and reciprocity

A key aspect of Indigenous peoples' cultural and political self-determination is that they have their own sciences for protecting ecosystems that matter to their societies' health, economy, and cultural vitality (Shilling and Nelson, 2018). **Ecosystem** is a term that has one of its origins in the field of ecology. The term often refers to a community of relationships of organisms that are connected through their sharing of nutrients and energy. Indigenous peoples have sought to articulate their own traditions about ecosystems to create dialogue with ecologists (Kimmerer, 2002, Ford and Martinez, 2000).

Indigenous peoples often emphasize that the community of relationships can be understood through kinship. The sharing of nutrients and energy is like mutual responsibility whereby the members of an ecosystem are interdependent and, in a way, are reciprocally giving and receiving gifts. The quality of reciprocity across members of an ecosystem is vital. Dennis Martinez writes that "the health of the land is an important requisite for the survival of the people, and the community well-being of the people is an important requisite for the survival and restoration of the ecosystems. It is a totally reciprocal relationship" (Martinez, 2008, p. 5). Consider some different reciprocal relationships that Indigenous peoples have discussed themselves or shared with others.

For Potawatomi people, we have longstanding traditions of relationships to corn. We have traditions of growing diverse varieties and see corn as related to other plants, such as beans and squash, to create sustainable ecosystems (Citizen Potawatomi Information Office, 2019). Quechua peoples in the Andes region, such as the Paru Paru, Chawaytiri, Sacaca, Pampallacta, Amaru, and Kuyo Grande communities, have created the Potato Park, a biodiversity conservation zone protecting more than 900 varieties of native potato (Argumedo, 2008, Huambachano, 2018). In the Arctic region, Sámi peoples have practices for maintaining sustainable relationships with reindeer through stewarding grazing pasture and protecting the health of herds living in dynamic landscapes (Ahrén, 2004, Labba, 2015).

The relationships between these Indigenous peoples with corn, potatoes, and reindeer are among the critical dimensions of each society's maintaining its own independent sources of food, having access to good nutrition, or being involved in healthy physical activity required to steward and harvest plants and animals. At a deeper level, Potawatomi, Quechua, and Sámi peoples understand that they are in **relationships** of reciprocity with the animals, plants, insects, and ecosystems where they live. Robin Kimmerer, speaking of ecosystems as "the living world," says that "the living world is understood, not as a collection of exploitable resources, but as a set of relationships and responsibilities. We inhabit a landscape of gifts peopled by nonhuman relatives, the sovereign beings who sustain us, including the plants" (Kimmerer, 2018, p. 27).

Reciprocity is focal. One Potawatomi heritage seed garden, called *gtegemen* (we grow it), involves bringing together humans with a plant combination called the three sisters (corn, beans, squash), whereby each being involved is considered as a relative (DeerInWater, 2019, Citizen Potawatomi Information Office, 2019). The relatives are deemed to express their independent contributions when they are in relations of interdependence or, as Robin Kimmerer calls it, a covenant of reciprocity (Kimmerer, 2017). Kaya DeerInWater says that human relationships to corn gift to people nutrition and stress reducing exercise (to care for corn). DeerInWater says "Those seeds are our relatives and ancestors, and you don't want to be an absent relative. It's a responsibility to plant the heritage seeds, and not taking care of that responsibility is not showing those seeds respect" (CPN Information Office, 2019).

The relationship of *papa arariwa*, for some Quechua peoples, means that humans are guardians of potatoes, which supports human and potato interdependence by engendering motivation in humans to care for potato habitat reciprocally (Stephenson, 2012). The conservation system of the Potato Park is importantly based on the *ayllu* system of conservation, which seeks to attain well-being of all members of an ecosystem. Argumedo and Wong interpret well-being, or *sumaq kawsay*, as "[being] in equilibrium with one's natural and social surroundings and to maintain reciprocity between all beings" (Argumedo and Wong, 2010, p. 84; see also Huambachano, 2018).

For Sámi, reindeer herders maintain particular family groupings with one another called *siida*, including sibling relationships and other familiar bonds. Siida allow them to combine and separate herds and change locations in order to avoid overgrazing pastures. It is a kin-based system of mutual responsibility and involves multiple levels of collective interaction, including individuals, households, groups of households, and recognized leaders (Labba, 2015). Kristina Labba writes that "Sami reindeer herding is characterized by a close connection with nature; the herding represents a complex coupled system of interchange among animals, ecology and reindeer herds" (Labba, 2015, p. 142).

For Potawatomi, Quechua, and Sámi peoples, *gtegemen*, *papa arariwa*, and *siida* are kinship relationships that serve as ethical systems that motivate humans to be responsible for protecting the environment. That is, they are relationships with qualities of reciprocity. Dialogue across Indigenous peoples on reciprocity is a global conversation that we have with each other.

Victoria Tauli-Corpuz, speaking of Igorot people of the Philippines, describes the practice of *ug-ugbo* during planting and harvesting of rice. She says that:

> we practiced, a traditional form of mutual labor exchange [and] collectively planted or harvested one field, moving on to other fields until the sun set. In the evening we all gathered to celebrate finishing the work. Our cultural rituals are linked with the phases of the agricultural cycle and the life cycle (birth, weddings, and deaths).
>
> (Tauli-Corpuz, 2001)

For Tauli-Corpuz, this traditional practice of mutual labor exchange involves reciprocity with the gifts of the land. She says that "The Igorots do not consider themselves the owners but the stewards or trustees of ancestral lands" (Tauli-Corpuz, 2001).

Workinah Kelbessa, in a paper on African **environmental ethics**, writes that for a number of African peoples there are beliefs and practices of "kinship . . . with different animals, plants, or a piece of land" (Kelbessa, 2014, p. 400). In Kelbessa's research, he describes how the *ukama* system of Shona people in Zimbabwe "embraces cosmological or ecological relationality as inseparable to what it means to be human" (p. 400). He cites the work of Munyaradzi Felix Mulrove, who discusses ukama is an "ethic of the interdependence of individuals," as it means "relatedness." "Human well-being" is considered to be "indispensable from our dependence on, and interdependence with, all that exists and particularly with the immediate environment on which all humanity depends" (Kelbessa, 2014, p. 400, citing Murove, 2009, p. 315). Concepts of dependence and interdependence resonate strongly with reciprocity.

These brief examples of Indigenous kinship traditions, if taken in their own right and in their unique contexts, are very different on kinship and reciprocity. What I am trying to highlight is that, in addition to respecting our differences, we also have broader conversations with one another about how our traditions may be related. These kin-making traditions involve a particular orientation toward environmental justice, which I will take up now.

Kinship and anti-colonial action

Kinship is an important way to understand environmental justice. For hundreds of years, invasion, exploitation, and colonization have greatly harmed Indigenous peoples (see Chapter 7). These acts of violence and oppression are sponsored by nation-states, corporations, non-profit entities, and discriminatory persons in their daily behavior. One way of understanding environmental injustice is as an assault on kinship relationships. Reciprocity is one of the kinship relationships that has been threatened most in cases of environmental injustice against Indigenous peoples.

In the work of Marama Muru-Lanning, she describes the significance of the Waikato River for Māori peoples, especially the Waikato-Tainui people. In genealogical terms, she explains: "The river is my ancestor, my *Tupuna Aawa*. For *hapū* (clans) and iwi of the lands that border its 425-kilometre length, the Waikato River is an ancestor, a *taonga* (treasure) and a source of *mauri* (life-essence)" (Muru-Lanning, 2018). Muru-Lanning's writing emphasizes the reciprocal exchange between people and the river. She describes one of the goals of the Waikato Māori is to "[emphasize their] duty to 'care for their ancestor'" and "the reciprocal relationship that Waikato- Māori have with the river" (Muru-Lanning, 2016, p. 161).

Her work goes on to document how in the middle of the 19th century, the New Zealand colonial government seized the river from Māori peoples (Muru-Lanning, 2016). Today the relationship between Māori and the river is disrupted by nearby farming that relies on fertilizer, wastewater due to the development of urban areas in the region, eight hydroelectric power stations, and a thermal power station that warms the water to be pumped into irrigating streams. She writes that "For Māori, another major desecration of the river occurs when its waters are diverted and mixed with waters from other sources so that it may be drunk by people living in Auckland" (Muru-Lanning, 2016, p. 42).

Muru-Lanning's work describes environmental injustice that is an affront to genealogical relationships—whereby genealogy relates to what I am calling kinship. The land use change imposed on Māori people by the New Zealand settlers undermines kinship. The undermining of kinship is exercised through the destruction of self-determination and reciprocity. The environmental injustice described by Muru-Lanning is a form of colonialism, in that the New Zealand settlers seek to dispossess Māori of their ancestral lands—where ancestral refers to generations of established reciprocal relationships—to make way for economic, cultural, social, and political systems that support New Zealand's self-determination.

At the same time, the Waikato-Tanui people have engaged in negotiations with New Zealand to establish co-management of the river. Many of the projects associated with this work involve actions aimed at protecting the river and restoring respect for its ancestral significance. While the river is facing so many issues that threaten Māori culture and health, the Waikato-Tanui are undaunted in their efforts to exercise **self-determination** and affirm kinship ties to the river (Muru-Lanning, 2016). Their website describes actions they are taking to create sustainable fishing practices, support traditional Māori relationships to the river, protect sacred places, and generate opportunities for Māori to have environmental-related employment (www.waikatotanui.com/services/taiao).

In a different place, The Tla-o-qui-aht First Nation and the Ahousaht First Nation are Nuu-chah-nulth peoples who have always lived in what is currently called the Clayoquot Sound region in British Colombia, Canada. Nuu-chah-nulth peoples have generations of reciprocal relationships with whales, salmon, and numerous other trees, animals, fish, and insects of the marine and forest environment. Nuu-chah-nulth peoples developed complex philosophies of reciprocal governance, including *hishuk ish ts'awalk* (oneness between humans and the environment) and *hahuulhi* (a responsibility- and family-based land tenure system) (Atleo, 2006, p. 2).

Marlene Atleo has shown that the land tenure system operated by ensuring that people understood their responsibilities to each other and non-humans (Atleo, 2006). There is a great emphasis on respect for equality. Richard Atleo writes that in the Nuu-chah-nulth governance system there is a "discourse" that people are "encouraged to see other species, as well as other peoples, as equals." He writes that this discourse is exemplified in "protocols such as the ceremony to acknowledge the first salmon of the season, or the ceremony to take down a great cedar for a great canoe." Atleo claims that "one of these demands is not to be disrespectful toward any part of creation" (Atleo, 2002, p. 200).

While the Nuu-chah-nulth governance system emphasized a respect for reciprocity, Richard Atleo writes that "Then came the discourse of colonialization." For Atleo, "The nature of this discourse is unilateral, evolutionary, linear, hierarchical, and presumptuous. It completely overshadowed and ignored the first discourse" (Atleo, 2002, p. 200). Indeed, the increase in direct settlement after 1850 ushered in disrespect for Nuu-chah-nulth kinship to land. Non-Indigenous investors began flooding the region in the 1980s and 1990s to profit from logging the forests. Indigenous persons, and numerous allies and environmentalists, fought against logging using blockades and other direct-action tactics, leading to global media attention being paid to the arrest of approximately 900 land protectors. In response, a mediation process was set up (Tindall et al., 2013).

A scientific panel was put together. The Forests Minister of British Columbia accepted the 127 recommendations of the panel, leading to commitments to "ecologically sound" logging and opportunities for First Nations to invest in major portions of the industry and take leadership in the regional decision-making. In 1994, an Interim Measures Agreement for Clayoquot Sound was developed between the province and the five Nuu-chah-nulth peoples (Tindall et al., 2013). Due importantly to Indigenous environmental justice advocacy, in 2000, the region became a Biosphere Reserve under the auspices of the United Nations Educational, Scientific and Cultural Organization (UNESCO). The First Nations have their own Tribal parks, for biodiversity conservation and recreation, such as the "The Haa'uukimun Tribal Park" (Carroll, 2014). A review of the public materials on any of these parks demonstrates how the park managers seek to provide experiences for visitors to understand reciprocity (see, for example, https://youtu.be/Sw4BToE48Ew).

Kinship and anti-colonial action

Reflecting on the work of the Waikato-Tanui and Nuu-chah-nulth peoples discussed in the previous section, environmental *injustice* can be understood through the idea of kinship relationships. Environmental injustice can be defined abstractly by thinking about how different societies treat each other in **colonialism**. Environmental injustice involves at least one society, a colonizing society, moving in somehow and disrupting the kinship relationships that matter to the sustainability of the societies whose homelands are being colonized. That is, environmental injustice occurs when at least one society dominates one or more other societies by disturbing their kinship relationships. The agents that carry out injustice for the colonizing society are the military, businesses and corporations, educational institutions, religious organizations, and private citizens. They engage in acts of terror and discrimination or take leadership in manipulating and managing people to support colonial goals.

The Great Lakes region in North America has been, and continues to be, a major site of environmental injustices against Indigenous peoples, especially related to mining and access to fish and plant harvesting. This is the region where I work. Anishinaabe, Oneida, Dakota, Menominee, and other peoples of the northern Great Lakes region have longstanding ancestral kinship traditions. Species like walleye, deer, wolves, and wild rice, among others, are integral to ecosystems, as well as entities and flows like water, lakes, forests, and prairies. In this section, I will review a number of examples that readers may or may not be familiar with. While briefly covered here, I hope readers will follow up on these cases on their own to learn more about them.

In many of the cultures of the region, people see themselves as relatives of non-humans. They have special kinship relationships with them, including ties of reciprocity. For these species, entities, and flows provide many gifts to humans in terms of supplies, food, clean water, forest health, and cultural identity. Indigenous persons feel compelled to honor these gifts and take actions that, in turn, give gifts back to these species in terms of habitat protection. The Anishinaabe water ceremony is one example. The water ceremony involves people sharing and drinking water, and engaging in practices, such as singing, that honor the gifts of life and health that water provides. The ceremony also solidifies humans' gifts back, as water protectors. Elder Josephine Mandamin, in an interview, says that "That's our responsibility, our role, and our duty, to pass on the knowledge and understanding of water, to all people, not just Anishinaabe people, but people of all colors" (Indigenous Environmental Network, 2014).

Indigenous peoples of the Great Lakes region have longstanding traditions of treaty-making across societies to create kinship bonds of reciprocity. The Dish with One Spoon Treaty between Anishinaabe and Haudenosaunee peoples in the Great Lakes region is an example. Leanne Simpson writes that:

> Gdoo-naaganinaa [the dish] acknowledged that both the Nishnaabeg and the Haudenosaunee were eating out of the same dish through shared hunting territory and the ecological connections between their territories . . . both parties were to be responsible for taking care of the dish. . . . All of the nations involved had particular responsibilities to live up to in order to enjoy the rights of the agreement. Part of those responsibilities was taking care of the dish.
>
> (Simpson, 2008, p. 37)

This treaty is an early example of how environmental *justice* was established through forging kinship bonds. When faced with the threat of colonization, Indigenous peoples of the Great Lakes attempted to forge bonds of kinship with European, American, and Canadian societies (Witgen, 2011, Lytwyn, 1997). Heidi Stark, speaking of treaties of Anishinaabe peoples with the US and Canada, claims that "Anishinaabe saw the treaties as vehicles for building relationships vested in reciprocal responsibilities" (Stark, 2010, p. 153).

Susan Hill, speaking of treaties and agreements of Haudenosaunee people and colonists, writes that the:

> relationship was to be as two vessels travelling down a river—the river of life—side by side, never crossing paths, never interfering in the other's internal matters. However, the path between them, symbolized by three rows of white wampum beads in the treaty belt, was to be a constant of respect, trust, and friendship. . . . Without those three principles, the two vessels could drift apart and potentially be washed onto the bank (or crash into the rocks).
>
> (Hill, 2008)

Hill's description of this treaty tradition, called the *Kaswentha*, emphasizes reciprocity as interdependence.

Despite Indigenous peoples' intentions to maintain kinship relationships through treaties, their treaty rights have been repeatedly violated. Treaty violations have threatened Great Lakes fish, wild rice, and wolves—disrupting the gift-giving and gift-receiving relationships that furnish the basis of interdependence. For example, in the 1980s and 1990s, in places currently known as Wisconsin and Michigan, some treaty Tribes had to fight to protect their responsibilities to fish. By treaty Tribes, I mean Indigenous peoples who have signed treaties with the US that protect their responsibilities to kin, like the aforementioned plants, animals, and fishes. They were harassed by white fishers and officials of the two states, experiencing both direct violence and the use of racial epithets. States like Michigan have entertained recreational and other types of wolf killing after wolf populations rebounded in the region that would interfere with Indigenous practices of relating to wolves. Commercial rice production, mining, dams, and genetic research have all threatened wild rice populations. Dams and logging have had deleterious impacts on anadromous fish, like sturgeon. Industrial pollution has for decades affected the capacity of Haudenosaunee people to harvest fish and medicinal plants that are safe to eat and use (Nesper, 2002, Andow et al., 2009, Fletcher and Reo, 2013, Holtgren, 2013, Arquette et al., 2002).

Zoltan Grossman has documented numerous struggles of Anishinaabe people against a threat to many non-human relatives: mines. A coalition of Potawatomi, Ojibwe, and Menominee peoples worked for years to stop the water pollution risks of the proposed Crandon zinc and copper mine in northeast Wisconsin, a mine seeking to boost the settler economy at the expense of Indigenous peoples' health and ways of life, including impacting fishing and wild rice. The Lac Courte Oreilles Tribes fought the Ladysmith copper-gold mine, and the Lac du Flambeau Tribe opposed Noranda's proposed zinc-silver mine in Oneida County (Wisconsin) (Grossman, 2005). Recently, the Back 40 zinc-copper mine is threatening Menominee ancestral territories. If built, the mine will affect the environmental quality of areas central to the Menominee's origin story (Reiter, 2017).

For decades, the Prairie Island Indian Community (Minnesota) has struggled with its reservation lands being lost to make way for a nuclear power plant and an Army Corps of Engineers dam system, both of which are meant to support settler energy and commercial transportation systems. The dam, for example, causes flooding, which recently forced children to cancel their maple sugar harvesting—a major tradition for them. The Tribe is now having to acquire new land farther away to mitigate the risks (Richert, 2019). Or, in cases where wilderness areas have been established by the US, Indigenous peoples have been excluded from culturally based and economically critical harvesting activities, such as fishing. The Boundary Waters Canoe Area, in what is currently Minnesota, excluded all motorboats from entering, which limited the access of nearby Anishinaabe peoples, especially the Boise Forte Band of Chippewa, who have treaty rights to continue their subsistence and other relationships to the area (Freedman, 2002).

A brief glance at the news will show that Anishinaabe peoples in the Great Lakes continue to struggle against pipelines and mines. The Enbridge Line 5 pipeline carries oil and gas through an area of Lake Michigan that is significant for biodiversity for five tribes. The Tribes, including the Little Traverse Bay Bands of Odawa Indians, have treaty rights that guarantee high water quality for fish harvesting and religious practices (Balaskovitz, 2018). Another aging pipeline, Enbridge Line 3 (finished in 1960), threatens wild ricing areas across Minnesota and Wisconsin. In what is currently Ontario, the Eabamatoong First Nation declared a state of emergency due to bad drinking water. Anishinaabe people have been widely affected by Canada's negligent drinking-water policies (McKenna, 2018; McLaughlin, 2019).

Anishinaabe peoples' **anti-colonial action** emphasizes kinship relationships as the solution to environmental justice, whether threats from mines and pipelines or from other problems referenced earlier in this section. Josephine Mandamin began the Mother Earth Water Walk to motivate people to take responsibility for clean water in the Great Lakes, honoring water's role as a sacred life-giver. Deborah McGregor, speaking of ceremony, writes that "We have a responsibility to take care of the water and this ceremony reminds us to do it" (McGregor and Whitaker, 2001, p. 24).

In terms of the issue of wolf hunting, The White Earth Land Recovery Project, in Minnesota, quotes elder Joe Rose, who says "We see the wolf as a predictor of our future. And what happens to wolf happens to Anishinaabe . . . whether other people see it or not, the same will happen to them" (White Earth Land Recovery Project, 2019). Regarding sturgeon, Jay Sam refers to it as "The grandfather fish," and explains how it would "sacrifice" itself "so the people would have food until the other crops were available" (Holtgren et al., 2014). In an Anishinaabe white paper on wild rice and treaty rights requested by the White Earth Nation chairwoman, Dr. Erma Vizenor, the authors state that:

> [wild rice] is a sacred plant. Manoomin is a living entity that has its own unique spirit. Anishinaabeg have a responsibility to respect that spirit and to care for it.... 'We consider it to be sacred, because it's a gift from the creator,' said White Earth elder Earl Hoaglund.
> (Andow et al., 2009, p. 2)

Given how many issues Anishinaabe continue to face, they nonetheless emphasize the importance of reciprocity in terms of solutions, whether they express this through discussing spiritual relationships or treaty rights. Environmental justice is about kinship and is anti-colonial, for environmental injustice is about the disruption of kinship. At least, this is one way of understanding justice and injustice from the perspectives of Anishinaabe and related peoples of the Great Lakes.

Indigenous environmental justice advocacy through kinship

Indigenous peoples' environmental justice movements emphasize different conceptions of kinship in humans' abilities to live respectfully within diverse ecosystems. Again, I am understanding kinship ethics as an approach that emphasizes moral bonds like reciprocity. Outside of the contexts of Indigenous people that I am most familiar with, I am sometimes struck by a particular division. The division is as follows. The activity of loving the environment, and therefore seeking to conserve nature, is separate from the activity of tackling environmental injustice. Through kinship ethics, Indigenous peoples show that these activities are not separate, for loving the environment occurs through establishing and honoring kin relationships. And environmental injustice is precisely an affront against kinship. In this way, realizing more reciprocal bonds with ecosystems and the plants, animals, insects, and humans who dwell in them *is a precise way to express environmental justice*.

In the Indigenous environmental justice movement, then, kinship with the environment cannot ever exclude responsibilities for humans to be in solidarity with each other toward stopping harms and the threats they face. Indigenous treaty-making is an example of this. There is no such thing as reciprocity with non-human relatives in the environment that would not, at the same time, entail human kinship across societies.

Indigenous environmental justice provides a trenchant critique of capitalism, colonialism, and industrialization as disruptions of kinship. At the same time, Indigenous peoples offer solutions, whether for clean energy, conservation zones, or rigorous science, among many others. But Indigenous intellectual traditions often propose that these solutions cannot be exercised at the expense of kinship bonds. Indeed, they can only be considered solutions if they support kinship relations moving into the future (Whyte, 2020).

Follow-up questions

- What would it mean to approach your actions as an environmentalist through kinship?
- What are the different ways in which reciprocal relationships support interdependence?

- Are there similarities or differences between the environmental justice struggles of Indigenous peoples, communities of color in industrial countries, and people of the Global South

References for further reading

Dhillon, J. 2018. Introduction: Indigenous Resurgence, Decolonization, and Movements for Environmental Justice. *Environment and Society*, 9, 1–5.

Estes, N. 2019. *Our History is the Future: Standing Rock Versus the Dakota Access Pipeline, and the Long Tradition of Indigenous Resistance*. Verso.

Kimmerer, R. 2017. The Covenant of Reciprocity. In: Hart, J. (ed.), *The Wiley Blackwell Companion to Religion and Ecology*. New York, NY, USA: John Wiley and Sons.

Yazzie, M. K. & Baldy, C. R. 2018. Introduction: Indigenous Peoples and the Politics of Water. *Decolonization: Indigeneity, Education, and Society*.

References used in this chapter

Ahrén, M. 2004. Indigenous Peoples' Culture, Customs, and Traditions and Customary Law-the Saami People's Perspective. *Arizona Journal of International & Comparative Law*, 21, 63.

Anaya, J. 2004. *Indigenous Peoples in International Law*. New York, NY, USA: Oxford University Press.

Andow, D., Bauer, T., Belcourt, M., Bloom, P., Child, B., Doerfler, J., Eule-Nashoba, A., Heidel, T., Kokotovich, A., Lodge, A., Lagarde, J., Lorenz, K., Mendoza, L., Mohl, E., Osborne, J., Prescott, K., Schultz, P., Smith, D., Solarz, S. & Walker, R. 2009. *Wild Rice White Paper: Preserving the Integrity of Manoomin in Minnesota. People Protecting Manoomin: Manoomin Protecting People: A Symposium Bridging Opposing Worldviews*. Mahnomen, MN: White Earth Reservation.

Argumedo, A. 2008. The Potato Park, Peru: Conserving Agrobiodiversity in an Andean Indigenous Biocultural Heritage Area. In: Amend, T., Brown, J., Kothari, A., Phillips, A. & Stolton, S. (eds.), *Protected Landscapes and Agrobiodiversity Values*. Protected Landscapes Task Force of IUCN's World Commission on Protected Areas.

Argumedo, A. & Wong, B. Y. L. W. 2010. The Ayllu System of the Potato Park (Peru). In: Belaire, C., Ichikawa, K., Wong, B. Y. L. & Mulongoy, K. J. (eds.), *Sustainable Use of Biological Diversity in Socio-Ecological Production Landscapes*. Montreal, PQ, Canada: Secretariat of the Convention on Biological Diversity.

Arquette, M., Cole, M., Cook, K., Lafrance, B., Peters, M., Ransom, J., Sargent, E., Smoke, V. & Stairs, A. 2002. Holistic Risk-Based Environmental Decision-making: A Native Perspective. *Environmental Health Perspectives*, 110, 259–264.

Atleo, E. R. 2002. Discourses in and about the Clayoquot Sound: A First Nations Perspective. In: Magnusson, W. & Shaw, K. (eds.), *A Political Space: Reading the Global through Clayoquot Sound*. Montreal, PQ, CA: MQUP.

Atleo, M. R. 2006. The Ancient Nuu-chah-nulth Strategy of Hahuulthi: Education for Indigenous Cultural Survivance. *International Journal of Environmental, Cultural, Economic and Social Sustainability*, 2, 153–162.

Balaskovitz, A. 2018. 'We were here First': Tribes say Line 5 Pipeline Tunnel Ignores Treaty Rights. *Energy News*, October 8.

Borrows, J. 1997. Living Between Water and Rocks: First Nations, Environmental Planning and Democracy. *The University of Toronto Law Journal*, 47, 417–468.

Carroll, C. 2014. Native Enclosures: Tribal National Parks and the Progressive Politics of Environmental Stewardship in Indian Country. *Geoforum*, 53, 31–40.

Citizen Potawatomi Information Office 2019. Heritage Seed Projects Help Decolonize Potawatomi Food Systems. *Hownikan*, July 10.
CPN Information Office. 2019. Heritage Seed Projects Help Decolonize Potawatomi Food Systems. *HowNiKan*.
DeerInWater, K. 2019. Harvesting Guide. *Hownikan*, October 18.
Fletcher, M. L. & Reo, N. J. 2013. Response to Sanders: Ma'Iingan as Property. *Wis. L. Rev. Online*, 69.
Ford, J. & Martinez, D. 2000. Traditional Ecological Knowledge, Ecosystem Science, and Environmental Management. *Ecological Applications*, 10, 1249–1250.
Freedman, E. 2002. When Indigenous Rights and Wilderness Collide: Prosecution of Native Americans for Using Motors in Minnesota's Boundary Waters Canoe Wilderness Area. *American Indian Quarterly*, 26, 378–392.
Grossman, Z. 2005. Unlikely Alliances: Treaty Conflicts and Environmental Cooperation between Native American and Rural White Communities. *American Indian Culture and Research Journal*, 29, 21–43.
Hill, S. 2008. Travelling Down the River of Life Together in Peace and Friendship, Forever: Haudenosaunee Land Ethics and Treaty Agreements as the Basis for Restructuring the Relationship with the British Crown. *Lighting the Eighth Fire: The Liberation, Resurgence, and Protection of Indigenous Nations*, 23–45.
Holtgren, M. 2013. Bringing Us Back to the River. In: Auer, N. & Dempsey, D. (eds.), *The Great Lake Sturgeon*. East Lansing, MI, USA: Michigan State University Press.
Holtgren, M., Ogren, S. & Whyte, K. P. 2014. Renewing Relatives: Nmé Stewardship in a Shared Watershed. *Tales of Hope and Caution in Environmental Justice*. Available from: https://hfe-observatories.org/stories/renewing-relatives-nme-stewardship-in-a-shared-watershed/. (Accessed March 31, 2020).
Huambachano, M. 2018. Enacting Food Sovereignty in Aotearoa New Zealand and Peru: Revitalizing Indigenous Knowledge, Food Practices and Ecological Philosophies. *Agroecology and Sustainable Food Systems*, 42, 1003–1028.
Indigenous Environmental Network 2014. Meet Josephine Mandamin (Anishinaabekwe), the "Water Walker". *indigenousrising.org*, September 25.
International Labour Organization 1989. Convention on Indigenous and Tribal Peoples. In: International Labour Organization (ILO), *Indigenous and Tribal Peoples Convention, C169*, June 27, 1989. Available from: https://www.refworld.org/docid/3ddb6d514.html. (Accessed March 31, 2020).
Kelbessa, W. 2014. Can African Environmental Ethics Contribute to Environmental Policy in Africa? *Environmental Ethics*, 36, 31–61.
Kimmerer, R. 2002. Weaving Traditional Ecological Knowledge into Biological Education: A Call to Action. *BioScience*, 52, 432–438.
Kimmerer, R. 2013. Returning the Gift. *What does Earth Ask of Us?* Available from: www.humansandnature.org/earth-ethic—robin-kimmerer-response-80.php.
Kimmerer, R. 2017. The Covenant of Reciprocity. In: Hart, J. (ed.), *The Wiley Blackwell Companion to Religion and Ecology*. New York, NY, USA: John Wiley and Sons.
Kimmerer, R. 2018. *Mishkos Kenomagwen*, the Lessons of Grass: Restoring Reciprocity with the Good Green Earth. In: Shilling, D. & Nelson, M. (eds.), *Traditional Ecological Knowledge: Learning from Indigenous Methods for Environmental Sustainability*. Cambridge, UK: Cambridge University Press.
Labba, K. 2015. The Legal Orginazation of Sami Reindeer Herding and the Role of the Siida. In: Allard, C. & Skogvang, S. F. (eds.), *Indigenous Rights in Scandanavia*. New York, NY, USA: Routledge.
Lytwyn, V. P. 1997. A Dish with One Spoon: The Shared Hunting Grounds Agreement in the Great Lakes and St. Lawrence Valley Region. *Algonquian Papers-Archive*, 28.
Manuel, G. 1974 *The Fourth World*. Toronto, ON, Canada: Collier-Macmillan Canada.
Martinez, D. 2008. Native Perspectives on Sustainability: Dennis Martinez. In: Hall, D. E. (ed.), *Native Perspectives on Sustainability: Voices from Salmon Nation*. nativeperspectives.net.
McGregor, D., and S. Whitaker. 2001. *Water quality in the province of Ontario: An aboriginal knowledge perspective*. Toronto: Chiefs of Ontario

McKenna, P. 2018. Controversial Enbridge Line 3 Oil Pipeline Approved in Minnesota Wild Rice Region. *Inside Climate News*. June 8.

McLaughlin, A. 2019. 18-year Water Crisis in Eabametoong First Nation 'Would Not Be Tolerated' Anywhere Else, Chief Says. *CBC Radio*. July 26.

Murove, M. F. 2009. An African Environmental Ethic Based on the Concepts of Ukama and Ubuntu. In: Murove, M. F. (ed.) *African Ethics: An Anthology of Comparative and Applied Ethics*. Scottsville, South Africa: University of KwaZulu-atal Press.

Muru-Lanning, M. 2016. *Tupuna Awa: People and Politics of the Waikato River*. Auckland, New Zealand: Auckland University Press.

Muru-Lanning, M. 2018. Interviewed by Peter Debaere. "Q&A With Marama Muru-Lanning About the Wakaito River" Global Water Blog, University of Virginia. June 15. Available from: https://blogs.darden.virginia.edu/globalwater/2018/06/15/qa-with-marama-muru-lanning-about-the-wakaito-river/#_ftnt_MML-1. (Accessed March 31, 2020).

Napoleon, V. 2013. *Thinking about Indigenous Legal Orders*. Dordrecht, Netherlands: Springer.

Nesper, L. 2002. *The Walleye War: The Struggle for Ojibwe Spearfishing and Treaty Rights*. Lincoln, NB, USA: University of Nebraska Press.

Reiter, G. A. 2017. Understanding the Menominee Tribal Perspective on the Back 40 Mine. *American Rivers Guest Blog*. Edited by Jessie Thomas-Blate. Available from: Americanrivers.org/2017/07/menominee-tribal-perspective-back-40-mine/. (Accessed March 31, 2020).

Richert, C. 2019. Environmental, Nuclear Worries Froce Pariarie Island Tribe to Seek New Lands. *Minnesota Public Radio News*, December 13.

Sanders, D. E. 1977. *The Formation of the World Council of Indigenous Peoples*. Copenhagen, Denmark: The International Secretariat of International Work Group for Indigenous Affairs.

Shilling, D. & Nelson, M. (eds.) 2018. *Traditional Ecological Knowledge: Learning from Indigenous Methods for Environmental Sustainability*. Cambridge, UK: Cambridge University Press.

Simpson, L. 2008. Looking after Gdoo-naaganinaa: Precolonial Nishnaabeg Diplomatic and Treaty Relationships. *Wicazo Sa Review*, 23, 29–42.

Sobrevila, C. 2008. The Role of Indigenous Peoples in Biodiversity Conservation: The Natural but Often Forgotten Partners. Washington, DC: The World Bank.

Stark, H. K. 2010. Respect, Responsibility, and Renewal: The Foundations of Anishinaabe Treaty Making with the United States and Canada. *American Indian Culture and Research Journal*, 34, 145–164.

Stephenson, A. 2012. The Quechua: Guardians of the Potato. *Cultural Survival Quarterly Magazine*, March.

TallBear, K. 2019. Caretaking Relations, Not American Dreaming. *Kalfou*, 6, 24–41.

Tauli-Corpuz, V. 2001. Innayan* Just Don't Do It! *Yes! Magazine*. September 30. Retrieved February 2, 2020. Available from: http://www.yesmagazine.org/issues/technology-who-chooses/innayan-just-dont-do-it.

Tauli-Corpuz, V. & Mander, J. 2006. *Paradigm Wars: Indigenous Peoples' Resistance to Globalization*. San Francisco: Sierra Club Books.

Tindall, D. B., Trosper, R. & Perreault, P. (eds.). 2013. *Aboriginal Peoples and Forest Lands in Canada*. Vancouver: UBC Press.

Todd, Z. 2017. Fish, Kin and Hope: Tending to Water Violations in Amiskwaciwâskahikan and Treaty Six Territory. *Afterall: A Journal of Art, Context and Enquiry*, 43, 102–107.

United Nations General Assembly 2007. United Nations Declaration on the Rights of Indigenous Peoples. *UN: Washington*, 12.

White Earth Land Recovery Project. 2019. Ma'iingan (The Wolf) Our Brother. Available from: www.welrp.org/about-welrp/maiingan-the-wolf-our-brother/. (Accessed January 10, 2020).

Whyte, K. 2020. Too Late for Indigenous Climate Justice: Ecological and Relational Tipping Points. *Wiley Interdisciplinary Reviews: Climate Change*, e603.

Witgen, M. 2011. *An Infinity of Nations: How the Native New World Shaped Early North America*. Philadelphia, PA: University of Pennsylvania Press.

21 Justice beyond humanity

Steve Cooke

Learning outcomes

- Gain an understanding of the concept of justice.
- Understand the role justice-claims play in moral reasoning.
- Consider key arguments for the value of non-human animals and the environment.
- Become familiar with animal rights and eco-centric theories of justice.

Introduction

In this chapter, we will explore attempts to extend the concept of justice beyond humanity. Traditional theories of justice restrict their concern to relationships between humans, tending to say very little, if anything, about what we owe to non-humans. These traditional theories are **anthropocentric**, which means that they give humans sole or preferential consideration. Anthropocentric theories of justice generally view non-human animals and the environment as resources, valuable only insofar as they contribute to the good of humans. This does not mean these theories assign no value to non-human animals or the environment. On the contrary, even if humans are the only beings owed justice, then we still have reason to preserve the environment. One reason is that a sufficiently hospitable environment is a necessary condition of human life. Anthropocentric approaches tend to see the environment as something to be preserved for the reason just given, and as a resource to be distributed according to whatever principles the particular theory rests upon. In other words, these theories claim that we may have duties *regarding* non-human animals or the environment, but we do not have duties *to* them. The Enlightenment philosopher Immanuel Kant made an argument along these lines about non-human animals. Kant claimed that duties to non-human animals are indirectly owed to humans. Animals, he argued, have no direct value of their own, but harming them makes a person more likely to treat fellow humans badly (Kant, 1963, pp. 239–241).

Anthropocentric theories have been challenged both by theorists who think non-human animals are also owed justice and by those who wish to bring the further environment under the scope of justice. The focus of this chapter is to introduce you to important examples of these two different approaches: the animal rights approach and the eco-centric approach. At the heart of animal rights and eco-centric theories are claims about what matters from a moral point of view.

Each argues that we have moral reasons to treat non-human animals or the environment better than we currently do, and that this moral concern can give rise to justice-claims.

Before we can get stuck into the matter of justice, however, we first need to get some background work out of the way. In order to make claims about what non-human animals and the environment are owed as a matter of justice, animal rights and eco-centric theorists first have to show that the objects of their concern have the right kind of moral status. Much of the thinking about animal rights and environmental ethics has been devoted to questions of moral value or moral status, so the first part of this chapter will explore where this value might come from. Once we have done this, we can then move on to looking at what moral value means for how agents should act. We will do this in quite abstract terms, and this will help us to get to grips with the concept of justice and its relationship with wider morality. By the end of the chapter, you should have a good understanding of the concept of justice, the role justice-claims play in moral reasoning, and of how animal rights and eco-centric theories have been developed to extend the concept of justice beyond its traditional boundaries.

The moral status of non-human animals

Let us begin with the question of the moral status or value of non-human animals, before moving on to perform the same exercise with the environment. To do this, I will start by outlining a sentience-based argument for the value of non-human animals. There are other arguments for the value of non-human animals besides the sentience-based one, but sentience is the most commonly argued-for grounding of value or status. It also has the virtue of being relatively straightforward to explain and understand.

These days, there is widespread agreement that non-human animals are worthy of moral concern for their own sakes, and that their moral status rules out certain ways of treating them. People tend to react in horror at stories of cruelty and mistreatment, and there are laws in many countries against causing unnecessary suffering or harm to non-humans. A key reason for this is that non-human animals are sentient. **Sentience** is a kind of consciousness that describes the capacity to feel and have emotions. Sentient beings have a perspective from which they experience life. In other words, sentient beings have subjective experiences. Different species of sentient beings are sentient to different degrees; they do not all experience the same range and intensity of feelings and emotions, but they do all experience pleasure and pain. An animal's capacity to feel means that the things that happen to it matter to it. A kick matters to a dog because it experiences the unpleasant feeling of pain. As a result, dogs have a preference for not feeling pain. The fact that animals feel pain and pleasure means that their lives can go better or worse for them depending upon whether those lives are painful or pleasurable. Another way to express this is that non-human animals have preference-based interests in the direction their lives take. It is for this reason that the utilitarian philosopher Peter Singer argued for the equal consideration of human and non-human animal interests, and it is also the usual justification for granting non-human animals legal protections (Singer, 1986, pp. 215–228; see Box 21.1).

Box 21.1 Utilitarianism and non-human animals

Utilitarianism is a consequentialist ethical theory. Utilitarians determine goodness and badness by looking at the consequences of actions rather than, for example, the intentions or motivations behind them. They argue that the best states of affairs are those where aggregate or average utility is maximised. Depending on the type of utilitarian, utility can be defined in terms of happiness, pleasure, or the satisfaction of preferences. Utilitarians argue that the morally obligatory thing to do in any situation is to maximise utility, such as by trying to achieve the greatest balance of happiness over unhappiness. For utilitarians, the utility of each counts for the same and ought to be considered impartially. Important utilitarian thinkers, like Jeremy Bentham and Peter Singer, have sought to include the happiness and suffering of non-human animals in their overall calculations of utility.

The fact that the lives of animals matter to those animals themselves means that they are not merely valuable as the means to human ends; they have value for their own sakes. The fact of sentience makes moral claims upon moral agents (see Box 21.2): if we think others ought to take our own suffering seriously and constrain their actions because of it, then we must concede that we ought to constrain our actions where another's suffering matters to them. Animal rights theorists argue that the moral status of non-human animals means that we cannot just treat them how we wish simply because it benefits us in some way. The sentience of non-humans means that we are required by morality to take their interests seriously when we act. In fact, say these theorists, the interests of sentient beings are sometimes sufficiently important to them that they generate rights (more on rights later). The most commonly claimed rights for non-human animals are those against suffering and being killed.[1] If non-human animals have rights against being made to suffer or being killed, then moral agents are duty-bound not to cause them suffering or kill them except in exceptional circumstances (such as in self- or other-defence, or when it is in the interests of the animal itself). These rights have some very significant consequences for how humans ought to treat non-humans and for the way our existing political communities are organised.

Box 21.2 Moral agents and moral patients

A **moral agent** is someone who is capable of choosing to act morally (or not). Non-human animals, like young children, lack the capacity to understand and act upon moral reasons, so they are not blameworthy or praiseworthy for their actions. Because they lack moral agency, they cannot be duty-bearers. Beings owed moral consideration for their own sakes but lacking the capacity to make moral choices are known as *moral patients*. Unlike *moral patients*, most adult humans are *moral agents*; they are responsible for their actions because they can choose how to act. Another way of saying this is that *moral agents* are *morally autonomous*. The capacity for moral choice makes *moral agents* capable of being duty-bearers.

The value of the environment

Just as animal rights theorists have argued that non-human animals matter for their own sakes, so too have some environmental ethicists argued that the environment is valuable for its own sake. Others have sought to extend interest-based accounts of value to encompass all living beings whether sentient or not. These *eco-* or *bio-centric* approaches claim that our reasons to treat the environment well are not necessarily because doing so benefits humans. Rather, they argue, it is because the environment is valuable independently of any instrumental **value** we gain from it. One possible explanation for this value might be that life itself, whether sentient or not, has intrinsic value. Environmental philosopher Robin Attfield uses a 'last man' thought experiment to test our intuitions on the subject, and then tries to work out principles that might account for those intuitions. Attfield asks us to imagine that all sentient life on earth has been wiped out save for one remaining human (Attfield, 1981). Would it be wrong, asks Attfield, for that man to cut down the last surviving elm tree, which would otherwise go on to propagate itself? The last man has no need of its wood; he chops it down merely as a kind of symbolic protest. Attfield thinks that the last man makes the world worse by reducing the amount of value in the world. If this is so, then it cannot be because the tree is only instrumentally valuable to humans or other sentient life. It must also be valuable in and of itself. Other environmental philosophers have taken similar approaches. Robin Elliot, for example, discusses whether something valuable is lost when a feature of the environment is destroyed, even if a restoration project is carried out somewhere else to offset the damage. Elliot uses an analogy between original artworks and forged artworks, suggesting that the naturalness of an environmental feature has value in the same way that an original artwork does (Elliot, 1982). The same analogy helps explain why people regard a convincing but artificial environment as less valuable than a natural one. These thought experiments suggest that perhaps the environment is intrinsically valuable rather than merely instrumentally valuable.

Other approaches to the value of the environment liken ecosystems to communities, seeing value in interconnected wholes and in maintaining their integrity. For example, Baird Callicott has argued that the value of individual living beings arises out of their contribution to the biotic community they are part of. This ecologically-based ethical holism gives plants, humans, and non-human animals instrumental value dependent upon their contribution to preserving and maintaining the biotic community (Callicot, 1992). Critics of holism have raised the concern that it implies the worryingly authoritarian view that individual members of a community, including humans, may be sacrificed for the sake of environmental goals (Regan, 2004, pp. 361–363). In response, some bio-centric theorists have claimed that members of organisms within a bio-centric community have both value for their own sakes and value insofar as they contribute to the community (Callicot, 1999; Katz, 1996). This modified **bio-centrism** places limits on trading off the well-being of individuals for the sake of the whole. However, there are still many unanswered questions to be asked about the ranking of values and the principles governing conflicts and trade-offs.

Reasons and rights

The attribution of the types of value described to non-human animals or the environment has some important consequences for human action. If we harm something that matters morally, then, other things being equal, we are doing something wrong. When we harm something that matters for its own sake, then we wrong that being itself. If non-human animals, individual non-sentient organisms, or the environment matter morally, then that means we have reasons to treat them in particular kinds of ways; we cannot simply discount their well-being or value when deliberating on courses of action that affect them.

The role played by morality in deliberation is an interesting one, and it is worth exploring it a little further because it is a topic that students often have difficulty getting to grips with. A brief discussion of the role of morality in guiding how we should act will hopefully therefore prove helpful. Let us begin with reasons and then briefly discuss rights and duties.

Moral reasons ordinarily have the special feature of trumping or defeating other reasons in the way that an ace trumps a king in a game of cards. It is part of the nature of morality that moral claims are more pressing on us than other sorts of considerations. I could benefit myself by taking bribes in return for giving some of my students higher marks than the quality of their work merits. I could use the money to buy a fine hat for myself. However, the fact that it would be morally wrong to do so trumps the selfish reason I have for taking bribes. Morality requires us to sometimes act in ways that conflict with our desires or our interests.

Moral reasons can also sometimes exclude other sorts of reasons even from being considered (Raz, 1975). A good example of an exclusionary reason comes from a thought experiment used by Bernard Williams to critique **utilitarianism**. Williams asks us to imagine a case whereby he must choose between saving the life of his wife or that of a complete stranger. Williams says that in such a circumstance, the right answer is 'his wife', and the reason is 'because she is his wife'. If we have to think about which life will produce the most utility before we decide, then we have had 'one thought too many'. We should not have to think about it: it is the nature of the relationship of marriage itself that excludes having to make such a judgement from consideration altogether (Williams, 1981, pp. 17–18).

Earlier we encountered the claim that the interest **non-human animals** have in not suffering is strong enough to ground a duty in moral agents, and this means that they have a right against being made to suffer. We can think of rights as particularly strong kinds of moral reasons. Much in the same way that moral reasons can trump or exclude non-moral reasons, rights can be thought of as trumping or excluding other sorts of moral as well as non-moral reasons. By way of example, let us imagine that I have promised to pay my friend some money that I owe him by next Wednesday. Because I have promised, my friend has a right to that money, and I have a duty to fulfil my promise. On my way to his house, I am confronted by someone collecting for charity. I have just enough money to repay my friend and no more. It would be morally good for me to give to this charity, and so I have reason to do so, but my promise to my friend rules out doing so. My friend's right to the money trumps the good I can bring about by giving it to someone else. If non-human animals have rights, then any social benefit that might be derived from causing them to suffer is ruled out by their right against suffering. Similarly, if a collective—such as a human community or a biotic community—has rights, then it will be protected against being changed or broken up in order to bring about a wider good. The right my friend gains in the example of promising arises out of the act of me making the promise, but, according to *interest* theorists, many of the rights we have are ours simply because we have very strong interests that must be protected in order for us to have any kind of decent life. Examples of these fundamental rights in humans include things like life, bodily security, and liberty.

The concept of justice

Now that we have explored some potential sources of value for non-human animals and the environment, and gained an understanding of the roles played by value and rights in moral deliberation, we can move on to think about justice in more detail. Justice is an extremely important concept, particularly in political morality, and the way we think about justice has an enormous influence on the way our political communities are structured. This section of the chapter will help explain why that is by clarifying the meaning of the concept and describing its features.

If non-human animals or the environment are not owed justice, then the way we treat them will be a matter of individual preference or merely of concern for humans who may be harmed as a result of our actions (recall Kant's argument in the introduction section of this chapter). Where stronger protections are granted, then these will have to arise out of an agreement amongst humans and remain contingent upon this. But, if the moral status of non-humans or the environment is higher than is commonly assumed in mainstream theories—if they are owed justice—then the principles by which we live and regulate our societies should change. Perhaps non-human animals or the environment are not merely things we ought to treat well because we benefit from doing so, or because it is good to be kind; perhaps they are also owed justice? Let us begin by considering the concept itself.

Justice is a moral concept; it is a way of thinking about and evaluating features of our social lives to determine the rightness or wrongness of them. There is more than one way to define justice, but when I use the term in this chapter, I mean social justice rather than **retributive justice**. Social justice is about social, economic, and political relations within and between political communities. Retributive justice, on the other hand, concerns the determination of innocence or guilt, or sentencing and punishment in the criminal justice system.

In ordinary language, we often use several moral concepts interchangeably: right, wrong, good, bad, fair, unfair, just, and unjust. In moral philosophy, the words refer to distinct concepts, and we need to be careful about how we use them. To moral philosophers, something can be bad but not wrongful, unfair but not unjust, good but not just, and so on. We can illustrate this with an example. Imagine a rockslide occurs, killing a group of hikers. The rockslide is a perfectly natural occurrence, but it causes enormous harm. In this example, we would be correct to say that something bad has occurred, but we would be mistaken to say that the hikers had been wronged. The rockslide is bad because it causes loss of life. Those lost lives were valuable: they contained projects that have been cut short prematurely, and friends and loved ones will be left unhappy and perhaps also worse off. A world in which the hikers are not killed is better than one where they are. What separates a bad occurrence like the rockslide from a wrongful one is the role played by **human agency**. The deaths of the hikers are not wrong because they do not result from any intentional human act; nobody is morally responsible for them. For the hikers to have been wronged, someone would have to have been to blame for the rockslide, either out of negligence or malice. Had someone intentionally set off the rockslide, then the deaths of the hikers would have been both bad and wrongful. Not only would the world have been made worse by the killing, but the people killed would also have been wronged.

The concept of justice is one that concerns right and wrong, rather than good and bad. This means that justice and injustice can only occur as a result of the actions or inactions of moral agents. As the political philosopher Thomas Scanlon describes it: justice is about what we owe to each other (Scanlon, 1998). This sets justice apart from moral acts that might be praiseworthy but are not necessarily required. For example, charity is a virtue that is usually considered distinct from justice.[2] Charitable giving is, by nature of the concept, giving that cannot be demanded of us. If we give to others because we owe them, then we cannot by definition be giving charitably. If I owe something to someone as a matter of justice, then I am under a strict duty to give them their due, and I wrong them if I do not. Justice can be rightfully demanded of another person or an institution; charity cannot.

Another important feature of the idea of justice is that it is strongly connected with political morality, so much so that Scanlon describes it as the area of morality governing social institutions (Scanlon, 1998, p. 13). Political morality governs the moral principles and rules governing our political institutions and relationships. Another influential political philosopher, John Rawls, went as far as to refer to justice as the first virtue of political institutions (Rawls, 1999, p. 3). By this, he meant that justice should be the virtue exemplified in the institutions that set and enforce the laws governing our lives. If our social arrangements are just as a first principle, then requirements of justice will come before other sorts of considerations in importance. For example, if we must make a choice between an efficient benefits system and a just benefits system, we ought to prefer the just one. Efficiency can be sacrificed for the sake of justice (recall the discussion on moral reasons trumping other reasons). One reason why justice and political morality are intertwined is because justice is often considered to be enforceable. This means that it can be permissible to use coercion to enforce the demands of justice when people fail to do as they ought. One of the key puzzles for political philosophers concerns not only determining what justice requires, but also who has legitimate authority to decide when its requirements have been violated and to enforce those requirements. It is because of this that there is a strong connection between justice and the role and purpose of government and the state. As a result, it can be useful to think about justice in terms of collectively enforceable morality.

The enforceable nature of justice-claims and the connection between justice and political morality helps explain why the language of rights is so well-suited to describing the requirements of justice. Theories of justice lay out principles for protecting the good of those owed justice, for resolving conflicts between them, and for distributing burdens and benefits between members of the community of justice.[3] Rights describe strict rules; if I have something by right, then I am entitled to it, and whoever has the correlative duty to fulfil my right may not simply decide to not fulfil their duty. Breaching a right counts as a serious wrong to the right-holder. This makes rights-talk a good way of describing requirements of justice. Once we describe what is owed to each as a matter of justice, it is then the task of social institutions to ensure compliance with rights and fulfilment of duties. Those concerned with protecting the vital interests of non-human animals have thus drawn heavily on rights discourses in pursuit of their aims. More recently, these rights discourses have begun to be incorporated into fleshed-out theories of justice for non-human animals. Ecological accounts have found it rather more difficult to do to so, but have nevertheless sought to make non-anthropocentric justice-based arguments for preserving the environment.

Non-anthropocentric theories of justice

If justice is owed to non-human animals or the environment, then what would this mean in practice? We have already seen that anthropocentric theories can offer protections to non-human life. If protecting non-human animals or the environment is necessary for achieving justice for humans, then even an instrumental account of the value of non-human animals, biotic communities, or the environment more broadly can achieve many of the goals of eco-centric and animal rights theories. This would not, however, represent the inclusion of non-humans or the environment within the community of justice. Animal rights and eco-centric theories of justice are based on claims that the moral status of non-human animals, bio-centric communities, living organisms, or ecosystems means that they too ought to be recognised as members of the community of justice. This means that members of the categories listed should not be treated as mere resources to be shared out: they ought to be recipients of justice rather than the currency of justice.

If non-human animals are owed justice, then their well-being should be protected, and their interests considered as a matter of right. For interest-based theorists, re-conceiving justice to encompass non-human animals has involved identifying the interests sufficiently strong as to ground duties in moral agents. From this, they have explored how political institutions and constitutional arrangements ought to change so as to account for those interests. For example, Alasdair Cochrane's sentientist theory of justice begins from the premise that non-human animals matter morally for their own sakes because of their sentience. This fact also means that they matter politically, and that the harmful treatment of non-human animals ought to be forbidden by law, even if those harms benefit humans. Cochrane defends what he calls a **sentientist cosmopolitan democracy**. This theory of justice involves an institutional framework designed to protect the basic rights of all sentient beings, starting with, but not limited to, a right to life and a right against suffering. Central to his theory is an argument that the basic rights non-humans have are held independently of any special relationships they might have: they are universal in nature, whether the non-human is domesticated or wild. A key plank of his account of non-human justice involves having the interests of non-humans represented in political deliberations and procedures (Cochrane, 2018). Another influential account, that of Sue Donaldson and Will Kymlicka, argues that domesticated non-human animals are owed citizenship rights, along with the same right as humans to enjoy substantively equal opportunities to realise their interests within their community. Because domesticated animals are granted citizenship rights, they are entitled, as a matter of justice, to political representation and to having a whole range of their interests given serious consideration in political decision-making.[4] Alongside this, Donaldson and Kymlicka develop principles governing the interaction between political communities and undomesticated animals. These they divide into liminal animals that move in and out of our communities without being a full part of them, and wild animals that live almost wholly apart. The former are treated as akin to migrant humans and entitled to similar sorts of consideration and hospitality, and the latter are regarded as sovereign over their own territories (Donaldson and Kymlicka, 2011).

Eco-centric theories of justice have found it more difficult to develop clear and intuitively plausible accounts. One reason for this has been that much theorising about justice has concerned the protection of individuals though rights, against being used as the means to achieve social goods. Another has been the focus in theories of justice upon achieving an equitable distribution of the burdens and benefits of social cooperation, which includes a fair distribution of resources between individuals. These features of theories of justice are difficult to apply to collectives or systems. How, for example, can we meaningfully talk of distributing natural resources to the environment itself? As Marcel Wissenberg has pointed out, in order to fit within a distributive paradigm, eco-centric theories need to consider the parts that make up the whole of nature as recipients of justice rather than the whole itself. Such a theory must, at the same time, acknowledge that the environment is made up of the resources to be distributed (Wissenburg, 1993). This is a difficult—perhaps impossible—task to manage. In response, some theorists have tried to move away from thinking about justice in terms of distributing goods. David Schlosberg attempts instead to focus on what is needed for ecological systems to function well and to flourish, making a claim along the way that systems 'are living entities with their own integrity' and that 'atomising nature into isolated animals devalues a form of life, and the way that this form of life flourishes' (Schlosberg, 2009, p. 148). Unfortunately, this argument suffers both from that fact that the claim that systems are living entities is false and the fact that shifting the focus away from substantive issues of distributive justice does not make those issues disappear. Nor does this kind of thinking do much to alleviate the worry raised earlier about the potential for totalitarianism in some strands of environmental philosophy.

A more convincing tack is taken by Brian Baxter. Baxter's approach is similar to the interest-based account used by animal rights theorists. However, Baxter adopts a broader conception of interests, including all things that affect the course of an organism's life, whether they have conscious preferences concerning them or not. This, according to Baxter, is because living beings are self-directed and possess a *telos*—an end of their own. Because they have a *telos*, they cannot be considered merely instrumentally valuable. Baxter contends that all living organisms, sentient or otherwise, are owed justice (Baxter, 2004, p. 9) and that human interests do not automatically trump those of members of other species, or non-human populations. Rather, when distributing environmental burdens and benefits, such as those from resource extraction or climate change, the interests of all living organisms affected ought to be considered from an impartial perspective. He writes:

> A fundamental aim of the theory of ecological justice, therefore, is to defend the claim that viable populations of organisms which are 'merely living' have a prima facie right to environmental resources necessary for those populations to continue to exist in a way which permits the flourishing of their individual members.
>
> (Baxter, 2004, p. 131)

Only under very stringent constitutional provisions may human interests automatically trump those of other beings. Where organisms lack meaningful individuality, he applies his distributive principles at the species level and at the level of viable populations (Baxter, 2004, ch. 9). In practice, this requires political communities to develop constitutional provisions representing the interests of the inarticulate via mechanisms such as the use of proxies or trustees (see Box 21.3).

Box 21.3   Including non-human animals: from theory to practice

Because of the role played by rights-talk in theorising about justice, attempts to get political communities to include non-human animals or the environment within the scope of justice are often conducted in the courts. In a number of notable cases, activists have attempted to extend the protections offered to humans by fundamental legal rights so that they also encompass non-human entities. For example, the Nonhuman Rights Project has, for many years, fought to have limited personhood rights granted to chimpanzees so that they have a right against being held in captivity and being considered as property. More recently, it has also begun filing lawsuits on behalf of captive elephants on the same basis. By petitioning on behalf of particular animals, the project aims to establish a legal precedent for the granting of liberty rights to all animals of the same species.

Similarly, in 2019, the High Court of Bangladesh recognised a river as a living entity with legal standing in its own right, and ordered a government commission to act as its guardian. The river's guardian is required to act in its interests, along with those of other rivers in the country. This model of court-appointed guardianship has also been applied to environmental entities such as rivers and mountains in India, Columbia, and New Zealand.

Justice and the social contract

Many of those working on mainstream theories of justice are deeply sceptical about non-anthropocentric visions of justice. A key reason for this has been the influence of the **social contract** model on thinking about justice. The contractualist model is one that supposedly creates serious, perhaps insurmountable, problems for including non-humans within the community of justice. Social contract theories have defined the community of justice, along with the principles governing it, by reference to moral agency, particularly in terms of voluntarism and reciprocity. In contractualist thinking, justice is a domain of morality that encompasses those capable of reciprocating with others and abiding by a set of principles on the basis that others will do likewise. A political community is envisioned to be akin to an organisation whose members have collectively agreed the rules that govern them. Because of this, contractualist thinking appears to exclude moral patients from the community of justice. In response, it has been pointed out that such thinking excludes many humans from consideration, including children and those with severe cognitive impairments. As the great Mary Midgley pointed out, this element of contractualist thought loses sight of the ordinary meaning of justice: 'in ordinary life we think that duties of justice become *more* pressing, not less so, when we are dealing with the weak and inarticulate' (Midgley, 1995, p. 96, original emphasis). Animal rights and eco-centric theorists have responded to these problems by either rejecting contractualism as a solid foundation for developing principles of justice, or by drawing upon alternative models of contractualism. For example, Donaldson and Kymlicka develop the contract model by drawing upon inter-species trust rather than collective agreement (Donaldson and Kymlicka, 2011, p. 106).

Conclusion

If non-human animals or other elements of the non-human world are owed justice, then this has profound consequences for the organisation of our political communities and for how we live our day-to-day lives. Extending justice beyond humanity means taking on a set of enforceable duties, duties that ought to be enshrined in the constitutional and institutional structures of political communities. Justice for non-humans means ensuring that their interests are accounted for to a much greater degree in political decision-making. It also means granting them an equitable share in the distribution of environmental goods and of the burdens and benefits arising out of social cooperation. Doing so raises a number of interesting problems for theorists to overcome, many of which are addressed in the recommended readings listed at the end of this chapter.

Follow-up questions

- If non-human animals have rights against being killed or made to suffer, then how should we address issues of predation, disease, and natural disasters in the wild?
- Does the rareness of a species make individuals of that species more valuable than similarly sentient or more sentient members of more abundant species?
- In what ways do animal rights and eco-centric theories of justice seem to be in conflict?

Notes

1 For a very good early articulation of an interest-based theory of rights for non-human animals, see Feinberg (1986).
2 For an excellent challenge to this view, see Buchanan (1987).
3 On distributive justice, see Chapter 3 of this volume.
4 On participation and procedural justice, see also Chapter 4 of this volume.

References for further reading

Armstrong, Adrian. 2012. *Ethics and Justice for the Environment*. London: Routledge.
Baxter, Brian. 2004. *A Theory of Ecological Justice*. London and New York: Routledge.
Cochrane, Alasdair. 2010. *An Introduction to Animals and Political Theory*. Basingstoke: Palgrave Macmillan.
Cochrane, Alasdair. 2018. *Sentientist Politics: A Theory of Global Inter-Species Justice*. Oxford: Oxford University Press.
Donaldson, Sue & Kymlicka, Will. 2016. *Zoopolis: A Political Theory of Animal Rights*. Oxford: Oxford University Press.
Garner, Robert. 2013. *A Theory of Justice for Animals: Animal Rights in a Nonideal World*. Oxford: Oxford University Press.
Midgley, Mary. 1983. *Animals and Why They Matter*. Athens: The University of Georgia Press.
Nussbaum Martha, C. 2006. *Frontiers of Justice: Disability, Nationality, Species Membership*. Cambridge, MA: Belknap Press of Harvard University Press.
Schlosberg, David. 2009. *Defining Environmental Justice: Theories, Movements, and Nature*. Oxford: Oxford University Press.

References used in this chapter

Attfield, R. 1981. The Good of Trees. *The Journal of Value Inquiry* 15, 35–54.
Baxter, B. 2004. *A Theory of Ecological Justice*, 1st ed. London and New York: Routledge.
Buchanan, A. 1987. Justice and Charity. *Ethics* 97, 558–575.
Callicot, B., 1992. Animal Liberation: A Triangular Affair. In: Hargrove, E.C. (Ed.), *The Animal Rights, Environmental Ethics Debate: The Environmental Perspective*. New York: SUNY Press.
Callicot, B. 1999. *Beyond the Land Ethic: More Essays in Environmental Philosophy*. Albany: State University of New York Press.
Cochrane, A. 2018. *Sentientist Politics: A Theory of Global Inter-Species Justice*. Oxford and New York: Oxford University Press.
Donaldson, S. & Kymlicka, W. 2011. *Zoopolis: A Political Theory of Animal Rights*. Oxford and New York: Oxford University Press.
Elliot, R. 1982. Faking Nature. *Inquiry* 25, 81–93. https://doi.org/10.1080/00201748208601955
Feinberg, J. 1986. The Rights of Animals and Unborn Generations. In: Werhane, P., Gini, A.R. & Ozar, D. (Eds.), *Philosophical Issues in Human Rights: Theories and Applications*. New York: Random House, pp. 164–173.
Kant, I. 1963. *Lectures on Ethics*. New York: Harper & Row.
Katz, E. 1996. *Nature as Subject*. Lanham: Rowman & Littlefield Publishers.
Midgley, M. 1995. Duties Concerning Islands. In: Elliot, R. (Ed.), *Environmental Ethics*. Oxford, USA and New York: Oxford University Press, pp. 89–103.
Rawls, J. 1999. *A Theory of Justice*, Revised ed. Cambridge, MA: Harvard University Press.
Raz, J. 1975. Reasons for Action, Decisions and Norms. *Mind* 84, 481–499.
Regan, T. 2004. *The Case for Animal Rights*, 2nd ed. Berkeley and Los Angeles: University of California Press.

Scanlon, T. 1998. *What We Owe to Each Other*. Cambridge, MA: The Belknap Press of Harvard University Press.

Schlosberg, D. 2009. *Defining Environmental Justice: Theories, Movements and Nature*. Oxford: Oxford University Press.

Singer, P. (Ed.) 1986. *Applied Ethics, Oxford Readings in Philosophy*. Oxford: Oxford University Press.

Williams, B. 1981. *Moral Luck: Philosophical Papers 1973–1980*, Reissue ed. Cambridge and New York: Cambridge University Press.

Wissenburg, M. 1993. The Idea of Nature and the Nature of Distributive Justice. In: Dobson, A., Lucardie, P. (Eds.), *The Politics of Nature: Explorations in Green Political Theory*. London and New York: Routledge.

Part IV
Future directions of environmental justice

22 Critical environmental justice studies

David N. Pellow

Learning outcomes

- Have a basic understanding of the concept of environmental justice.
- Develop a critical approach to environmental justice that pushes beyond the traditional boundaries of the field.
- Be able to demonstrate how prisons and environmental justice concerns are linked.
- Learn to discuss how environmental justice movements also include prisoners who are leaders in this effort.

This chapter considers new developments in the field of environmental justice studies and applies those ideas to the case of the US prison system in order to develop a stronger understanding of the nature of incarceration and its range of impacts. The historical and contemporary character of prisons examined here offers the reader an innovative way of thinking about both the criminal legal system and environmental issues.

The field of environmental justice studies has moved us toward a clear understanding that, where we find social inequalities by race and class, we tend to also find **environmental inequalities** in the form of marginalized groups being exposed to greater levels of pollution, toxics, "natural" disasters and the effects of climate change/disruption, as well as their exclusion from policy-making bodies and government agencies that influence those outcomes (Harrison, 2019). These populations are more likely to live in neighbourhoods polluted by industry and government facilities (such as toxic waste dumps, landfills, incinerators, etc.) than other groups because they lack sufficient political power to prevent such discriminatory outcomes. Thus, environmental risks disproportionately affect poor communities, communities of colour, immigrants, Indigenous peoples, and other marginalized communities around the globe (Bullard, 2000; Bullard and Wright, 2012; Whyte, 2017).

Critical environmental justice studies is a perspective put forth by a number of scholars, intended to address some limitations and tensions within environmental justice studies (Adamson, 2011; Pellow, 2018b; Pellow and Brulle, 2005). These include four major concerns. The first concern is the need to examine *multiple* forms of inequality and their intersections (Malin and Ryder, 2018). Environmental justice scholars have a tendency to limit our analyses to race or class, while a small but growing group of scholars has explored the role of gender and sexuality in environmental justice studies (Buckingham and Kulcur, 2010; Gaard, 2017; see also Chapter 18 of this volume). Moreover, the category *species* remains, at best, at the margins of the field (see Chapter 21 of this volume). So, critical environmental justice insists on a focus across a broad range of categories of difference and is inclusive of more-than-human beings and things. This framework also seeks to challenge the forms of inequality and institutional violence that correspond with these social categories, which would include **heteropatriarchy**, all forms of racism (not just white supremacy), ableism, speciesism/dominionism, transphobia, and colonialism, so this project is firmly aimed at confronting dominance in all forms.

The second concern of critical environmental justice is that the environmental justice studies literature could be improved through greater attention to *multi-scalar* approaches. **Scale** is of critical importance because it allows us to understand how environmental injustices are facilitated by decision-makers who behave as if places where hazards are produced "out of sight and out of mind" are somehow irrelevant to the health of people and ecosystems at the original sites of decision-making power and sites of consumption (Voyles, 2015). Attention to scale also assists us in observing how grassroots *responses* to environmental injustices draw on spatial frameworks, networks, and knowledge to make the connections between hazards in one place and harm in another. Critical environmental justice studies thus advocate multiscalar methodological and theoretical approaches to studying environmental justice issues in order to better comprehend the complex spatial and *temporal* causes, consequences, and possible resolutions of environmental justice struggles.

The third concern of critical environmental justice is the degree to which various forms of social inequality and power—including **state power**—are viewed as entrenched and embedded in society. Environmental justice scholarship generally looks to the state and its legal systems to deliver justice. But, as numerous studies have demonstrated, the track record of state-based regulation and enforcement of environmental and civil rights legislation in communities of colour has not been promising. In its more than a quarter-century history of processing environmental discrimination complaints, the US Environmental Protection Agency's (EPA) Office of Civil Rights has reviewed some 300 complaints filed by communities living in the shadows of polluting industry and has never once made a formal finding of a civil-rights violation (Lombardi et al., 2015). Yet, a number of scholars contend that the environmental justice movement continues to seek justice through a system that was arguably never intended to provide justice for marginalized peoples and nonhuman natures (Benford, 2005; Pulido et al., 2016). The concern here is that such an approach runs the risk of leaving intact the very power structures that produced environmental injustice in the first place. My review of the environmental justice studies literature concludes that, while scholars are generally critical of the abuse of power by dominant institutions with respect to the perpetuation of environmental injustices, they generally do not question the existence of those institutions. In other words, the approach is reformist rather than abolitionist, and this is concerning for me and many other scholars who view certain institutions (for example, the military) as an existential threat to visions of social and environmental justice (Aikau, 2015; Heynen, 2016). Critical environmental justice studies seeks to push our analyses and actions beyond the state, beyond capital, and beyond the human via a broad anti-authoritarian perspective.

Finally, the fourth concern of critical environmental justice is the largely unexamined question of the **expendability** of human and nonhuman populations facing threats from states, industries, and other political economic forces. Environmental justice studies suggests that various marginalized human populations are treated—if not viewed—as inferior and less valuable to society than others. In the book *Black and Brown Solidarity*, John Márquez introduces the concept of "racial expendability" to argue that black and brown bodies are, in the eyes of the state and its constituent legal system, generally viewed as criminal, deficient, threatening, and deserving of violent discipline and even obliteration (Márquez, 2014). Laura Pulido (2017) describes this as the process whereby producing social difference becomes central to creating value, wherein some groups are defined as disposable, a process that Ananya Roy argues leads to civil death (the loss of basic rights and privileges associated with full membership in a society) and social death (the condition of being defined as less than fully human) (Roy, 2019). Critical environmental justice studies makes this theme explicit by arguing that these populations are marked for erasure and early death, and that ideological and institutional othering is linked to the more-than-human world, as well. Critical environmental justice counters that perspective by viewing these threatened bodies, populations, and spaces as *indispensable* to building socially and environmentally just and resilient futures for us all.

Box 22.1 Case study: US prisons and critical environmental justice

Bryant Arroyo is a Puerto Rican man serving time at a prison in Pennsylvania. He recently organized 900 of his fellow inmates to write letters to the town supervisors protesting the planned construction of an $800 million coal gasification plant next door. The plant would have been located just 300 feet away from the prison, placing the inmates in clear danger of exposure to significant air pollution. Arroyo's group of letter writers included LGBTQ prisoners, rival prison gangs, and Black and racist white prisoners; in other words, a highly unlikely and extremely diverse coalition. The gas plant project was defeated, earning Arroyo the title of "jail house environmentalist." Arroyo is one of many persons in the US who has fought against environmental injustices under the unimaginable conditions of brutality and repression of the prison system.

Source: Pellow, 2018b

In this portion of the chapter, I discuss conditions facing people in prisons in order to consider the utility of a critical environmental justice studies framework. The US prison system is unique in that it is the single largest prison system on earth, housing more people and more inmates per capita than any other nation. The more than 2 million inmates behind bars in the US include adults, juveniles, and immigrants, and the vast majority of prisoners are people of colour and low-income persons (Alexander, 2012).

I recently began a project exploring the myriad intersections between the US prison industrial complex and environmental justice concerns. There are numerous ways that prisons and environmental threats intersect to produce harms to the bodies of prisoners and corrections officers and to nearby ecosystems. For example, there are confirmed reports of water contaminated with arsenic, lead, and other pollutants at prisons in dozens of states. In Flint, Michigan, the Genesee County jail's inmates—including pregnant women—were forced to drink toxic water while prison guards drank filtered water out of bottles. This example raises key questions about what Katsi Cook and Elizabeth Hoover call **environmental reproductive justice**—a vision of ensuring that a community's physical and cultural reproductive capabilities are not inhibited by environmental contamination (Hoover, 2018). Another investigation found that nine of 11 juvenile detention facilities in the state of Oregon had exorbitantly high levels of lead in their water systems (Loew, 2017). When Hurricane Harvey hit Texas in August of 2017, prisoners at the Federal Correctional Complex in Beaumont were placed on lockdown and were not evacuated, despite the fact that the storm dropped 35 inches of rain and caused massive flooding, creating a toxic stew of contaminants from nearby petrochemical industries, including a large sulphur dioxide leak from a nearby Exxon refinery (Feltz, 2017), and as temperatures neared 100 degrees, the prison officials cut off the water system because it was visibly contaminated, and many prisoners experienced dehydration as a result. A traditional environmental justice studies approach to this topic would likely place a primary emphasis on the degree to which there may be a geographic concentration of prisons and jails in communities of colour and low-income neighbourhoods and what remedies might be sought through state-based policy mechanisms. A critical environmental justice studies approach builds on that perspective but goes further.

The first concern of critical environmental justice studies expands our analytical reach to consider how multiple social categories of difference are intertwined in the making of (and in resistance to) environmental injustice. Let us consider race and gender. The US prison system is overwhelmingly made up of people who are poor and non-white. As the National Association for the Advancement of Colored People (NAACP) recently concluded that: "Though African Americans and Hispanics make up approximately 32% of the US population, they comprised 56% of all incarcerated people in 2015"; and that "the imprisonment rate for African American women is twice that of white women" (NAACP, 2017). Women overall constitute the single fastest growing group of prisoners in the US (Alexander, 2012). All of these groups face great risks to human health due to a range of threats inside prisons. But we must go beyond categories of difference restricted to the human community. With respect to the category of species and the broader ecosystems in which they dwell, the impacts of prisons include the fact that sewage and other water discharges, chemical toxins, fossil fuels, air pollution, and hazardous waste produced from within carceral institutions affect nonhuman species in waterways, ambient air, and nearby land bases. Letcher County, Kentucky, is the site of a proposed federal prison project that would be placed on land where mountain-top removal (MTR) coal-mining has occurred, a process whereby mountains have been destroyed to extract coal, reducing ecosystems to poisonous rubble and dust. The prison is also to be placed near an area that is largely low income, raising clear environmental justice concerns about multiple and layered uses over time that will produce harm to local ecosystems and human health. The Letcher County site is also home to second-growth forests that serve as habitat for the Indiana bat and grey bats, both of which are endangered species whose fate is further placed in jeopardy as a result of this proposal. In June of 2019, after years of battling this proposed prison, environmental and social justice activists (including inmates at other prisons who protested this proposal) were rewarded with a victory when the Bureau of Prisons dropped its efforts to construct the facility (Cufone, 2019).

The second concern of critical environmental justice studies advances multiscalar analyses. **Temporal scales** figure quite strongly in this discussion about prisons, and that means attention to history. An early example of a prison where environmental justice issues were evident was the concentration camp at Fort Snelling in Minnesota, which was set up by the US government in the 1860s to neutralize the resistance movement by Dakota peoples who were pushing back against the conquest of their lands (Waziyatawin, 2008). This concentration camp was a site of deep mistreatment of the Dakota, resulting in malnutrition, mass starvation and death among hundreds—a profound instance of environmental injustice and racism because inmates were unable to feed themselves from their traditional land base. The wounds of that era remain with us as Fort Snelling today is a monument to US empire that thousands of white Minnesotans celebrate every year.

During World War II, President Franklin Delano Roosevelt issued an executive order to send thousands of Japanese-American citizens to concentration camps in the Western US in order to contain the perceived threat that they posed to the nation (see Chapter 17 in this volume), since the government suspected many of them of being sympathizers with the Japanese empire (a claim that was proven false). During that time of Japanese-American mass incarceration, the prisoners faced serious water contamination because the US military delivered drinking water into the camps through old pipes that had previously been used to store oil and gas. As a result, the water was laced with hydrocarbons and toxins, placing the Japanese-American community's health at significant risk. In many other cases—such as the camps at Gila River and Poston, Arizona—numerous prisoners at these sites experienced extreme heat, dehydration, and many stillbirths as a result of the harsh climate and social conditions imposed on them.

Let us consider the microscale: an emphasis on the body and embodiment reflects a core focus in feminist scholarship in general and on feminist environmental justice in particular; for example, scholars have argued that when we redefine our bodies as "lands" and "homes" and "environments," then we can more effectively personalize and politicize environmental justice for people whose personal well-being is often at risk from a range of threats (Gaard, 2017; Moraga, 1993). At the microscale, study after study finds that women and LGBTQ folk face extraordinarily high threats of sexual abuse and violence in the prison system. Moreover, consider the deliberate use of toxins on prisoners' bodies. In many US jails and prisons, inmates with *mental disabilities* are frequently subjected to punishment and pain-compliance techniques that include the use of toxic chemical sprays. So sexual abuse and chemical attacks are examples of environmental injustice in prisons, revealing intersections of gender, sexuality, and disability, and it is precisely those same persons who have experienced those forms of violence who are resisting them.

At a *broader scale*, we find that many prisons are located on or near current and former waste dumps and toxic waste sites. For example, the largest immigrant prison in the nation is in Dilley, Texas, a place often called "the waste epicentre" (Bernd, 2017) of the Eagle Ford Shale, a 30-county region where 20 saltwater disposal wells pump fracked wastewater back into the ground. The Karnes County Civil Detention Center is another immigrant prison in the region and has been described as the "drilling epicentre" of the Eagle Ford Shale, a region with several well blowouts, local evacuations, and the majority of residents reporting major environmental health problems. The water at the prison is heavily chlorinated and immigrant prisoners report that it smells and tastes foul (Bernd, 2017). As is the case at many immigrant prisons around the nation, inmates routinely resist through hunger strikes, sit-ins, and other means of protest and defiance against this brutal system. The International Energy Agency forecasts that growth in US oil production will meet 80% of new *global* demand for oil in the coming years, with much of that growth coming directly from oil produced by fracking in Texas (Worland, 2018). Thus, international migrants, refugees, and asylum seekers are caged, with children often separated from their families, and placed at great risk of exposure to contaminated water as an indirect result of repressive, militaristic US foreign policies in Latin America, as well as climate change-induced drought, and as a direct result of the intersection of oppressive anti-immigrant policies and an intensified commitment to global fossil fuel production, amplifying nativism and global climate injustice.

The third concern of critical environmental justice studies signals an opportunity to think and act in ways that question our reliance on state dominance to imagine and achieve environmental justice. A particularly pronounced way in which prisons and environmental injustice converge is through the methods the state deploys to criminalize, control, and incarcerate activists within revolutionary movements dedicated to radical forms of environmental justice, and in this category I include the Black Panther Party, the Black Liberation Army, the American Indian Movement, the Puerto Rican *independentistas*, the Weather Underground, MOVE, the Young Lords Party, and revolutionary social movements because they, among their primary concerns, were combatting state and corporate power, racism, colonialism, and militarism against people of colour and Indigenous peoples in favour of community-based solutions that would result in peoples exercising democratic control over land, territory, and space (Pellow, 2018a). The prison is a vicious site of state repression, and the fact that activists working to protect and defend vulnerable communities and environments have been targeted by the state and imprisoned suggests to radical movements that embracing and reinforcing state power may be counterproductive for efforts to achieve environmental justice. Joy James's analysis of slain politicized prisoner and black liberation activist George Jackson's writings makes this plain. She argues that Jackson "calls into question the very right of the state (as master) to exist. In abolitionist insurrectionary narratives, such as those offered by Jackson, what is sought is not the mere abolition of penal captivity or slavery, but the abolition of all masters" (James, 2005, p. xxii).

Other examples of prisoner resistance against the violence of state dominance and captivity abound (see Box 22.1). They include hunger strikes and labour strikes, commissary strikes and boycotts, lawsuits, demanding clean water and access to showers, demanding healthy food, demanding health care, and demanding educational programmes with a particular emphasis on courses in the fields of ethnic studies, women's studies, and sociology (Law, 2012). The fact that prisoners have to stage protests and risk further harm to themselves in order to have access to basic human and environmental rights reflects the need to imagine and plan for a world in which we do not rely on the institutions perpetrating mass violence to provide us with security and other fundamental needs. In other words, meaningful environmental justice will rarely be achieved through state-sanctioned means. In my view, this evidence suggests that we must build social relations that are sustainable, healthy, and just, and that meet needs that states cannot or will not fully provide—like non-militarized security, peace, socially just relationships, and ecological sustainability.

The fourth concern of the critical environmental justice studies framework counters the discourses and practices of expendability with indispensability. It is hard to imagine a more appropriate site for such a discussion than the prison system, which is premised on the assumption that its captives are expendable and disposable. A critical environmental justice perspective challenges that view with an embrace of prison abolition. This is not a vision focused on simply removing prisons from society, but instead is predicated on undertaking the more difficult work of making prisons irrelevant and obsolete by transforming the social relationships *outside* of prisons throughout society that fuel the prison system (Saed, 2012). It is also a vision and practice that organizes around an "all or none" orientation—a refusal to leave anyone behind, which is a powerful declaration of solidarity and of indispensability.

This is an important aim because the prisoner rights movement has often minimized the needs of women, trans, queer, disabled, and immigrant prisoners, although this is changing. Ultimately, the prison system is a particularly brutal example of enslavement and forced migration in the 21st century, insofar as incarcerated people are literally kidnapped by the state and siphoned away from their communities. What this means is that prisoners are also *migrants* and therefore *mass incarceration is a form of forced mass migration*. And given the fact that prisoners in the US are legally and juridically enslaved via the 13th Amendment, when prisoners face ecological risks, this is a powerful example of where slavery and environmental racism converge. Let me take this a bold step further: I would argue that since prisoners are enslaved people, then the many forms of prisoner resistance against environmental injustice mean that the prison environmental justice movement is, in fact, nothing short of a slave rebellion!

Discussion and conclusion: why environmental justice, prisoners, and prisons?

A focus on prisons and socioecological concerns advances the scholarship in environmental justice studies in several ways.

First, in broad terms, a focus on prisons and environmental concerns generally provides a new angle on environmental justice struggles and politics that has thus far been largely overlooked. Given the emphasis within the field of environmental justice studies on the relationships among race, space, mobility, the regulatory state, and risk, focusing those concepts on the prison system facilitates a transformation of our understanding of how they are interconnected, how they relate to the depths of racism and institutional violence, and how they suggest solutions that exceed the typical reformist orientation of the field and the environmental justice movement. Relatedly, the environmental justice reframing of the environment as those spaces where we "live, work, play, learn, and pray" is particularly true in the prison system since, unlike "free persons," prisoners do all of those things in a single place. So, we might revise and expand that definition of the environment so that it reads, where we "live, work, play, learn, pray . . . *and do time.*"

Second, the links among space, race, and environment are much starker given the immobilizing effects of prison (insofar as inmates' mobility is almost entirely determined by prison authorities). This is particularly relevant considering the longstanding debate in environmental justice studies about the relative freedom of choice that people of colour have to move in and out of contaminated neighbourhoods. Scholars like Paul Mohai and Robin Saha (2015a,2015b) have presented strong evidence that the **"minority move-in"** hypothesis (the claim that people of colour move into polluted neighbourhoods because rents are cheaper, rather than polluters targeting communities of colour) is misguided, since the vast majority of cases of environmental racism occur when polluting facilities follow residents of colour, not the other way around. In the case of the prison system, this is essentially a moot point since prisoners have virtually *no* say in where they serve time and are therefore entirely at the mercy of prison authorities.

Third, if we extend our understanding of environmental justice activism into the carceral system, we find that prisoners are some of the most influential activists and leaders of the environmental justice movement today. Sacoby Wilson (2016) uses the term "environmental slavery" to underscore the violent and deeply historical and racist roots of environmental injustice. Theorizing **environmental racism** in the prison system allows one to go much deeper on this question because the 13th Amendment to the US Constitution allows prison slavery as an exception to the Emancipation Proclamation. This also means that the environmental justice movement can now be thought of more properly as part of the broader abolitionist movement because many prisoners fighting for environmental justice are also fighting for their freedom and for the abolition of slavery, as we see in the nationwide prison strikes occurring each summer against prison slavery. What this means is that the environmental justice movement can now be directly linked to the long historical movement to abolish slavery in the US.

Let me state clearly just one reason why thinking about social inequality and the environment is of utmost importance to me: the United Nations Convention on Genocide defines genocide in five ways: a) killing members of the group; b) causing serious bodily or mental harm to members of the group; c) deliberately inflicting on the group conditions of life calculated to bring about its physical destruction in whole or in part; d) imposing measures intended to prevent births within the group; e) forcibly transferring children of the group to another group (United Nations, 1948). That is precisely what environmental racism is, whether we are talking about the proliferation of toxic-waste landfills in communities of colour or the prevalence of environmental threats within the prison system in the US. This is precisely what people are experiencing every day. So, let us call it what it is. It is **genocide**, by the legal definition articulated at the highest levels of international law and at the most basic level of common sense (it also reflects Principle 10 from the 1991 Principles of Environmental Justice: "environmental justice considers governmental acts of environmental injustice a violation of international law, the Universal Declaration on Human Rights, and the United Nations Convention on Genocide" [EJnet.org, 1991]). So, the next time someone asks me "Why should I care about environmental issues when there's so much going on in the world in terms of social justice concerns?" my response will be: "Do you care about genocide? Do you? Because that's what this is."

Some might caution that it would take a miracle to make real the changes needed to address these ills. But as George Lipsitz (2011) reminds us, we witness miracles all the time, including the fact that people who have been the targets of genocide and relentless dehumanization have survived and endured. So, let us recommit ourselves to making the miracle of environmental justice commonplace and an everyday reality. If, from the bowels of a prison in Pennsylvania, Bryant Arroyo can uplift himself and hundreds of his fellow inmates to achieve a victory for environmental justice against a massive polluting power plant, then surely those of us in the "free" world can exercise far more power than we ever imagined. What could be more important to the field of environmental justice studies than deepening our understanding of systems of oppression, inequality, and of effective ways to confront those social problems? Put more bluntly, what could be more important to scholars than strengthening our grasp on the ways in which brutal systems of mass violence—such as slavery and the prison system—function, and how we might reform, or even abolish them? A critical environmental justice lens on the US prison system does exactly that.

Follow-up questions

- Discuss how and why "environmental" issues are inseparable from "social" issues and how social justice concerns have been articulated through the environmental justice movement.
- Have you ever thought of prisons as an "environmental" issue? Why, or why not? How has this chapter shaped your thinking about this topic?

- If prisoners can be leaders in the environmental justice movement, how might you rethink your own level of freedom to make social change on this issue and other causes (assuming you are not imprisoned at the moment)?
- In what ways can you and your friends improve the environment in your community directly, without waiting on or demanding this of elected officials?

References for further reading

Àlvarez, L. and Coolsaet, B. (2018). Decolonizing environmental justice studies: A Latin American perspective. *Capitalism Nature Socialism*, 31:2, 50–69., doi:10.1080/10455752.2018.1558272

Bales, K. (2016). *Blood and earth: Modern slavery, ecocide, and the secret to saving the world*. New York: Spiegel & Grau.

Dillon, L. and Sze, J. (2016). Police power and particulate matters: Environmental justice and the spatialities of in/securities in U.S. cities. *English Language Notes*, Fall/Winter, 13–23.

Shakur, A. (1987). *Assata: An autobiography*. Chicago: Lawrence Hill Books.

References used in this chapter

Adamson, J. (2011). Medicine food: Critical environmental justice studies, Native North American literature, and the movement for food sovereignty. *Environmental Justice*, 4(4), 213–219.

Aikau, H. (2015). Following the alaloa kīpapa of our ancestors: A trans-Indigenous futurity without the state (United States or otherwise). *American Quarterly*, 67(3), 653–661, September.

Alexander, M. (2012). *The New Jim Crow: Mass incarceration in the age of colorblindness*. New York, NY: The New Press.

Benford, R. (2005). The half-life of the environmental justice frame: Innovation, diffusion, and stagnation. In Pellow, D.N. and Brulle, R.J. (eds.), *Power, justice, and the environment: A critical appraisal of the environmental justice movement*. Cambridge, MA: The MIT Press, 37–53.

Bernd, C. (2017). U.S. is locking immigrants in toxic detention centers. *Earth Island Journal*, July 31.

Buckingham, S. and Kulcur, R. (2010). Gendered geographies of environmental justice. In Holifield, R., Porter, M. and Walker, G. (eds.), *Spaces of environmental justice*. New York: Wiley-Blackwell, Chapter 3.

Bullard, R.D. (2000). *Dumping in Dixie: Race, class, and environmental quality* (3rd ed.). Boulder, CO: Westview Press.

Bullard, R.D. and Wright, B. (2012). *The wrong complexion for protection: How the government response to disaster endangers African American communities*. New York City: New York University Press.

Cufone, M. (2019). Prisoners and activists stop new prison on coal mine site in Kentucky. *Press Release by Green Justice*, June 13.

EJnet.org. (1991). *The principles of environmental justice*. Washington, DC, October 24–27. www.ejnct.org/ej/principles.html.

Feltz, R. (2017). Texas prisoners are facing horrid conditions after hurricane Harvey & retaliation for reporting them. *Democracy Now*, September 8.

Gaard, G. (2017). *Critical ecofeminism*. New York: Lexington Books.

Harrison, J.L. (2019). *From the inside out: The fight for environmental justice within government agencies*. Cambridge, MA: The MIT Press.

Heynen, N. (2016). Urban political ecology II: The abolitionist century. *Progress in Human Geography*, 40(6), 839–845.

Hoover, E. (2018). Environmental reproductive justice: Intersections in an American Indian community impacted by environmental contamination. *Environmental Sociology*, 4(1), 8–21.

James, J. (ed.). (2005). *The new abolitionists: (Neo)Slave narratives and contemporary prison writings.* Albany, New York: SUNY Press.

Law, V. (2012). *Resistance behind bars: Struggles of incarcerated women* (2nd ed.). Oakland, CA: PM Press.

Lipsitz, G. (2011). *How racism takes place.* Philadelphia: Temple University Press.

Loew, T. (2017). High lead levels found in 154 taps in 9 Oregon youth correctional facilities. *Statesman Journal*, Salem, Oregon, April 11.

Lombardi, K., Buford, T. and Greene, R. (2015). Environmental justice denied. *Center for Public Integrity*, August 3.

Malin, S.A. and Ryder, S.S. (2018). Developing deeply intersectional environmental justice scholarship. *Environmental Sociology*, 4(1), 1–7.

Márquez, J. (2014). *Black-brown solidarity: Racial politics in the new gulf south.* Austin, TX: University of Texas Press.

Mohai, P. and Saha, R. (2015a). Which came first, people or pollution? A review of theory and evidence from longitudinal environmental justice studies. *Environmental Research Letters*, 10, 1–9.

Mohai, P. and Saha, R. (2015b). Which came first, people or pollution? Assessing the disparate siting and post-siting demographic change hypotheses of environmental injustice. *Environmental Research Letters*, 10, 1–17.

Moraga, C. (1993). *The last generation: Prose and poetry.* Toronto: Canadian Scholars Press.

NAACP. (2017). Criminal justice fact sheet. *NAACP*, www.naacp.org/criminal-justice-fact-sheet/ (Accessed on July 22, 2019).

Pellow, D.N. (2018a). Political prisoners and environmental justice. *Capitalism Nature Socialism*, 29(4), 1–20.

Pellow, D.N. (2018b). *What is critical environmental justice?* Cambridge, UK: Polity Press.

Pellow, D.N. and Brulle, R.J. (eds.). (2005). *Power, justice, and the environment: A critical appraisal of the environmental justice movement.* Cambridge, MA: The MIT Press.

Pulido, L. (2017). Geographies of race and ethnicity II: Environmental racism, racial capitalism and state-sanctioned violence. *Progress in Human Geography*, 41(4), 524–533.

Pulido, L., Kohl, E. and Cotton, N. (2016). State regulation and environmental justice: The need for strategy reassessment. *Capitalism Nature Socialism*, 27(2), 12–31.

Roy, A. (2019). Racial banishment. In Antipode Editorial Collective (eds.), *Keywords in radical geography: Antipode at 50.* New York, NY: Wiley.

Saed. (2012). Prison abolition as an ecosocialist struggle. *Capitalism Nature Socialism*, 23(1), 1–5.

United Nations. (1948). *Convention on the prevention and punishment of the crime of genocide.* UN General Assembly. December 9.

Voyles, T.B. (2015). *Wastelanding: Legacies of uranium mining in Navajo country.* Minneapolis, MN: University of Minnesota Press.

Waziyatawin. (2008). *What does justice look like? The struggle for liberation in Dakota homeland.* St. Paul, MN: Living Justice Press.

Whyte, K. (2017). The Dakota access pipeline, environmental Injustice, and U.S. colonialism. *RED INK: An International Journal of Indigenous Literature, Arts, and Humanities*, 19(1), 154–169.

Wilson, S. (2016). Environmental justice and health disparities: Passion, partnerships, and progress. Presentation at University of California, Santa Barbara, May 24.

Worland, J. (2018). A 'Major second wave' of U.S. fracking is about to be unleashed upon the world. *Time*, March 6.

23 Sustainable materialism and environmental justice[1]

David Schlosberg

Learning Outcomes

- Gain an understanding of the meaning of 'sustainable materialism'.
- Explore the notions of justice that are important to sustainable materialist movements.
- Gain an understanding of how 'material participation' is a political act.
- Learn how health and community are central to activists' notion of justice.

Introduction to sustainable materialist action

There has been a growing interest in social movements focused on constructing more sustainable food, energy, and fashion systems—movements that I have been calling sustainable materialism (Schlosberg and Coles, 2016; Schlosberg, 2019; Schlosberg and Craven, 2019). When actors in these kinds of movements talk about justice, which they do a lot, they actually articulate a range of different conceptions of social and environmental justice. The focus of these activists' idea of justice is most often around three core things: the crucial nature of political participation, the importance of responding to power, and, most often, addressing a set of basic capabilities or needs. And in all of those, the role of community and attachment is central.

In addition to this pluralistic set of conceptions of the idea of justice, what is crucial is that most activists understand and engage the relationships and linkages between these different types of social and environmental justice—or injustice. And yet, even with the range of ideas of justice, and the connections between them, there are some important disconnects with past uses of environmental justice, and with recent arguments for the need for a more critical environmental justice approach.

Before getting into those conceptions of justice in a bit more depth, it is important to briefly explain a little about the movements themselves. I also want to introduce motivations of actors in these movements, in addition to justice itself.

What is sustainable materialism?

It seems clear that there is a growing focus, and an increasing number of people and groups who are working on reworking everyday environmental practices, and are attentive to the sustainability of materials that flow through their communities. These include movements focused on food systems, community energy, and sustainable crafting and fashion. While such practices have long been the norm in subsistence economies and across the Global South, this is now becoming a particular and growing focus in many industrialized (or declining/post-industrial) economies. Most recently, these movements have been growing out of a frustration with governmental progress on environmental issues and a real sense of a lack of accomplishment felt by actors who have engaged in movements for the implementation of broad environmental and sustainability policies (Schlosberg, 2019).

Participants active in these sustainable materialist movements in the US, UK, and Australia (Schlosberg and Craven, 2019) have often discussed their frustration with a sole focus on national and global environmental and climate policy, especially the failure of the various annual United Nations climate change conferences. More broadly, we also see a generational shift in priorities—with younger people quite interested in actually *doing* something about the food, energy, and other things in their everyday lives, and not solely protesting or lobbying for new environmental policies and politics.

Crucially, the idea of **sustainable materialism** does not just address individual ethical or political consumer action (such as that discussed by Micheletti, 2010)—it is really about collective action and self-identified movements around material flows. The attention is, for example, on food movements that have been focused on sustainable food *systems* and new relationships between production and consumption. This includes the slow food, ethicurian, anti-alienation focus on food—embracing growing, cooking, and gathering to eat. And it also encompasses movements specifically dedicated to food justice or food sovereignty (see Chapter 14 of this volume)—groups responding more directly to the unavailability of good food, to economic development potential, and to a re-growth of cultural ties to land, subsistence, and practice (Sbicca, 2018).

The **community energy** movement also illustrates sustainable materialism. Again, such movements are not simply about individual homeowners buying solar panels. It includes neighbourhood and cooperative solar, community energy corporations (see Chapter 13 of this volume), broader **just transition** movements (see Chapter 19 of this volume), and green city initiatives and their advocates (see Chapter 15 of this volume) (Walker et al., 2010). As with food, there are many types of actions and actor motivations, from a fear of peak oil to the desire to add zero emissions energy to climate adaptation planning. But, again, as with food, the focus is not just about individual consumption choices, but rather about a desire to be part of an alternative energy system and flow of power—electrical and political.

Another example of a movement examining the environmental and social impacts of material flows is sustainable fashion, primarily new producers of more sustainable clothing that is attentive to the realities of the full supply chain (Gordon and Hill, 2015). A concern for both social and environmental justice is absolutely central in this movement—there is this very self-conscious attention both to workers and to the natural environments harmed by the textile and clothing industries, from the land and workers where cotton is grown, to conditions in sewing factories, to chemicals in manufacturing, to the immense waste of the industry.

Broadly, I have been calling the focus of these kinds of movements sustainable materialism, or an environmental practice of everyday life (see also Meyer, 2015, Meyer and Kersten, 2016). Such a focus has also been seen as representing a new generation of **ecological democracy** (Eckersley, 2019). These initiatives also represent a version of what Agyeman (2013) has called "**just sustainability**" (see Chapter 9 of this volume), but with a clear focus on the actual sustainable flows of materials and practices in and across communities.

What motivates sustainable materialist activists?

In addition to a dedication to social and environmental justice, there are other important and key motivations of actors in these movements. One is a critique of the classic idea of postmaterialism—specifically the political argument long made by Inglehart (1977) that once your basic material needs are met, you can demand more value-based policies. The postmaterialist thesis argues that with political and economic development comes some level of comfort, when people can develop values around things like identity or environment. Politically, **postmaterialism** has a built-in assumption that liberal pluralism works—that once people have these 'developed' values, they can lobby to have those values brought into the political process and public policy.

It is clear, however, that both of these assumptions of postmaterialism—on values and on politics—are simply wrong. There is a longstanding literature—and the entire movement around environmental justice—that illustrates that poor people consistently have environmental concerns (Guha and Martinez-Alier, 1997). Those values do not only come to those with resources, as postmaterialism implies. And we know about the political frustration with the implementation gap—that such values are not being represented in political decision-making in most countries (again, especially English language countries like the US, UK, and Australia). Rather than postmaterial, many movements are clearly engaged with a concern with material life; this is more along the lines of the new materialism (Coole and Frost, 2010), including an attentiveness to connections, flows, entanglements, and attachments between human and nonhuman systems.

Another core motivation for these movements is their clear articulation as responses to power. There is a real Foucauldian aspect to these movements, as movement activists see themselves—their bodies and practices—as replicating or participating in systems of power they disagree with (Schlosberg and Coles, 2016). This bodily participation and replication of injustice is important to the articulation of justice concerns in these movements, so I will return to this later in the chapter.

Finally, another motivation in sustainable materialist movements is a concern with the human–nonhuman relationship, or its dysfunction. This is the sustainability part of the sustainable materialism—people are really concerned about the relationship between the everyday provision of basic needs—like food, energy, and clothing—and the impact on nonhuman beings and functioning ecosystems. When people in various movements focused on actual practices of the environmentalism of everyday life talk about sustainability, which they do a lot, their focus is on three core ideas and practices linked to a new materialism. First, there is a dedication to the crucial role of rebuilding connections with the vital materiality of the natural world. Second, such movement actors often discuss their attention to flows, systems, stages, or circulations of materials through entangled human and nonhuman communities. And third, movement actors clearly articulate the importance of ethically informed action—that sustainable material practice is a crucial form of political action. For example, food system activists understand soil as a living entity, analyze the way that 'waste' can be brought back into food systems, and articulate something as straightforward as establishing a farmers' market as a political act. Sustainable fashion activists are increasingly attentive to both the ecological and social impacts of every aspect of their supply chains, from pesticide runoff from cotton fields that decimates rivers, to the labor conditions of those sewing and packing clothes; they see demands for the sustainability of the industry as both ethical and political.

Methodology and movement-generated theory

There is one other question to address before we get to the ideas of justice articulated and used by sustainable materialist movements, and that is why we should pay any attention to movement definitions of social and environmental justice at all. What is the actual point of talking to activists about the concepts that motivate them? On the one hand, social theory is not just made up by academics; social movements make social theory. Interviews with activists are key; it is crucial to actually talk to folks about what matters to them, what motivates them, and how they describe the practices in which they are engaged. The discussion in this chapter is based on 100 interviews with activists in food, energy, and sustainable fashion movements in the US, UK, and Australia.

This is different than an alternative approach to academic engagement with social movements and activists, which tends to focus on a more top-down role of academics to inspire, inform, or direct these movements. We often encounter surprisingly paternalistic language from scholars who see theory moving in a single direction, from the experts to the activists. This is not my method or goal. The point here is to examine the political discourse of these movements—in this chapter specifically around their motivating idea of justice. This is not to inspire or direct movements, but to understand the theorizing that goes on in movements, and to bring that movement theorizing into broader political reflection and discussion.

It is only through such intensive interviews with community activists that we can learn how integral notions of social justice are to the sustainable materialism being developed and practiced. In interviews with activists in food, energy, and sustainable fashion movements, the concept of justice kept coming up. As many of these groups identify as **food justice** movements (see Chapter 14 of this volume) or are concerned with **energy justice** (see Chapter 13 of this volume), it is not surprising to find a range of ideas about social and environmental justice. It obviously central to the self-identity of these groups and actors. In an important way, this focus really demonstrates the discursive power of the idea of environmental justice in environmental movements dedicated to new material flows and systems.

Three notions of justice in sustainable materialist movements

Procedural justice, political participation, and material participation

What do activists in these movements mean when they say they are motivated by justice, food justice, energy justice, environmental justice, or social justice? Importantly, these new social movements are expressing a broad range of justice concerns. In particular, there are three different ideas or conceptions of justice articulated by these groups focused on environmentalism and everyday life. In some more depth, let's look at these three.

First, activists focus on political participation and **procedural justice**. We hear groups and interviewees repeatedly emphasize the importance of increasing community *involvement* in the *production* of food and energy relative to their other aims and objectives. This is not, then, simply about classic political participation, but also illustrates an insistence on a sense of *material* participation, or social inclusion in everyday practice. Crucially, movement actors note that this can only be achieved by mobilizing the community and creating a voice.

This dual notion of participation really gets to the core of the materialist aspect of this politics. Participation is not just about voting or being consulted about a policy. It is about *doing* things on the ground—literally, in the case of food movement activists, it is about getting one's own hands dirty. One interviewee[2] noted the relationship between material action and the necessary authenticity of democratic politics when they insisted that "rebuilding parts of democratic participation and people's ability to have a say and actually *do* is a fundamental need of theirs . . . to take back [control]." It is this mix of the political and the material that makes these movements unique and illustrates a reconstructive and democratic idea of procedural justice.

Theoretically, participation can be seen and understood as a distinct notion of justice, as discussed in Chapter 4 of this volume —procedural justice is clearly key in itself, in terms of democratic legitimacy, and key to just outcomes in terms of more fair distributions. Being impacted by political decisions without having a say—especially without a say in decisions that lead to environmental bads in one's community—has been a central focus of environmental justice movements.

But participation and procedural justice are also amongst the central components of a capabilities-based notion of justice. Nussbaum (2011, p. 34) includes the capability of 'control over one's environment' in her key list of capabilities (see also Chapter 6 of this volume) and is explicit about how it entails inclusion in political decision-making. Such participation and control over one's environment is just one aspect of a broad set of needs to be met if citizens are to be capable of building a functioning life for themselves.

We might understand participation as a separate distinct notion of justice, central to movement demands. Or we might see it as a highlighted area of a broader capabilities-based understanding of justice. Either way, the point here is simply that this sense of a participatory duty as an element of justice is very common, broad, and strong in the discourse of these kinds of material environmental movements of everyday life. It is also about more than just political inclusion, though that is crucial; a just process also includes the opportunity for material participation in the making of everyday life.

Justice and power

The second aspect of social justice articulated in these material movements—rather consistently—is the relationship between justice and power. Simply put, groups articulate the need for social justice to require new forms and circulations of power. Members of food groups, for example, consistently note the importance of the movement as a response to the industrialization and alienation of food systems, and to particular growers, corporations, and/or supermarket chains. The idea is to refrain from participating in such circulations of power, and to create new circulations that are more just. Activists in community energy groups clearly see things in a similar way, in terms of creating an energy system that is distinct from, and a threat to, the existing fossil fuel-based energy system.

Importantly, groups articulate this response as going beyond simple resistance or individual action, and, in a very Foucauldian sense, it includes the construction of new modes, circulations, or institutions of power that replace problematic industrialized and power-driven practices. The point, groups note, is to have participants step outside of the industrialized circulation of power and create decentralized, local alternatives and flows that are both collective and connective. Justice is about the construction of, and participation in, new flows of materials through everyday life.

The work of these groups is an attempt to be a counter-power that addresses both social and environmental justice in the construction of these new flows. While activists understand the small scale of their own efforts, and are certainly aware of the ongoing power of the industries they seek to replace, their desire is to step outside oppressive and unjust systems of production and distribution and create and embody alternative practices and flows of power. For sustainable materialist groups, this is how you build a just food system, a just energy system, a sustainable fashion supply chain.

Justice in these movements, then, is about empowerment and the creation of alter-flows of power. Those new flows of food or energy or stuff through communities is, in part, about empowerment—and is very strongly and directly related to a sense of both community and social justice. This conception of justice as power, then, is also directly related to the first notion of justice as procedural; material participation is about both political participation and resistance to unjust power.

Justice and capabilities

A capabilities approach to justice offers a means to bring together the way individuals and communities articulate a variety of related justice concerns, all focused on basic needs and broader community functioning. The general idea of a **capabilities approach** is that justice is not simply about equity of resources, but what people need to convert those resources into a functioning life they choose for themselves. These capabilities are generally discussed as basic political, economic, and social freedoms and rights, as well as basic material needs such as health, shelter, water, and food. And they are broadly used to define indices of well-being, including the UN's new Sustainable Development Goals.

In interviews with food system, community energy, and sustainable fashion activists, the importance of health comes up a lot. For these activists, the point is not simply about individual health, or just physical health, but also more about a general experience of well-being or flourishing for both individuals and communities as a whole.

This concern about health is more prominent in food and sustainable fashion groups than among community energy advocates. However, it is important to note the reality (though one which is not surprising, given work such as Guthman, 2008) that there seems to be much less of an explicit concern for the health of *workers* among food systems activists—and much *more* of that concern in the sustainable fashion interviews, where health of workers in both cotton fields and in production facilities was almost always a key issue. This points to a crucial reality about the concern for justice in the sustainable fashion movement—it is not just about the question "What are my clothes made of?" but more generally "Who has contact with my clothes, and what is the impact of their manufacture, along the supply chain?"—and are those impacts just and equitable? Here, the concern for health is not just about the health of one's own community, but also about the health of those all along the supply chain, even if those lives are far away.

Also, crucially, activists consistently make a connection between the notion of health as an element of justice and the health of the community as a whole. For food system activists, a concern for health is not just about you or me eating better and being healthier; it is about a healthy and functioning community. Many movement organizations refer to their mission as promoting social well-being and connectedness or nourishing people and building community. Activists see their work as creating a better quality of life—and health here is understood as an individual and community life that is more fulfilling, abundant, socially connected, and resilient.

For those in food movements, the attractiveness of working in food is its ability to contribute to improving quality of life. Again, this is not just about one's own quality of life; activists make a connection between changes in the production and distribution of food and local self-sufficiency, health, and well-being. Very similar justice and health connections are made in sustainable fashion. Here, the concern was not just about the individual workers or wearers, but also about communities all through the supply chain, from those in cotton fields to workers who do the sewing all the way to the consumer.

But health is not the only capability; community itself is seen as a capability. Importantly, when it comes to the idea of capabilities, activists tend to see the very vibrancy of community as a capability, in two distinct ways. **Community** is both a capability for each and every person within that community—it is something necessary for individual functioning, so it must be attended to. But community is also understood as an important *subject* of justice—it is one of the things that activists are trying to keep functioning, or to improve its functioning. There is much in activist discussion about the very health of the community and the improvement of community well-being. Such a focus is similar to the findings of Walker (2011), who examined the concept in the area of community energy governance. There, Walker saw community understood as an actor, a scale, a place, a network, a process, and an identity. In these kinds of materialist movements, my point is that community is articulated in a variety of ways, depending on a given context and the actors involved, but it is both an enabler of individual justice and a subject deserving of justice itself.

One way of thinking about this stated importance of community—or one way in which sustainable materialist activists are talking about community—is in how community is about creating connections, attachments, and ongoing relationships. Attachments can be to each other as individuals, or they can be attachments to specific places or neighbourhoods, between people and the local environment, and between people, places, and specific cultures in those places.

Attachment to place is a process and a relationship; attachments are an important and constitutive element of our identity and functioning (Manzo and Devine-Wright, 2014). Disruptions to, and conflicts around, place attachment have long generated key issues in environmental justice organizing and action (see, for example, Schlosberg et al., 2018). What this focus on attachments means is that the connections or attachments people have between their place, identity, and agency really need to be part of our understanding of a broad conception of social and environmental justice. Such a concern for place attachment is especially important in food movements, but comes up in all materialist movements; it is seen as an important contrast to industrial food, energy, and other material systems that do not illustrate any care for places, and instead tend to despoil and undermine them.

The idea here is that **attachment**—to both people and place—is a basic capability (see also Groves, 2015). Attachment is a basic need, essential for a functioning life. And in a capabilities approach, justice occurs with the availability of the basic capabilities of human life. To deny or disrupt those capabilities, then, is an injustice; therefore, the disruption of attachment is an injustice and the creation of attachments displays a dedication to justice.

This focus on preserving and constructing attachments as an element of justice offers a contrast to what we usually hear in environmental justice movements. In most environmental justice issues, it is the act of detachment—or the undermining of attachment to place—that is the issue. For example, with lead in drinking water in Flint, Michigan, in addition to being unequally distributed—certainly because of race—we hear about how the toxicity in the water alienates people from place (see also Chapter 16 of this volume). If the very place we live is toxic, it undermines our attachment and is a clear form of injustice.

But what we hear more in these sustainable materialist movements is that everyday life around the provision of material needs is currently alienating. The desire—for example, in the food movements—is to *reattach* in some way with each other, and with the nonhuman realm. It is the *lack* of attachment—rather than detachment—that is the focus. Yes, activists talk about how alienating current food systems are, and that they have created practices where we don't engage each other, or the farmers, or the land.

But, crucially, what activists articulate is not just about *d*etachment, but more about how food and energy justice movements are seeking to *r*eattach people to each other, to specific places, and to the material and natural worlds more broadly—to build attachments and community. Even in sustainable fashion, the justice concern is about how the current material practices of fashion are toxic to many in the supply chain, but that by bringing attention to these practices, sustainable fashion can both bring those practices out from the dark and make connections and attachments between producers, consumers, and environment.

This concern with attachment makes clear that activities focused on building community or rekindling place, through food and energy and everyday practices, can and should be thought of as environmental justice activism. As one activist put it:[3]

> Do we feel connected deeply to each other, to other human beings, not just our immediate tribe but to humans, all of us? Do we feel connected in a reverent way to the larger, natural world that we're a part of? All humans feel good when we do.

And that's the point—we feel good when we are attached, and these movements are attempting to achieve that as an element of the just lives they are trying to construct.

This concern about attachment to community is one of the threads that ties together the whole range of justice concerns in these movements—around participation, capabilities, and community. Rebuilding such attachments, communities, and sustainable relations with each other and the nonhuman realm are all absolutely part of what the achievement of environmental justice would look like.

Pluralistic justice

This illustrates one aspect of the conceptions of social and environmental justice in sustainable materialist movements. These conceptions are pluralistic in a simple sense, in that activists have multiple conceptions of justice and hold them simultaneously.

The importance of community and attachment, however, illustrates a second aspect of justice in these movements, as well. Similar to findings on environmental justice more generally, participants in these movements see these conceptions of justice not as divergent and incommensurable, but as interrelated, very practically linked, and mutually reinforcing. In other words, not only do we find arguments for participation, power, capabilities, and community attachment as elements of justice, but we also find arguments for the **relationship** between them. The interviews with activists illustrate that these different articulations of justice are understood as thoroughly and practically linked.

For example, one activist in a food movement states the following:[4]

> For me, it's about health, it's about supporting your local economy and supporting your community, which is the people who grow your food, produce your food, process your food. . . . Food is about community, it's about nourishment, and it's about community as well.

In that one statement, we see a concern for health, power, and a new system and community. These concerns about different aspects of justice are inherently linked. Another activist illustrates the same kind of pluralistic understanding:[5]

> I think again it's people feeling connected to the land again. . . . People taking their power back, really. . . . So I think that's helped people get into social justice once they start, because once they start to understand the connection between things . . . where food comes from . . . then it enhances their capacity to make connections in other ways and therefore they become more politically aware and just their sense of outrage or injustice actually—you know, they can actually hone that in to something.

Many activists make these same kinds of connections, linking social justice, equity, food security, community power, just and sustainable supply chains, community engagement, and material participation. As many communicated, all of these things and approaches are simply part of the activism that they do.

These movement groups illustrate that it is possible to articulate *qualitatively different* sorts of justice claims simultaneously. While the conceptions of justice differ, what we see in these groups—just as I have observed in the environmental justice movement (Schlosberg, 2007)—is a willingness, or an insistence, to articulate the commonalities, connections, and relationships between different experiences of injustice and arguments for just arrangements. Beyond the theoretical attractiveness of pluralism, any attempt to reduce group statements about justice to their parts ignores the reality that, for groups themselves, notions of justice are fundamentally interconnected and interdependent—and linked to the place and environment in which they live.

As I and others have argued, a capabilities approach to justice has a kind of '**internal pluralism**' and can incorporate a range of conceptions of justice without privileging any particular notion (Walker, 2009). But I think that these findings about sustainable materialist movement conceptualizations and motivations necessitate two key additions to this pluralist theoretical position.

First, it is clear that individual and collective functionings are bound together, relationally, and *often* in place. That is, participants understand and articulate the relationship of different capabilities, at different levels—from individual to community to ecosystem to the planetary. These linkages are key, and the broader normative project of a pluralistic understanding of environmental justice requires that we acknowledge and embrace the reality of the relational complexity between understandings of, and desires for, social and environmental justice.

The second point to make is that, in order to embrace the reality of the relationship between notions and experiences of justice, and to understand environmental injustice as the undermining and depriving of individuals and communities of basic capabilities, issues of power are central. We must understand and accept the reality of the role of the circulation of power in the construction of injustice, and the role of resistance to that power and the construction of alternative flows of power, as central to environmental justice activism. For participants in sustainable materialist movements, the act of reconstructing sustainable material flows as an explicit act of resistance is inseparable from the project of realizing a just world where the mutual flourishing of individuals, communities, and environments is possible.

The key point here is about the importance of the *relation* between power, interruption, and proper functioning of those individuals, communities, and environments. In order to properly understand environmental *injustice* as multiple, pluralistic, and—crucially—fought against, power needs a more prominent place in our theorizations. Movement activists certainly do not take on all of the components of what Pellow (2017) calls a critical environmental justice in Chapter 22 of this volume, but there is clearly an understanding of the role of a critique of power and the role of bodily resistance.

This is really an argument about the role of power in a capabilities approach to justice in general, and environmental justice in particular. As I laid out earlier, my interest here is not only in getting a sense of how these sustainable materialist movements actually create a theory of social and environmental justice for themselves, but also in what it is that they are articulating that can help identify the limits of existing academic theory. On community, attachments, plurality, and power, sustainable materialist movements offer much to the academic realm in terms of innovative and engaged ways of theorizing about justice.

The limits of sustainable material action

In closing, I want to address two important issues—first about the potential real political impact of sustainable materialist movements, and second to do with the limitations of the conceptions of justice they articulate. As for the first, it is always important to address the classic question that arises when we think about these small-scale, localist movements like sustainable materialist groups focused on community food, energy, and sustainable fashion. Can they, as small-scale alternative flows, really take on large-scale and old-school sovereign and corporate powers and injustices? Critics (for example, Blühdorn, 2017) argue that staying local and focused on practice, these movements may be seen as apolitical or postpolitical, avoiding the real policy realm, and the power and injustices of international regimes; they are, the argument goes, more a form of coping in an unjust world than a way of constructing social and environmental justice.

To be clear, none of the activists interviewed for this project claimed that their practices will simply and easily bring about a major transformation to social justice; nor do they argue that their materialist action should replace traditional political strategies and battles. Activists in these movements see them as supplements to—not replacements for—more traditional political battles and tactics. But they do thoroughly understand the construction and practice of these movements around more sustainable material flows as political acts, and they see making a difference on the ground as a crucial contribution to battles for justice.

Sustainable materialist movements can be seen as examples and forms of what are called prefigurative politics (Cornell, 2009, Maeckelbergh, 2011). Their activism is about constructing forms of organization and practices that are ready to scale up or, more likely, replicate and multiply horizontally. For the activists and organizations involved in sustainable materialism, the very crucial act of stepping out of flows of power and setting up counterflows in sustainable, productive, and vital ways is a form of doing politics, doing social and environmental justice, and of living and replicating just practices and flows in the face of larger power structures and injustices. Is this all that is necessary to achieve environmental justice and social justice more broadly? No. Do activists really believe that it is all that can be done politically? Again, no. Do they believe it is something important, a contribution to broader political battles for justice? Absolutely. Can such sustainable materialist movements, systems, and flows serve as examples of just practices that can be replicated? Yes, and that is one of the motivations articulated by activists.

Politically, then, there is potential and vibrancy. And yet, even with the breadth and plurality of conceptions of justice at play in these movements, there are some real disconnects between the notions of justice raised by activists and previous understandings of environmental justice such as those in the first part of this volume. It is important to note what is *absent* in the vast majority of the 100 interviews of sustainable materialist activists. Ideas of inequity or inequality were rarely mentioned, which means that more classic notions of **distributive injustice** (see Chapter 3 of this volume) were not really articulated. And there were rarely comments on, or definitions of, justice that address recognition or disrespect (see Chapter 5 of this volume) of communities as a whole.

In addition, and even with the pluralist conception and practice of environmental justice embodied by these movements, the issue, obviously, is that they do not articulate the kind of intersectional multidisciplinary understanding of a broad critical environmental justice offered by David N. Pellow (2017) or Laura Pulido (2015, 2017). On the one hand, these activists are clearly in tune with some of the ideas of a critical environmental justice in their understanding and response to *power*. There are certainly critiques of the power of corporations and of the state—and of their collusion in undermining both environment and community autonomy. But on the other hand, the focus is often quite limited—for example, the focus is rarely racialized or decolonized. Other than African American food justice groups in the US that referred to race and movements like Black Lives Matter, race was rarely mentioned. It was also rare to hear explicit commentary from sustainable materialist activists on feminism or gender, or on state violence. Comments on decolonizing or recognizing complicity in settler colonialism were also nearly absent. Overall, there was much more talk about species interdependence, and reconnecting with the nonhuman realm, than there was about race. While the interviews of activists were designed to match the demographics of the countries in which they are based, these are primarily white movements—and white movements that are not thinking as active allies with other movements.

There are, then, some key differences between current work on understandings of environmental justice and the discourses of the term being used in these sustainable materialist movements. This illustrates a real disconnect between the necessity and importance of critical environmental justice in theory, in the academy, and in our understanding of the structures that build and maintain a whole host of oppressions on the one hand, and, on the other hand, the way the idea and discourse of environmental justice is being used by an increasing number of activists outside the usual focus areas of the traditional environmental justice movement. One of the key questions, then, is whether all of these movements—these quite empowered and empowering movements creating new systems for the flow of materials through communities in just ways—can actually achieve what they hope to. Can we have just material systems without sustained public attention to a broader range of **systems of oppression**?

Conclusions

Summarizing, the central argument about environmental justice in sustainable materialist movements is twofold. First, these new social movements focused on sustainable food, energy, and fashion systems are articulating a broad range of justice concerns—around procedural justice, a response to power, a set of capabilities, and a desire for community and attachments.

Second, these various concepts of social and environmental justice are seen not as competing, but as plural, related, and interlinked. They are also seen as not only interlinked with each other, but also with a broader sense of sustainability and just treatment of the environment and ecological systems as well.

The broad point here is that some core conceptions of environmental justice are alive and well in movements around the basic needs of everyday life. The conception of justice in sustainable materialist movements is rich, engaged, linked, and vibrant—illustrating the strength and influence of the idea and discourse of social justice generally, and environmental justice specifically. And yet, the broad and pluralistic conception and discourse of environmental justice in such movements does not address the full range of concepts and issues in environmental justice, in particular around racial justice (see Chapter 17 of this volume). Overall, while there has been an embrace of much of the environmental justice movement discourse in the sustainable materialist movement, that discourse has taken on a different set of frames and targets, and illustrates key differences between the movements.

Follow-up questions

- What are similarities and differences between sustainable materialist movements and other groups dedicated to environmental justice?
- Is the activist focus on the material flows of things like food and energy through communities really 'political', or is it just consumerist?
- Can we address crucial environmental justice issues of race and class through a focus on material flows?

Note

1. The ideas in this chapter are drawn from and more fully developed in David Schlosberg and Luke Craven, *Sustainable Materialism: Environmental Movements and the Politics of Everyday Life* (Oxford University Press, 2019).
2. Food Activist, Australia, 2017.
3. Food Activist, USA, 2017.
4. Food Activist, Australia, 2017.
5. Food Activist, USA, 2017.

References for further reading

Agyeman, Julian, David Schlosberg, Luke Craven and Caitlin Matthews. 2016. 'Trends and directions in environmental justice: from inequity to everyday life, community, and just sustainabilities', *The Annual Review of Environment and Resources*, 41, 6.1–6.20.

Middlemiss, Lucie. 2018. *Sustainable Consumption: Key Issues*. Routledge.

Schlosberg, David and Romand Coles. 2016. 'The New Environmentalism of Everyday Life: Sustainability, Material Flows, and Movements', *Contemporary Political Theory*, 15(2), 160–181.

Schlosberg, David and Luke Craven. 2019. *Sustainable Materialism: Environmental Movements and the Politics of Everyday Life*. Oxford: Oxford University Press.

References used in this chapter

Agyeman, J. 2013. *Introducing Just Sustainability: Policy, Planning, and Practice*. London: Zed Books.

Blühdorn, I. 2017. 'Post-capitalism, post-growth, post-consumerism? Eco-political hopes beyond sustainability', *Global Discourse*, 7, 42–61.

Coole, D. and S. Frost. 2010. *New Materialisms: Ontology, Agency, and Politics*. Durham: North Carolina Duke University Press.

Cornell, A. 2009. 'Anarchism and the movement for a new society: direct action and prefigurative community in the 1970s and 80s', *Perspectives on Anarchist Theory*, 12, 36–69.

Eckersley, R. 2019. 'Ecological democracy and the rise and decline of liberal democracy: looking back, looking forward', *Environmental Politics*, doi:10.1080/09644016.2019.1594536

Gordon, J.A. and C. Hill. 2015. *Sustainable Fashion: Past, Present, and Future*. New York: Bloomsbury.

Groves, C. 2015. 'The bomb in my backyard, the serpent in my house: environmental justice, risk, and the colonisation of attachment', *Environmental Politics*, 24, 853–873.

Guha, R. and J. Martinez-Alier (eds.). 1997. *Varieties of Environmentalism*. London: Earthscan.

Guthman, J. 2008. 'Bringing good food to others: investigating the subjects of alternative food practice', *Cultural Geographies*, 15, 431–447.

Inglehart, R. 1977. *The Silent Revolution: Changing Values and Political Styles Among Western Publics*. Princeton, NJ: Princeton University Press.

Maeckelbergh, M. 2011. 'Doing is believing: prefiguration as strategic practice in the alterglobalization movement', *Social Movement Studies*, 10, 1–20.

Manzo, L.C. and P. Devine-Wright. 2014. *Place Attachment: Advances in Theory, Method, and Applications*. London: Routledge.

Meyer, J.M. 2015. *Engaging the Everyday: Environmental Social Criticism and the Resonance Dilemma*. Cambridge, MA: MIT Press.

Meyer, J.M. and J.M. Kersten. 2016. *The Greening of Everyday Life: Challenging Practices, Imagining Possibilities*. Oxford: Oxford University Press.

Micheletti, M. 2010. *Political Virtue and Shopping: Individuals, Consumerism, and Collective Action*. New York: Palgrave.

Nussbaum, M.C. 2011. *Creating Capabilities: The Human Development Approach*. Cambridge, MA: Harvard University Press.

Pellow, D.N. 2017. *What is Critical Environmental Justice?* Cambridge, UK: Polity Press.

Pulido, L. 2015. 'Geographies of race and ethnicity 1: white supremacy vs white privilege in environmental racism research', *Progress in Human Geography*, 39, 809–817.

Pulido, L. 2017. 'Geographies of race and ethnicity II: environmental racism, racial capitalism and state-sanctioned violence', *Progress in Human Geography*, 41(4), 524–533.

Sbicca, J. 2018. *Food Justice Now! Deepening the Roots of Social Struggle*. Minneapolis, MN: University of Minnesota Press.

Schlosberg, D. 2007. *Defining Environmental Justice: Theories, Movements, and Nature*. Oxford: Oxford University Press.

Schlosberg, D. 2019. 'From postmaterialism to sustainable materialism: everyday life and practice-based environmental movements', *Environmental Politics*, doi:10.1080/09644016.2019.1587215

Schlosberg, D. and R. Coles. 2016. 'The new environmentalism of everyday life: sustainability, material flows, and movements', *Contemporary Political Theory*, 15(2), 160–181.

Schlosberg, D. and L. Craven. 2019. *Sustainable Materialism: Environmental Practice and Everyday Life*. Oxford: Oxford University Press.

Schlosberg, D., L. Rickards and J. Byrne. 2018. 'Environmental justice and attachment to place: Australian cases', in R. Holifield, J. Chakraborty and G. Walker (eds.), *The Routledge Handbook of Environmental Justice*. London and New York: Routledge.

Walker, G. 2009. 'Beyond distribution and proximity: exploring the multiple spatialities of environmental justice,' *Antipode*, 41, 614–636.

Walker, G. 2011. 'The role for 'community' in carbon governance', *WIREs Climate Change*, 2, 777–782.

Walker, G., P. Devine-Wright, S. Hunter, H. High and B. Evans. 2010. 'Trust and community: exploring the meanings, contexts and dynamics of community renewable energy', *Energy Policy*, 38, 2655–2663.

24 Mobilizing 'intersectionality' in environmental justice research and action in a time of crisis

Giovanna Di Chiro

Learning outcomes

- Gain basic understanding of the concept of "intersectionality," charting its roots in Black feminist thought and social movement activism.
- Understand how intersectionality's focus on linking critical analysis and social justice praxis resonates with and strengthens the transformative power of environmental justice research and action.
- Develop deeper awareness of how multiple and mutually reinforcing systems of oppression and unequal relationships to power and privilege shape people's experiences of environmental and climate disruption.
- Gain broader appreciation of how the knowledge and agency of those most marginalized build diverse coalitions across social and geographic borders to address intersecting social, ecological, and climate crises.

Introduction

> There is no such thing as a single-issue struggle, because we do not live single-issue lives.
> (Lorde, 1984, p. 28)

> Let us wake up, humankind! We're out of time. We must shake our conscience free of the rapacious capitalism, racism, and patriarchy that will only assure our own self-destruction.
> (Berta Cáceres, 2015)[1]

The idea expressed in Audre Lorde's oft-quoted statement from her 1984 book *Sister Outsider*—that we inhabit complex, multi-dimensional lives, shaped by a multiplicity of identities, a diversity of experiences, and unequal relationships to power, privilege, and oppression—is today generally accepted as axiomatic within feminist thinking and activism around the world. More than thirty years later, the powerful words spoken by Indigenous Honduran environmental justice activist Berta Cáceres embody this idea. Lorde's quote would presage the emergence of the now widely used word "intersectionality," a term that has achieved global dispersion as a "travelling theory" and that has been mobilized in many forms across a broad spectrum of geographic, disciplinary, and social movement contexts

(Cho et al., 2013). Additionally, as illustrated in Cáceres' quote and her environmental justice activism, intersectionality has been widely adopted by social activists as a model for guiding community organizing practices committed to coalition building among diverse social movements, including movements dedicated to advancing environmental and climate justice. How have the concepts underlying the idea of intersectionality called forth by Lorde "travelled" into environmental justice activist framings expressed, for example, in Cáceres' call for humankind to "wake up"?

Intersectionality's rapid ascendancy as a word that could represent the inclusion and intermingling of multiple identities and diverse political standpoints has garnered a substantial digital presence, "going viral," while also sparking heated debates. In 2011, the meme "My feminism will be intersectional or it will be bullshit," coined by Flavia Dzodan (2011), a young feminist blogger bemoaning the racism that all too often passes unchallenged by white feminists whose politics largely focus on a liberal agenda promoting equal rights among women, generated considerable internet traffic, and was appropriated as a marketable commodity, printed and sold on T-shirts and coffee mugs with no credit or remuneration given to the author (Romano, 2016). The unapologetic brassiness of the quote amplified intersectionality's proliferation in and beyond feminist circles. On conservative websites, intersectionality has been decried as "the most dangerous problem in America," perpetrating an extreme identity politics and creating a "hierarchy of victimhood" where "white men would be at the bottom" (Coaston, 2019). Intersectionality surfaced on liberal feminist websites during the 2017 Women's March on Washington proclaiming women's unity against misogyny, and it showed up in Senator Kirsten Gillibrand's (D-N.Y.) presidential primary campaign when she tweeted "the future is female [and] intersectional."[2] Some left-leaning environmental and sustainability websites deploy intersectionality as shorthand signalling an environmentalism that embraces a socially critical edge: "My environmentalism will be intersectional or it will be bullshit" (Ramsey, 2014). Or, more directly, "If your environmental activism isn't intersectional, you aren't actually doing good in the world. You're just helping those who already hold privilege" (Willow, 2018). It would seem that citing intersectionality—whether as peril or virtue—becomes a yardstick measuring the breadth and depth of one's politics.

Like many viral vocabularies that travel beyond the context and content of their provenance (c.f. **diversity, inclusion, sustainability**) and that cross diverse academic, political, and social movement domains, intersectionality has been mainstreamed, leading to what some argue is a "shallow[er] apprehension" (Carastathis, 2016, p. 3) of its original meaning. Having achieved the status of a cliché, even a buzzword, how can intersectionality be reclaimed as the powerful critical tool and valuable guide for coalitional praxis in support of social justice that many argue encompasses its enduring legacy? I am particularly interested in the travel or viral nature of intersectionality into the realm of environmentalism, and especially into academic and activist circles focusing on environmental justice. In this chapter, I offer a brief discussion of the use, appropriation, and mobility of this now ubiquitous "keyword," and reflect on how in my environmental justice research and activist collaborations I draw on and mobilize the idea in the current moment of global social and ecological crisis.

The mobility and mobilization of a keyword

While the *term* **intersectionality** (based on the use of the metaphor of a "traffic intersection" developed in a 1989 essay by feminist legal scholar Kimberlé Crenshaw)[3] would catch fire in the late 1980s and 1990s, the *analysis* underlying intersectionality's core understanding—that people's experiences of social inequality and vulnerability are shaped not by a single factor, such as their gender or racial identity, but by multiple and mutually reinforcing axes of social division buttressed by unequal relations of power and privilege—can be traced through a much longer genealogical path. Many Black, Indigenous, and other feminist scholars of colour have pointed out that the underpinnings of intersectionality are grounded in diverse conceptual and political lineages pre-dating the moment it was named (Collins and Bilge, 2016). These lineages include 19th-century Black women's abolitionist writing and politics, such as Sojourner Truth's critical analysis of the irreducibility of multiple oppressions in her "Ain't I a Woman?" speech (Painter, 1997); the 1960s Black feminist critiques of racism and patriarchy, such as Frances Beal's (1969) elucidation of the "double jeopardy" Black women face under white supremacist US capitalism; the 1970s Combahee River Collective's naming the "interlocking" and "simultaneous" **systems of oppression** impacting the lives of Black queer women (Taylor, 2017); and the heterogeneous coalition politics forged through social movement activism among women and men from Black, Chican@/Latinx, Indigenous, and Asian American communities revealing the complexity of the **matrix of domination** that affected the lives of minoritized women and men in the US and beyond (Collins and Bilge, 2016).

Intersectionality is a constant struggle

Explaining and recovering the origins of intersectionality (either as a concept, a word, or an academic field) became a growth industry in the early decades of the 2000s with the production of an assortment of critical feminist analyses underlined by a consistent tenet: intersectionality fundamentally calls for a synergistic relationship between critical inquiry and critical praxis. For many feminist critics, intersectionality must 1) provide *both* more complex and comprehensive analyses of how and why social inequalities exist *and* the critical tools to aid in the empowerment of marginalized communities and individuals (Collins and Bilge, 2016); 2) advance the goals of social justice and be grounded in and support liberatory politics (Cho et al., 2013); and 3) generate critical analyses of multiple and interlocking identities and oppressions, *not* as an "additive" junction of essentialized categories, but as critical praxis enabling the possibilities of coalition building to dismantle all systems of oppression across a wide spectrum of social justice movements and decolonial projects (Spade, 2013; Carastathis, 2016; Taylor, 2017).

For some critics, the mainstreaming and incorporation of intersectionality into academic institutions and policy arenas such as UN human rights frameworks erased its historical origins in **Black feminist** activism and evacuated it of its socially critical muscle (Hancock, 2016; Carastathis, 2016). As Collins and Bilge (2016) argue, the institutionalization and naming of a concept that originated in concrete campaigns for social justice serve to make it more compatible with the academic norms of abstraction and objectivity, and often lead to the defanging of its core critiques of inequality and domination. The incorporation of intersectionality within academia, even in disciplines that offer a critique of the exclusionary histories and practices of the university, such as women/gender/sexuality studies, ethnic studies, or Black studies, they write, "seemingly suppressed the transformative and potentially disruptive dimensions" of these critical interdisciplinary interventions (p. 85).

Many feminist critics' qualms about intersectionality's mobility point to the tendency to reduce the critical focus on the multiplicity of mutually reinforcing oppressions to a shallow, individualized form of identity politics that serves as a "guarantor of political or theoretical inclusion, or . . . as a unified theory of 'multiple oppressions'" (Carastathis, 2016, p. 8). Intersectionality zeros in on the conceptual limitations of single-issue analyses, revealing how they render invisible, or absent, those individuals and groups who inhabit more than one subjugated and "othered" category, thus making them illegible in legal or other institutional structures (governmental, medical, educational, political), and therefore having few or no options for legal or other remedies to combat the harms of prejudice and discrimination (Crenshaw, 1989). The insistence instead on the need to focus on *multiple* axes of oppression has, according to many critics, been misread as connoting an additive approach suggesting that "adding more to the list" increases the visibility of oppressed identities. But the brainwave of intersectionality as a concept is that liberal structures of representation, by only recognizing single categories of difference, are insufficient to remedy the injustices and discrimination faced by Black women or others who are dominated by multiple categories of difference. Intersectionality, therefore, is *not* about inclusion, it is about **absence**; it helps to make the dangers of multiply interacting exclusions visible. The remedy is not about including and reasserting essentialized identities; it is about recognizing that identity categories are socially and historically situated and are not merely about individuals: they are *coalitional*, "constituted by internal differences as much as by commonalities" (Carastathis, 2016, p. 184). Furthermore, understanding identity categories not structured by a single axis but through multiply intersecting axes helps to imagine potential cross-identity and cross-border coalitions by illuminating "interconnections and interrelations, as well as grounds for solidarity that reach across and reveal differences within categories of identity" (p. 185).

Environmental justice and the politics of intersectionality

Political theorist María José Méndez (2018) invokes the story of the late Indigenous Honduran environmental justice activist Berta Cáceres to illustrate how erroneously equating intersectionality with "inclusion" evades the fundamental truth that "intersectionality kills" (p. 8). In stark contrast to the liberal feminist framing of intersectionality as a cheery metaphor for inclusion, the intersectional feminist politics of Cáceres was embedded in a critical, **decolonial environmental justice**[4] project to resist the perpetual violence, death, and destruction wrought by the mutually reinforcing, oppressive, and controlling forces of "rapacious capitalism, racism, and patriarchy."[5] Cáceres, the renowned Lenca activist, demonstrated a profound understanding of the core tenets of intersectionality, though she never used the term. To prevent our own and the earth's demise, she warned, humankind must "wake up" to the current global crisis and together break free from the interconnected, violent systems of oppression that subjugate our bodies and psyches.

Along with hundreds of grassroots activists and allies, Cáceres and the Council of Popular and Indigenous Organizations of Honduras (COPINH), the organization she co-founded, led a decade-long battle opposing the 2009 US-backed military coup in Honduras that ousted its democratically elected president and put into power a repressive government working hand in glove with global capital to sell off the country's land and natural resources, and exploit its predominantly peasant and Indigenous population (Frank, 2018). In 2015, Cáceres and COPINH were awarded the prestigious Goldman Environmental Prize to commemorate their successful organizing against the construction of the Agua Zarca Dam on the Gualcarque River, a joint project of the Honduran company Desarrollos Energéticos SA (DESA) and Chinese state-owned Sinohydro, the world's largest dam developer. The Gualcarque River is sacred to the Lenca and Afro-Indigenous Garífuna peoples' lifeways and cosmovisions: "In our worldviews," Cáceres explained in her awards speech, "we are beings who come from the Earth, from the water, and the corn."⁶ Cáceres' speech displayed her affinity with an intersectional, **multi-species environmental justice** perspective founded on the interconnectedness between human and non-human nature. She affirmed that we must act in coalition with the Gualcarque River, and many other "gravely threatened rivers" who need our support and solidarity to fight this existential planetary crisis. "Our mother Earth," she continued, which "has been militarized, fenced-in, and poisoned . . . demands that we take action."⁷

Cáceres' intersectionality grew from a systemic analysis of the historical and global entanglements of corporate and colonial power grounded in Lenca and Garífuna knowledge systems based on **natureculture** relationalities and the interdependence of human and non-human beings (Haraway, 2018). Recounting an occasion when she observed Cáceres' grounded intersectionality, Méndez (2018) describes a community assembly in 2014 at which Cáceres and other activists co-created a web-like power map:

> The map looked like a spider web that tried to systematize the invisible connections between development projects, criminal gangs, private security companies, paramilitary groups and U.S.-funded military forces; relations with severe consequences for the bodies and territories of the various peasant, indigenous, feminist, union and LGBTQ groups gathered at the assembly. Unlike the formal spatial imaginary of rigid state and non-state boundaries, this map revealed the blurrier and trans-nationally embedded infrastructure of power in Honduras. Standing against this chalkboard, Berta insisted that this multi-layered killing machine of bodies and worlds could only be disassembled through cross-border solidarities that drew strength from spiritual and ancestral wisdom in the hopes of nurturing alternative designs of relating to nature and living beings.
>
> (p. 8)

By providing a spatial representation of the globally interlinked power systems that enable new extractions and land grabs, this complex map helped the community envision the systems it would need to resist and the cross-border, trans-national coalitions it would need to organize. For Méndez, this activist-generated map reveals the kind of intersectionality that "destroys bodies and worlds," the intersectionality of structures of subordination and violence "that kills" (p. 8). Tragically, the intersectional, systemic linkages Cáceres and others had mapped out delivered on their deadly promise; Cáceres was assassinated in her home on March 3, 2016, becoming one of more than 127 land and water protectors killed in Honduras since 2010 (Global Witness, 2019).

"Berta no murió, se multiplicó!" (Berta did not die, she multiplied!) was the rallying cry that reverberated across the globe in demonstrations from Tegucigalpa to Washington DC, signalling that the global coalitions Cáceres helped create would not be intimidated into silence and inaction (Albaladejo, 2016). Thousands of her *comadres* came together to demand that her killers be brought to justice and to reassure people across the world that the struggle for an intersectional, anti-racist, feminist, Indigenous, LGBTQ-supportive, anti/decolonial, naturecultural, and care-based environmental justice would live on. Not only did Cáceres' memory and presence grow and multiply in movements linking feminist, Indigenous, and environmental justice struggles, but her critical analysis of *multiplicity*—that reveals the interlinking of structures of social and ecological domination and exploitation—would also multiply (Frank, 2018). A documentary film produced a year after her death, taking the rallying cry for its title, tells the stories of the hundreds of Indigenous and other marginalized communities from all corners of the earth who have strengthened their resolve to defend "madre tierra, their territories, their rivers, and their rights to **Buen Vivir**,[8] that is, their own indigenous forms of sustainable development" (Vinal, 2017).

Mobilizing intersectionality for environmental justice

Like Cáceres' understanding of the interconnected axes of **power** that harm bodies and lands, the environmental justice movement has long embraced the paired axioms of intersectionality: that the multiple and intermeshed oppressions experienced by marginalized communities and environments around the world *kill*, and that the interconnections brought to light by the critical analysis of these multiplicities open up possibilities for trans-border collaborations and political coalitions. These coalitions might not resemble the cosiness of "home," as Reagon (1983) has made clear, but would acknowledge their common purpose and "join together" to defend and care for the Earth—our *collective home*. This conjuring of social and ecological "commonality" situated in a context of intersectional "difference" similarly animates many other Indigenous, feminist, and **anti-colonial environmental justice** activists around the world. (e.g. Desmarais, 2007; Pellow, 2018; Estes and Dhillon, 2019; Sze, 2020). Yet, in our current moment of intersecting social and environmental crises, we see the resurgence of appeals to a shallower "commonality" in, for example, the calls for "inclusion" in the women's movement as noted earlier, and in the exclamation that "our house is on fire," the global metaphor voiced by teenaged climate justice activist Greta Thunberg (2019) beseeching world leaders to unite and take action to avert catastrophic climate disruption.

Whose house is on fire? Coalition and the commons in a time of crisis

Invoking a shallow conception of the commons is not new. As the 1980s were racking up multiple "hottest years" in one decade, the legendary Brundtland Report, *Our Common Future*, an interdisciplinary social and economic policy analysis commissioned in 1987 by the World Commission on Environment and Development, would raise the alarm of a coming global social and ecological breakdown (World Commission, 1987). *Our Common Future* called for a renewed vision of a "global commons" rooted in the conviction that global (particularly North–South) cooperation was now vitally necessary to stave off certain catastrophe and an impending, worldwide tragedy of the commons. This 1980s-style commons discourse of one-worldism, signified by the universal "we" (Di Chiro, 2003) was roundly criticized as a tone-deaf pretence of the international environmental establishment's attempt to paint the UN's family of nations as if we were all paddling around in the same boat or comfortably seated on "Spaceship Earth." For many environmental justice critics, the anaemic response to the declarations of global crisis and the need for North–South unity as "sustainable development" or "sustainability" has amounted to a set of dismally minimalist reforms to slow down capitalist growth while leaving intact the interlocking and unjust global political and economic systems that led to worldwide crises in the first place.[9]

Thirty-some years later, the current global crisis narrative insists that we embrace our common danger and act, as Thunberg (2019) implores, as if our collective "house is on fire." Thunberg's house-on-fire metaphor is a plea for unity and commonality of purpose for saving the earth, a compelling and rational argument coming from a remarkable and gutsy teenager, although she was not the first courageous young person to reprimand the bigwigs of the international environmental establishment (Asmelash, 2019).[10] The burning house metaphor relies on a readiness to envisage the whole earth as our house or home, and a willingness to connect the global scale idea of the earth as a "house on fire" to the terrifying idea of your *own* house engulfed in flames. As a political message meant to mobilize unity, this metaphor presumes that everyone can relate to it. Many can, except, of course, those who do not have a house, or who are struggling with homelessness and gentrification, or who are displaced and forced to migrate from their homelands. Drawing on an intersectional environmental justice analysis of disproportionate impacts of ecological destruction and climate disasters, the question becomes: *whose* house is on fire?

The appeal to the universality of the commons has often plagued environmental and climate politics, and many environmental justice activists maintain that the **Anthropocene** story has emerged as yet another commons discourse bringing with it the colonialist assumption of a universal humanity and the erasure of differential human experiences and impacts (Di Chiro, 2018). Environmental justice scholars argue that not *all* humans have been responsible for or benefited from sucking the earth dry (Yusoff, 2018), and—like the Honduran Lenca people's **cosmovisions** interconnecting humans and nature—*some* humans have created societies based on models of interdependence, reciprocity, and responsibility to and with each other and with non-human relatives and ecosystems. For many Indigenous environmental justice scholars and activists, any collective human project that aims to address the urgency of climate disruption must first *urgently* dismantle and decolonize the various arms of the settler state, including the university and its corporate sustainability regime (Todd, 2015; McGregor, 2018; Whyte, 2018).

If the aim is to build a thriving, ongoing commons, the way to instigate more fruitful coalitions and collusions would be *first* to engage in a politics of refusal and repair: a refusal to reproduce the settler, imperial, and carceral logics of mainstream environmentalism or the corporate university, and a commitment to repairing and healing damaged relationships and broken lands (Grande, 2018; Tuck and Yang, 2018; Whyte et al., 2018). Any new "commons" proposal, these critics argue (whether from the UN, the scientific community, or university offices of sustainability), that hopes to spur unified action to save the earth, and that claims that we are *all* responsible and *all* in the same boat, will not readily succeed. The dream of collectively co-creating our common future remains a fantasy without a meaningful engagement with the principles of intersectionality.

Intersectionality does appear, however, in some new commons talk. Many responses to the current socio-ecological crisis focus on anti-capitalist, collectivist, environmental justice, and feminist critiques of global capitalism's assault on social reproduction (Hoover, 2017; MacGregor, 2017; Federici, 2019; Chapter 18). In eco-activist spaces, for example, we have seen the rise of movements for **degrowth** (see Chapter 8 of this volume), worker-owned cooperatives, and solidarity economies. Some **just transition** movements (see Chapter 19 of this volume), like the Climate Justice Alliance's "It Takes Roots" movement, have started to build alternative collective governance and economic structures (Gabriel, 2018; Patterson, 2018; Movement Generation, 2019; Chapter 19 of this volume). In national politics, we have seen a revival of eco-socialist proposals such as the Green New Deal emanating from the quasi left-leaning wing of the US Democratic party (Klein, 2019).

Diverse cultures of collective care, commoning, and mutuality, however, have always existed and persist today in many Indigenous land-based social and eco-systems and pedagogies (Simpson, 2017; Nelson and Shilling, 2018). **Collective care** work also exists in communities that have been historically excluded from mainstream economic activity, such as in the long history and current wave of African American cooperatives and community-based, mutual aid organizing for sustaining everyday life and collective survival (White, 2018). Extensive examples and creative models of Black collective endeavours in the colonies, in the Caribbean, in Africa, and in cooperatives such as the Free African Society starting in the late 19th century in Philadelphia, are richly documented in W.E.B. DuBois' (1907) early sociological research.

Commons/commoning

Many commons theorists pay homage to Nobel Prize-winning economist Elinor Ostrom's (1990) work on rational choice theory and the co-management of the commons by small-scale farmers, fishers, and peasant herders. Ecofeminist commons theorists, however, informed by intersectional feminist political ecology lenses (e.g. Gibson-Graham, 2013; Kawano, 2018; Clement et al., 2019; Federici, 2019), do not see "the commons" exclusively as resources or places to be managed, rather they refer to practices of **commoning** (for example, community food production, land restoration, cooperative housing, and cooperative childcare), understood as a complex set of social processes and relationships that support a shared sense of interdependence and resist neoliberalism's and the state's ongoing and *new* forms of enclosure.

Feminist critic Silvia Federici (2019) argues that contemporary commoning practices have arisen in response to a "new round of enclosures" that create ever higher rates of precarity and dispossession (p. 24). New enclosures include land grabs for fossil fuel extraction and new oil and gas pipelines, water grabs for industrial agriculture and hydroelectric power, roboticization and outsourcing leading to rising unemployment and mass incarceration, conservation programmes like REDD+ creating forest resource enclosures and the loss of Indigenous land rights, seed patenting and the loss of farm and food sovereignty, and austerity regimes calling for the privatization of social entitlements, which end up pricing out most people from access to decent healthcare, housing, and education. These new enclosures, fuelled by the acceleration and expansion of neoliberal and racial capitalism (Pulido and De Lara, 2018), represent an ongoing threat to social reproduction—that is, the capacity to maintain everyday life—for millions of people around the world. In contrast to the slow violence sanctioned by the new politics of enclosure, the politics of the commons, as defined by these ecofeminist thinkers, challenges the logics of extraction and possession and, instead, cultivates relationships founded on reciprocity, mutuality, and, most importantly, *care*. As feminist scholars have long shown, care work, or the work of social reproduction, is a necessity for sustaining everyday life, as well as for social and ecological well-being, but it remains invisible, underpaid, or done for free by women, people of colour, immigrants, and other marginalized groups (MacGregor, 2017; Raghuram, 2019; Clement et al., 2019).

Ecofeminists draw heavily on the many commons that pre-existed or have survived outside or beneath settler colonial contexts in the Americas, including the commoning systems of Native American tribes and the Maroon/African commons that fugitive slaves constructed in the heart of plantation economies (Steady, 2009). They also highlight current examples, including the Zapatista movement, the Standing Rock encampment in North Dakota, and the traditions of *Buen Vivir* or "living well together" (the commons practices created by Indigenous groups in Latin America) as examples of commoning systems that collectively reproduce everyday life, create new social institutions and mutualist safety nets, and build new visions of self-government demonstrating that the world can be organized and lived otherwise (Walsh, 2010; Estes and Dhillon, 2019). But the commons, historically or presently, do not always embody egalitarian or inclusive forms of organization. Despite the claims by some new commons theorists that commoning is about difference—*not* commonality—and that the ideal commons would constantly expand the scope of its participants, the new, self-identified commons movement tends to be very homogeneous and very white (Velicu and Garcia-Lopez, 2018).

Decolonizing the commons and carework: intersectionality and the undercommons

As Cáceres urged, we need to "wake up" and build diverse coalitions to protect our lands, rivers, and homes on Mother Earth. How do we co-create a decolonized idea of the commons and shape a new collective sphere committed to transformative eco-politics in this time of climate urgency? How might intersectional and anti/decolonial approaches to the commons that bridge community and university contexts take shape? In contemplating these questions, I have found helpful the interventions of critical theorists Stefano Harney and Fred Moten (2013), whose widely read open access book, *The Undercommons: Fugitive Planning and Black Study*, zeros in on and interrogates the university as a commons.

Harney and Moten disavow the very idea that the university is a progressive, enlightened place where meaningful answers or solutions to the world's most pressing problems can be found. In fact, they argue, the university in its current form is complicit in all manner of destruction and suffering and has been associated with a host of violent practices (McLeod, 2019). Far from its self-promotional claims that it is producing global leaders who will tackle the world's social and ecological crises, the university, they write, is "dedicated largely to enhancing professionalization, to legitimating the myth of meritocracy, to techno-scientific efficiency, to individualizing and privatizing the pursuit of knowledge . . . and to war" (Harney and Moten, 2013, p. 40). In their blistering critique, Harney and Moten advocate for taking refuge in and engaging in anti/decolonial practices from the location of the **undercommons**, what they describe as a subaltern, subversive way of "being in but not of" the neoliberal university. They argue that "the undercommons is a fugitive network whose escapees are committed to collectivity rather than a university culture bent on creating socially isolated individuals whose conformity to practices of academic distance and claims of objectivity leave the world-as-it-is intact" (p. 35).

Harney and Moten call for the development of subversive intellectual practices that engage a politics of *refusal* committed both to the abolition of oppressive institutions and to the creation of a collective commons practice for research, study, and action. Abolition, they argue, is a standpoint, not just about abolishing oppressive institutions, like prisons, for example, but the abolition of a society forged by racial capitalism that could *have* prisons (p. 42). Likewise, an abolitionist position on academia is not about abolishing the university, but rather about the dismantling of a university system that, for example, could be financed by fossil-fuel–heavy endowments and fails to effectively challenge "the world as it is," recklessly putting at risk the common futures of its *own* students. Scholar/activist members of the emerging interdisciplinary field of Abolitionist University Studies (Boggs et al., 2019) are interested not simply in "getting rid of" the oppressive elements of an institution, but in the healing and repair of the traumatized, damaged relationships and lands brought about by colonialism, in order to prefigure a new university and a new commoning approach to research and action.

Engaged scholars from a wide variety of disciplines propose strategies for "undercommoning," or refusing the settler logics of the university, to co-create anti/decolonial practices for environmental justice research and collaborations dedicated to building our common future grounded in intersectionality (e.g. Grande, 2018; Tuck and Yang, 2018). Included in these strategies is, first, a commitment to collectivity by refusing the individualistic logic underlying the university's incentives system that creates hierarchies of individual worth and labour; and, second, abolitionist/undercommoning advocates endorse a commitment to *reciprocity* and *mutuality* by being accountable to each other and to the communities we claim to serve by co-creating relationships not contingent upon the productionist, single-voice imperatives of academia or the non-reciprocal, extractivist logic of university research branded as "high" impact (Grande, 2018). The critical (not the shallow) conceptualizations of intersectionality—including those being shaped and re-shaped in diverse environmental justice, ecofeminist, anti-racist, anti-ableist, decolonial, LGBTQ, abolitionist assemblages—are building transformative coalitions and envisioning regenerative possibilities for thinking and doing otherwise (e.g. Kafer, 2013; Gaard, 2017; Ray and Sibara, 2017; Haraway, 2018; Jampel, 2018).

Decolonizing our common future: undercommoning in Philadelphia

Intersectionality's power lies in its insistence that our analyses have the capacity to shift meanings and to influence actions in the world. I am interested in strategies for constructing intersectional understandings of both the *commons* (working across different social experiences and identities to challenge racial and fossil capitalism and to support collective well-being of all human and non-human communities) and the *undercommons* (refusing to be an arm of the carceral, imperial, settler state by building spaces outside of its enabling institutions—including the settler academy—to engage in anti/decolonial, mutualistic co-resistance) (Grande, 2018, p. 60). This includes challenging and/or refusing academia-as-usual and the universalizing logic of single-axis environmentalism, and embracing forms of critical, active, and collaborative social praxis armed with the insight that intersectionality kills.

The environmental justice collaborations I work with in Philadelphia embody critical approaches to intersectionality, generating from the ground up incisive analyses of how the city's social, economic, and environmental problems are interwoven. Our work explores and participates in diverse commons-building or commoning/undercommoning practices carried out by community organizations and networks in Philadelphia. This work makes visible the "matrix of domination" (Collins and Bilge, 2016) and complex of intersectional factors that have resulted in the city's ranking as the poorest major city in the US, and contests the standard, racist assumption that Philadelphia's problems stem from its own people's ignorance, lack of development, and incapacity for self-governance (Pew, 2019). Instead, an intersectional, multi-scalar environmental justice analysis helps reveal how global capitalism's "new enclosures" and the history of exploitation and land dispossession that made possible the plantation and industrial economies in the Americas—including the theft of the Indigenous Lenni Lenape people's homelands to establish the Pennsylvania colony—is connected to Philadelphia's current socio-environmental conditions (Black and Chiarappa, 2012). Likewise, an intersectional, historical perspective illustrates that marginalized communities in Philadelphia have always resisted the colonial and environmental violence enacted through these global practices of dispossession and exploitation; histories of *resistance* come to light by tracing the rich abolitionist background and long presence of commoning and cooperative movements in the city (Nembhard, 2014).

Collectivist and solidarity economies are proliferating in Philadelphia as evidenced by the hundreds of local, democratically organized cooperatives focusing on food, housing, energy, healthcare, childcare, credit unions, and schooling, supported and underwritten by the Philadelphia Area Cooperative Alliance (PACA).[11] One such cooperative is Serenity Soular, a collaborative I co-founded in 2012 with community leaders in North Philadelphia (Di Chiro and Rigell, 2018).

Serenity Soular is a campus–community collaboration between students and faculty at Swarthmore College and community members from North Philadelphia with the mission to bring solar technology, sustainable community development, and green jobs training opportunities to local residents in this section of the city.[12] We spell *solar* s-o-u-l-ar to emphasize our commitment to keeping the *soul*—or our connection to the people and to the community—at the forefront of the solar energy transition, which includes ensuring that a just transition to the renewable energy economy prioritizes those communities historically excluded and trampled over in the interests of profit and accumulation. Moreover, we want to keep the *soul* in solar energy to hold our sustainability initiatives true to an intersectional analysis of racial, economic, and environmental injustice, and true to supporting the community's efforts to address the ongoing problems of poverty, unemployment, gentrification, and environmental racism.

Another community commoning and undercommoning strategy in Philadelphia focuses on the city's housing crisis and rampant gentrification leading to the displacement of historically Black neighbourhoods and pricing residents out of their homes. Community Futures Lab, embracing the empowered, future-enacting imaginaries of *Afrofuturism* (Imarisha and brown, 2015), documents stories of African American environmental histories and resilience in North Philadelphia to augment community organizing calling for fair housing policy, neighbourhood security, and the commitment to sustainable *futures* for vulnerable and neglected communities in the city (Kim, 2016).[13] Echoing the determination to fight for "Black futures" (as well as documenting Black histories) that underlies the intersectional organizing of the broader Movement for Black Lives, the Lab adopts the theory of abolition as "a desire to heal, transform and ultimately change the power structures that damaged our communities and families. That is what the next 10 years will call for; that is the answer for our future" (Cullors, 2020).

Inspired by the Lab's approach to documenting histories and imagining thriving Black futures in North Philly, Serenity Soular draws on these storytelling practices to organize "soul-ar stories workshops": community gatherings inviting residents to share their experiences of environmental and climate disruption and visions for a more just and sustainable city.

Other creative methods for cultivating commons and engaging in undercommoning sprout from the network of community gardens and community land trust movements promoting local food sovereignty and reclaiming urban agricultural land rights for the predominantly low-income Black neighbourhoods of North Philadelphia. My campus–community environmental justice action research collaborations join together with Philadelphia Urban Creators, a local undercommoning organization working to foster regenerative urban agriculture and community empowerment through food and agriculture-based initiatives. In a commons-building environmental justice convergence, the Serenity Soular cooperative installed a 14-panel solar array on a building at Urban Creator's "Life Do Grow Farm" that will become a neighbourhood-based, solar-powered storefront selling and distributing organic produce in the heart of North Philadelphia.[14]

Undercommoning is evident in the work of South Philadelphia's Philly Thrive, an environmental justice organization (co-founded by a student who took my environmental justice course in 2013) spearheading the "Right to Breathe" campaign, an alliance among the city's environmental and climate justice groups that joined together to shut down and repurpose the 1,300-acre site of Philadelphia Energy Solutions (PES), the largest oil refinery on the east coast of the US.[15] A classic case of environmental racism, the PES refinery was located in South Philadelphia, directly adjacent to neighbourhoods that are 70% African

American and that struggle with high rates of poverty, substandard housing, and disproportionately high instances of asthma, lung cancer, and other environmental illnesses. In June 2019, the ageing PES refinery experienced a massive explosion, destroying a significant section of the facility and releasing harmful chemicals into the air (Ramirez, 2019). Soon thereafter, the facility was shut down and put up for auction. In the fall of 2019, Philly Thrive led a coalition of environmental and climate justice allies, including the students in my environmental justice course, to demand a voice in the city's decisions about the site's redevelopment, and to advocate for a just transition to a healthier and more sustainable development model. Deploying a coalitional "commons" discourse in their direct action organizing at Philadelphia's City Hall, activists and students held signs saying "No Polluters on the People's Land," decrying the industry and state collusion that has for too long defiled the community's air, water, and land and violated the people's right to a healthy future (Glovas, 2020). At a community meeting with my class after one of the actions, Miss Sylvia, a longtime Philly Thrive member, commented that she was "so proud of the students. They stepped up to support us. We're all in this together; the polluted air from the oil refinery endangers all of us."[16]

I include in this chronicle of the coalitional, anti/decolonial, and care-based strands weaving together an intersectional web of a new commons/undercommons framework emerging in Philadelphia, the new *sanctuary movement*, a movement that at first glance might not seem to fall within the domain of environmental justice. The sanctuary movement in Philadelphia and nationwide has multiplied since 2017 in the wake of the illegal and violent crackdown on immigration by the Trump administration. Pushing against the racist and xenophobic policies implemented by the US Department of Homeland Security and enforced by US Immigration and Customs Enforcement (ICE), the new visions of sanctuary counter with a strong discourse of *belonging*. A designated sanctuary city since 2014, Philadelphia—the City of Brotherly Love—boasts one of the country's highest rates of arrests and deportations by ICE agents (Sontag and Russakoff, 2018). In defiance of these actions, the sanctuary movement aims to ensure that those who seek refuge are welcomed as belonging in the city and belonging as members of the greater human commons. In the spirit of belonging, the movement challenges *all* forms of exclusion, defining "sanctuary" as a big tent inclusive of immigrant and refugee communities, the city's Black and Brown residents, queer and trans youth, people with disabilities, Indigenous communities, and people experiencing homelessness.[17] This decolonized vision of "home" (and, indeed, of "homeland security") prefigures a more capacious, coalitional understanding of what it would mean to protect the commons and our common future.

Philadelphia-based organizations such as Juntos and the New Sanctuary Movement develop intersectional constructs of home and the commons through their work engaging in community defence and care. This includes creating systems of defence (rapid response actions, legal and immigration policy clinics), zones of community resistance (designated safe spaces and homes), mutual aid networks (community fundraisers, mentorship), and care services (childcare sharing, medical/healing skills, transportation sharing, food procurement/gardening, job training, and schooling).[18]

Blending an intersectional environmental justice analysis with the critical perspectives of the new sanctuary movement shows how the enormous increase in migrations of people from Central America to sanctuary cities like Philadelphia can be correlated to a long history of global scale environmental and climate disruptions and violence rooted in colonialism and military intervention in countries from this region (Blitzer, 2019). Climate change-linked food insecurity—triggered by water shortages, severe drought, and a fungal blight that has destroyed many agricultural crops—has, since 2017, forced over 100,000 Guatemalans alone, along with a sizeable number of Hondurans, to migrate north due to malnutrition and starvation, adding to the growing numbers of people from Central America that are projected to become climate refugees in the coming decades (Gustin and Henninger, 2019). Here we return to Cáceres' intersectional environmental justice battle against global capitalism, racism, and patriarchy, the unholy matrix of domination that sought to destroy the homelands, livelihoods, and futures of the Indigenous and peasant farming communities in her country. By mobilizing intersectionality, it becomes clear how the need for sanctuary and the practices of healing, repair, and care are intertwined in the struggle for environmental justice.

I am inspired by the flourishing cultures of solidarity uniting the new sanctuary movement in Philadelphia with the city's many environmental justice organizations. These intersectional coalitions are challenging the interlocking global and local forces that are putting peoples' everyday lives and the very future of the planet at risk by joining together to oppose the fossil fuel industry's presence in the city, decades of economic disinvestment, gentrification decimating the city's historically Black neighbourhoods, and land displacement forcing more and more people to migrate. Moreover, these coalitions are imagining alternatives and co-creating prefigurative economies of restoration, regeneration, and care. While our earth burns out of control, fuelled by global warming, these activists and movements are fired up. They are imagining and practicing commoning/undercommoning, creatively mobilizing the many critical resources needed to build a vibrantly intersectional global commons movement with the power to act with urgency.

Follow-up questions

- In what contexts have you encountered the term intersectionality in your own experience?
- How has your definition of intersectionality changed (if at all) after having read this chapter?
- Why do some feminist and environmental justice critics argue that "intersectionality kills"?
- Describe the challenges you see to building "global unity" to work towards protecting "our common future"? Where do you see some promising opportunities for environmental justice alliance building?
- How does a critical, intersectional environmental justice perspective help build diverse social movement coalitions? What examples of these kinds of coalitions can you think of from your own community(ies)?

Notes

1 From Berta Cáceres' speech at the 2015 Goldman Environmental Prize ceremony in honor of her environmental justice activism: www.goldmanprize.org/recipient/berta-caceres.
2 https://twitter.com/sengillibrand/status/1070106980298186753?lang=en.
3 Examining discrimination suits brought against corporate employers by Black women plaintiffs, Crenshaw (1989) argues that because anti-discrimination law considers only a "single-axis" lens for adjudicating racial or gender subordination, it implicitly imagines, in the case of anti-racist analysis,

the subject of discrimination as male (Black men as the subjects of racism), or the subject as white, as in the case of single-axis anti-sexist/feminist analysis (white women as the subjects of sexism). Crenshaw explains, "the intersection of race and gender renders Black women's experiences invisible" (p. 143). Using the "traffic intersection" metaphor, the experience of discrimination can be envisioned as many different vehicles driving through an intersection from every direction and at some point causing an accident. If a Black woman crossing the intersection is harmed, Crenshaw explains, "her injury could result from sex discrimination or race discrimination" (p. 149). In assessing her injury, one cannot easily determine whether the "skid marks" from the car of sexism or the car of racism occurred separately or simultaneously, thereby making it difficult to determine which caused the harm. As Crenshaw argues, intersectionality, thus, is about *absence*; it is about the invisibility of those whose lives are marked by a multiplicity of structural harms.

4 On decolonial environmental justice, see Chapter 7 of this volume.
5 Cáceres (note 1 in this chapter).
6 Cáceres (note 1 of this chapter).
7 Cáceres (note 1 of this chapter).
8 On *Buen Vivir*, see Chapter 7 of this volume.
9 On sustainability and environmental justice, see Chapter 9 of this volume.
10 While the media spotlights Thunberg as the sole representative of the international youth climate movement (a position she rejects), there have been many earlier appeals by young activists of color to the international community demanding action on environmental and climate justice.
11 PACA: https://philadelphia.coop.
12 Serenity Soular: www.serenitysoular.org.
13 Community Futures Lab: www.facebook.com/communityfutureslab.
14 Philadelphia Urban Creators: www.phillyurbancreators.org.
15 Philly Thrive: www.phillythrive.org.
16 Personal communication, December 4, 2019, Philadelphia Friends Center.
17 Among the diverse sanctuary coalition partners in Philadelphia are the New Sanctuary Movement of Philadelphia: www.sanctuaryphiladelphia.org; and Juntos: https://vamosjuntos.org.
18 See note 17 of this chapter.

References for further reading

Carruthers, C.A. (2018). *Unapologetic: A Black, Queer, and Feminist Mandate for Radical Movements*. Boston, MA: Beacon Press.
Gaard, G. (2017). Posthumanism, Ecofeminism, and Inter-species Relations. In S. MacGregor (ed.), *Routledge Handbook of Gender and Environment*. London: Routledge, pp. 115–129.
Jampel, C. (2018). Intersectionality of Disability Justice, Racial Justice, and Environmental Justice. *Environmental Sociology*, 4(1), pp. 122–135.
Klein, N. (2019). *On Fire: The Burning Case for a Green New Deal*. New York: Simon & Schuster.
Pulido, L. and De Lara, J. (2018). Reimagining 'Justice' in Environmental Justice: Radical Ecologies, Decolonial Thought, and the Black Radical Tradition. *Environment and Planning E: Nature and Space*, 1(1–2), pp. 76–98.
Simpson, L.B. (2017). *As We Have Always Done: Indigenous Freedom Through Radical Resistance*. Minneapolis, MN: University of Minnesota Press.

References cited in this chapter

Albaladejo, A. (2016). Berta Did Not Die, She Multiplied. *The Advocate*, 41(1), Available at: www.lawg.org/wpcontent/uploads/storage/documents/The_Advocate_Spring_2016.pdf
Asmelash, L. (2019). Greta Thunburg Isn't Alone: Meet Some Other Young Activists Who Are Leading the Environmentalist Fight. *CNN*, September 29, Available at: www.cnn.com/2019/09/28/world/youth-environment-activists-greta-thunberg-trnd/index.html
Beal, F. (1969). *Black Women's Manifesto; Double Jeopardy: To Be Black and Female* (3rd ed.). New York: World Women's Alliance (Pamphlet). Available at: www.hartford-hwp.com/archives/45a/196.html

Black, B. and Chiarappa, M. (eds.). (2012). *Nature's Entrepot: Philadelphia's Urban Sphere and its Environmental Thresholds*. Pittsburgh: University of Pittsburg Press.

Blitzer, J. (2019). How Climate Change Is Fueling the U.S. Border Crisis. *The New Yorker*, April 3, Available at: www.newyorker.com/news/dispatch/how-climate-change-is-fuelling-the-us-border-crisis

Boggs, A., Meyerhoff, E. Mitchell, N. and Schwartz-Weinstein, Z. (2019). Abolitionist University Studies: An Invitation. Available at: https://abolition.university/wp-content/uploads/2019/08/Abolitionist-University-Studies_-An-Invitation-Release-1-version.pdf

Carastathis, A. (2016). *Intersectionality: Origins, Constestations, Horizons*. Lincoln: University of Nebraska Press.

Cho, S., Crenshaw, K. and McCall, L. (2013). Toward a Field of Intersectionality Studies: Theory, Applications, and Praxis. *Signs: Journal of Women in Culture and Society*, 38(4), pp. 785–803.

Clement, F., Harcourt, W., Joshi, D. and Sato, C. (2019). Feminist Political Ecologies of the Commons and Commoning. *International Journal of the Commons*, 13(1), pp. 1–15. Available at: http://doi.org/10.18352/ijc.972

Coaston, J. (2019). The Intersectionality Wars. *Vox*, Available at: www.vox.com/platform/amp/the-highlight/2019/5/20/18542843/intersectionality-conservatism-law-race-gender-discrimination?fbclid=IwAR0l7uqXi3AJbhIfRplafY72JIWiBc1DCp22RukIkAcHGuHnp5O08_rruPI&__twitter_impression=true

Collins, P.H. and Bilge, S. (2016). *Intersectionality*. Cambridge, UK: Polity Press.

Crenshaw, K. (1989). Demarginalizing the Intersection of Race and Sex: A Black Feminist Critique of Antidiscrimination Doctrine. Feminist Theory and Antiracist Politics. *University of Chicago Legal Forum*, 1989(1), pp. 139–67.

Cullors, P. (2020). Black Lives Matter Began After Trayvon Martin's Death. Ferguson Showed its Staying Power. *NBC News*, January 1, Available at: www.nbcnews.com/think/opinion/black-lives-matter-began-after-trayvon-martin-s-death-ferguson-ncna1106651

Desmarais, A. (2007). *La Via Campesina: Globalization and the Power of Peasants*. London: Pluto Press.

Di Chiro, G. (2003). Beyond Ecoliberal 'Common Futures': Toxic Touring, Environmental Justice, and a Transcommunal Politics of Place. In D. Moore, J. Kosek and A. Pandian (eds.), *Race, Nature, and the Politics of Difference*. Chapel Hill: Duke University Press, pp. 204–232.

Di Chiro, G. (2018). Canaries in the Anthropocene: Storytelling as Degentrification in Urban Community Sustainability. *Journal of Environmental Studies and Sciences*, 8, pp. 526–538. Available at: https://doi.org/10.1007/s13412-018-0494-

Di Chiro, G. and Rigell, L. (2018). Situating Sustainability Against Displacement: Building Campus-Community Collaboratives for Environmental Justice from the Ground Up. In J. Sze (ed.), *Sustainability: Approaches to Environmental Justice and Social Power*. New York: NYU Press.

DuBois, W.E.B. (1907). *Economic Co-Operation Among Negro Americans Free African Society)*. Available at: https://archive.org/stream/economiccooper00duborich#mode/1up

Dzodan, F. (2011). My Feminism will be Intersectional or it will be Bullshit! *Tiger Beatdown*. Available at: http://tigerbeatdown.com/2011/10/10/my-feminism-will-be-intersectional-or-it-will-be-bullshit/

Estes, N. and Dhillon, J. (eds.). (2019). *Standing with Standing Rock: Voices from the #NoDAPL Movement*. Minneapolis, MN: University of Minnesota Press.

Federici, S. (2019). *Re-enchanting the World: Feminism and the Politics of the Commons*. Oakland, CA: PM Press.

Frank, D. (2018). *The Long Honduran Night: Resistance, Terror, and the United States in the Aftermath of the Coup*. Chicago: Haymarket Books.

Gaard, G. (2017). Posthumanism, Ecofeminism, and Inter-species Relations. In S. MacGregor (ed.), *Routledge Handbook of Gender and Environment*. London: Routledge, pp. 115–129.

Gabriel, R. (2018). Interview with Elizabeth Yeampierre: Capitalism Is Going to Kill the Planet. *NACLA Report on the Americas*, Available at: www.tandfonline.com/doi/abs/10.1080/10714839.2018.1479484?af=R&journalCode=rnac20

Gibson-Graham, J.K. (2013). *Take Back the Economy: An Ethical Guide for Transforming Our Communities*. Minneapolis, MN: University of Minnesota Press.

Global Witness. (2019). *At What Cost?* Available at: www.globalwitness.org/en/campaigns/envi ronmental-activists/at-what-cost/?gclid=EAIaIQobChMIuKuApL3-5gIVgY3ICh1s8AdVEA AYASAAEgKvnPD_BwE

Glovas, K. (2020). Anti-refinery Protesters Want New PES Owner to End Pollution at Site. *KYW News*, January 10, Available at: https://kywnewsradio.radio.com/articles/news/anti-refinery-protesters-want-new-pes-owner-to-end-pollution

Grande, S. (2018). Refusing the University. In E. Tuck and K.W. Yang (eds.), *Toward What Justice?* New York: Routledge.

Gustin, G. and Henninger, M. (2019). Central America's Choice: Pray for Rain or Migrate. *NBC News*, July 9, Available at: www.nbcnews.com/news/latino/central-america-drying-farmers-face-choice-pray-rain-or-leave-n1027346

Hancock, A. (2016). *Intersectionality: An Intellectual History*. New York: Oxford University Press.

Haraway, D. (2018). Making Kin in the Chthulucene: Reproducing Multispecies Justice. In A. Clarke and D. Haraway (eds.), *Making Kin Not Population*. Chicago: Prickly Paradigm Press, pp. 67–99.

Harney, S. and Moten, F. (2013). *The Undercommons: Fugitive Planning and Black Study*. New York: Minor Compositions. Available at: www.minorcompositions.info/wp-content/uploads/2013/04/undercommons-web.pdf

Hoover, E. (2017). Environmental Reproductive Justice: Intersections in American Indian Community Impacted by Environmental Contamination. *Environmental Sociology*, 4(1), pp. 8–21.

Imarisha, W. and brown, a.m. (eds.). (2015). *Octavia's Brood: Science Fiction Stories from Social Justice Movements*. Chico, CA: AK Press.

Jampel, C. (2018). Intersectionality of Disability Justice, Racial Justice, and Environmental Justice. *Environmental Sociology*, 4(1), pp. 122–135.

Kafer, A. (2013). *Feminist, Queer, Crip*. Bloomington, IN: Indiana University Press.

Kawano, E. (2018). Solidarity Economy: Building an Economy for People and Planet. *The NextSystem Project*, Available at: https://thenextsystem.org/learn/stories/solidarity-economy-building-economy-people-planet

Kim, H.N. (2016). An Afrofuturist Community Center Targets Gentrification. *Hyperallergic*, June 22, Available at: https://hyperallergic.com/307013/an-afrofuturist-community-center-targets-gentrification/. Accessed November 30, 2019.

Klein, N. (2019). *On Fire: The Burning Case for a Green New Deal*. New York: Simon & Schuster.

Lorde, A. (1984/2007). *Sister Outsider: Essays & Speeches by Audre Lorde*. Berkeley, CA: Crossing Press, p. 138.

MacGregor, S. (ed.). (2017). *Routledge Handbook of Gender and Environment*. London: Routledge.

McGregor, J. (2018). Toward a Philosophical Understanding of TEK and Ecofeminism. In M.K. Nelson and D Shilling (eds.), *Traditional Ecological Knowledge*. Cambridge, UK: Cambridge University Press.

McLeod, A. (2019). Toward Abolition. *Critique & Praxis*, 13(13), Available at: http://blogs.law.columbia.edu/praxis1313/allegra-mcleod-toward-abolition/

Méndez, M.J. (2018). 'The River Told Me': Rethinking Intersectionality from the World of Berta Cáceres. *Capitalism Nature Socialism*, 29(1), pp. 7–24.

Movement Generation (2019). *From Banks to Tanks to Cooperation and Caring*. Available at: https://movementgeneration.org/wp-content/uploads/2016/11/JT_booklet_English_SPREADs_web.pdf

Nelson, M.K. and Shilling, D. (eds.). (2018). *Traditional Ecological Knowledge*. Cambridge, UK: Cambridge University Press.

Nembhard, J.G. (2014). *Collective Courage: A History of African American Cooperative Economic Thought and Practice*. State College, PA: Pennsylvania State University Press.

Ostrom, E. (1990). *Governing the Commons: The Evolution of Institutions for Collective Action*. Cambridge, UK: Cambridge University Press.

Painter, N.I. (1997). *Sojourner Truth: A Life, A Symbol*. New York: W.W. Norton.

Patterson, J. (2018). Intersectionality and Climate Justice. *MIT Climate Conversations*, February 8, Available at: www.youtube.com/watch?v=W08MGHhAkpk

Pellow, D.N. (2018). *What is Critical Environmental Justice?* Cambridge, UK: Polity Press.
Pew Charitable Trusts (2019). *The State of Philadelphians Living in Poverty.* Available at: www.pewtrusts.org/en/research-and-analysis/fact-sheets/2019/04/the-state-of-philadelphians-living-in-poverty-2019
Pulido, L. and De Lara, J. (2018). Reimagining 'Justice' in Environmental Justice: Radical Ecologies, Decolonial Thought, and the Black Radical Tradition. *Environment and Planning E: Nature and Space*, 1(1–2), pp. 76–98.
Raghuram, P. (2019). Race and Feminist Care Ethics: Intersectionality as Method. *Gender, Place & Culture*, 26(5), pp. 613–637.
Ramirez, R. (2019). Residents Say they've Already had Enough as Investigation Starts into Philadelphia Refinery Fire. *Grist*, Available at: https://grist.org/article/residents-say-theyve-already-had-enough-as-investigation-starts-into-philadelphia-refinery-fire/
Ramsey, A. (2014). My Environmentalism will be Intersectional or it will be Bullshit. *Open Democracy*, Available at: www.opendemocracy.net/en/opendemocracyuk/my-environmentalism-will-be-intersectional-or-it-will-be-bullshit/
Ray, S.J. and Sibara, J. (eds.). (2017). *Disability Studies and the Environmental Humanities.* Omaha, NB: University of Nebraska Press.
Reagon, B.J. (1983). Coalition Politics: Turning the Century. In B. Smith (ed.), *Home Girls: A Black Feminist Anthology.* New York: Kitchen Table Women of Color Press, pp. 356–368.
Romano, A. (2016). This Feminist's Most Famous Quote has been Sold all over the Internet. She hasn't Seen a Cent. *Vox*, Available at: www.vox.com/2016/8/12/12406648/flavia-dzodan-my-feminism-will-be-intersectional-merchandise
Simpson, L.B. (2017). *As We Have Always Done: Indigenous Freedom Through Radical Resistance.* Minneapolis, MN: University of Minnesota Press.
Sontag, D. and Russakoff, D. (2018). In Pennsylvania, It's Open Season on Undocumented Immigrants. *ProPublica*, April 12, Available at: www.propublica.org/article/pennsylvania-ice-undocumented-immigrants-immigration-enforcement
Spade, D. (2013). Intersectional Resistance and Law Reform. *Signs: Journal of Women in Culture and Society*, 38(4), pp. 1031–1055.
Steady, F. (2009). *Environmental Justice in the New Millennium: Global Perspectives on Race, Ethnicity, and Human Rights.* London: Palgrave MacMillan.
Sze, J. (2020). *Environmental Justice in a Moment of Danger.* Oakland, CA: University of California Press.
Taylor, K.Y. (ed.). (2017). *How We Get Free: Black Feminism and the Combahee River Collective.* Chicago: Haymarket Books.
Thunberg, G. (2019). Our House is On Fire. *The Guardian*, Available at: www.theguardian.com/environment/2019/jan/25/our-house-is-on-fire-greta-thunberg16-urges-leaders-to-act-on-climate
Todd, Z. (2015). Indigenizing the Anthropocene. In H. Davis and E. Turpin (eds.), *Art in the Anthropocene: Encounters Among Aesthetics, Politics, Environment and Epistemology.* Open Humanities Press, pp. 241–260. Available at: https://law.unimelb.edu.au/__data/assets/pdf_file/0005/3118244/7-Todd,-Zoe,-Indigenizing-the-Anthropocene.pdf
Tuck, E. and Yang, W.K. (2018). R-Words: Refusing Research. In E. Tuck and W.K. Yang (eds.), *Toward What Justice?* New York: Routledge.
Velicu, I. and Garcia-Lopez, G. (2018). Thinking the Commons Through Ostrom and Butler: Boundedness and Vulnerability. *Theory, Culture & Society*, 35(6), pp. 55–73.
Vinal, S. (2017). Berta Didn't Die, She Multiplied. *Documentary Short. Mutual Aid Media Productions*, June, Available at: www.imdb.com/title/tt7070686/
Walsh, C. (2010). Development as Buen Vivir: Institutional Arrangements and (De)Colonial Entanglements. *Development*, 53(1), pp. 15–21.
White, M. (2018). *Freedom Farmers: Agricultural Resistance and the Black Freedom Movement.* Chapel Hill: University of North Carolina Press.
Whyte, K.P. (2018). Indigenous Science (fiction) for the Anthropocene: Ancestral Dystopias and Fantasies of Climate Change Crises. *Nature and Space*, 1(1–2).

Whyte, K.P., Caldwell, C. and Schaefer, M. (2018). Indigenous Lessons About Sustainability are not just for 'All Humanity'. In J. Sze (ed.), *Sustainability: Approaches to Environmental Justice and Social Power*. New York: NYU Press.

Willow, F. (2018). Intersectionality is Important for Environmental Activism, Too. *Ethical Unicorn*, Available at: https://ethicalunicorn.com/2018/02/09/intersectionality-is-important-for-environmental-activism-too/

World Commission on Environment and Development (1987). *Our Common Future*. Oxford, UK: Oxford University Press.

Yusoff, K. (2018). *A Billion Black Anthropocenes or None*. Minneapolis, MN: University of Minnesota Press.

Index

Note: page numbers in *italic* indicate a figure and page numbers in **bold** indicate a table or box on the corresponding page.

academic-activist nexus 88
academic debates on 11–12
access 40; access patterns **210**
accumulation by dispossession 79–80
activism 176–177, 187–188; academic-activist nexus 88; anti-coal activism 3; romanticizing 242–243; urban agriculture as **185**; *see also* protests/protesters; solidarity activism; sustainable materialist action
actors 216–218
adaptation *see* climate change adaptation
Africa **200–201**
agency 284
agents, moral **281**
agriculture 4, 176–180; agro-export **210**; in the environmental justice movement 179; and food justice 180–184, 186–187; and food sovereignty 184–187 *see also* Black farmers; farmworker livelihoods; urban agriculture
a-growth 95
alignment 214–215
allocation, water 209, **210**, **211–212**, 212–214
alternative food initiatives (AFIs) 181
alternative food movement 181
alternative worldviews 143
ambition 251–253
androcentrism: anti-androcentrism principle **237**
animal rights approach 4
animals *see* animal rights approach; non-human animals
Anthropocene 114, 322
anthropocentrism 137–138, 279; non-anthropocentric theories of justice 285–287
anthropogenic climate change 238
anti-androcentrism principle **237**
anti-colonial action 266–275
anti-colonial environmental justice 321
anti-exploitation principle **237**
anti-marginalization principle **237**
anti-poverty principle **236**

anti-toxics movement 123; global 125–126; in the United States 122–124
anti-toxic struggles 15n2, 122–126; *see also* anti-toxics movement
Asia: rapid urbanisation in **200–201**
Aspen (Colorado): population control in 230–231
attachment 309

Baltimore: water quality in **110**
being: coloniality of 85
bio-centrism 282
biocultural diversity 143
biodiversity 3, 132–133, **136–137**, 143–144; and conservation injustices 137–139; and the extinction crisis 134–137; and just conservation 141–143; and protected-area conservation 139–141; and recognition **60–61**
biopiracy 100
biopower 111
Black farmers **180**
Black feminism 318
Black Love Canals 7, 126–128
Bolivia **82**
Buen Vivir 83, **83**, 100, 143, 321

California 26
Cancer Alley (Mississippi Chemical Corridor) 124, **124**
capabilities 3, 44, 64–65, **65**, 75, 79; capabilities approach 22, 71–74, 75n7, 162, 308–309, 311–312; and distributive environmental justice 71–73; and the environment 68–70; and environmental justice 71–74; and justice 308–310; Nussbaum's list of central human capabilities **67**; and procedural environmental justice 72–73; and recognition-based environmental justice 73–74; as well-being 65–68

capabilities approach 22, 71–74, 75n7, 162, 308–309, 311–312; *see also* minimum-capabilities approach
cap-and-trade 27
carework: collective care 322–323; decolonizing 324–325
case study: evictions in Chavez Ravine 229–230; Japanese-American internment 228–229; population control in Aspen, Colorado 230–231; US prisons and critical environmental justice **295**
Chavez Ravine: evictions in 229–230
choice 250–253
cities *see* urbanization
class domination 10
climate change, anthropogenic 238
climate change adaptation 149
climate justice 3, 100, 148–149, 158, 163; calling for **152–153**; climate justice pyramid **156–158**, *157*; demanding global climate justice 241–242; as an ethical question 149–152; as a political question 153–156
coalition 321–325
cognitive justice 87–88
collective care 322–323
colonialism 110, 272; *see also* anti-colonial action; coloniality; settler colonialism
coloniality 79; of the being 85; environmental injustices generated by **86**; of knowledge 85; of power 85; *see also* settler colonialism
coloniality of power 85
Colorado 227, 230–231, 257
commoning 323
common property resource 166
commons 103, 321–325; *see also* undercommons
community 309; *see also* community energy; community food security; frontline communities
community energy 304
community food security 181
complementarities: between environmental justice and degrowth 100–102, **102**; and water injustice 209–215
conflict 132–133, 143–144; and conservation injustices 137–139; and the extinction crisis 134–137; and just conservation 141–143; and protected-area conservation 139–141
conservation: biodiversity 60; debates about the future of **141–142**; towards just conservation 141–143; *see also* conservation injustices; conservation justice; fortress conservation; protected-area conservation
conservation injustices 137–139
conservation justice 132–133; towards just conservation 141–143
conserved areas 142–143
contexts 25–28

control 212–213
cosmopolitanism 151, 155, 229, 286
cosmovisions 322
crisis 132–133, 143–144; and conservation injustices 137–139; the extinction crisis 134–137; and just conservation 141–143; and protected-area conservation 139–141
critical environmental justice 5, 109; critical environmental justice studies 293–301; and prisons **295**
critical sustainabilities 112–115
cultural turn 54

Dakota Access Pipeline (DAPL) 60, 111; #NoDAPL 111
DAPL *see* Dakota Access Pipeline
Day, Rosie xvii, 3, 168, 170
decision-making 212–213
decolonial environmental justice 3, 78–79, 319; Latin American 78–89; roots of 79–84; theory of 84–89
decolonial perspective 114
decolonial theory 59
decolonial turn 78
decolonization 78, 324–325, 325–328
defending land and livelihoods 240
degrowth 3, 95, 322; and environmental justice 94–103, **102**
demand management 170–171
democracy *see* ecological democracy; participatory democracy; sentientist cosmopolitan democracy
development: resisting and mobilizing 80–81; *see also* sustainable development
difference 84
difference principle (John Rawls) 23
differing preferences 24
dignity 55
discursive power 136
disobedience 229
dispossession: accumulation by 79–80; historical process of 79–80; *see also* energy dispossessions
disrespect 55
distribution **57**, 138, 250; of risk or harm 30–31; water distributive inequities 210–211; *see also* ecological distribution conflicts
distributional injustice 28–29; causes of 31–32; responding to 32–33; *see also* distributive injustice
distributional justice 161; *see also* distributive justice
distributive environmental justice 21, 33; and capabilities 71–74; and distributive injustice 32–33; distributive justice and lived experience 25–32; theories of distributive justice 22–25
distributive injustice 28–29, 31–32, 313; responding to 32–33

Index

distributive justice 2, 12, 21, 78, 161, 202, 217, 235; and lived experience 25–32; theories of 22–25; *see also* distributive environmental justice
diversity 317; water normative 212–213
domination, matrix of 318

eco-centric approach 4, 286–287
ecocentrism 137–138; *see also* eco-centric approach
ecofeminism 4, 103, 234, 243–245, 323; environmental justice 243–244
ecogentrification 183
ecological catastrophism 128
ecological climax theory 140
ecological debt 100, 156
ecological democracy 304
ecological distribution conflicts (EDC) 95–96
ecological integrity 207
ecological justice 252; interconnected 143
economics *see* steady-state economics
ecosystem services 68, 111
ecosystems 268–270
EDC *see* ecological distribution conflicts
Ende Gelände ('Here and no further') protests 103
energy consumption 167–171
energy dispossessions 166
energy efficiency 170
energy justice 3, 161–162, 306; challenges and future directions 171; justice and energy consumption 167–171; justice in energy production 162–167
energy poverty 169
energy production 162–167
energy services **167–168**
energy sovereignty 155
engaged understanding: of water justice 216–218
environment, the: and capabilities 68–70; the value of 281–282
environmental conflicts 96
environmental enforcement 28
environmental ethics 270
environmental governance 37–39; moving towards procedural justice 46–48; and participation 40–43, **43**; when participation is not justice 44–46; why procedural justice matters 39–40, 48
environmental heritages 13–15
environmental histories 12, 13–15
environmental identities 13–15
environmental impacts: of urbanisation **196–197**
environmental inequality 10, 122, 136, 226, 293; drivers of **198–199**; and urbanisation 194, **196**, 203
environmental injustices: and drivers of urban environmental inequality **198–199**; and environmental impacts of urbanization **196–197**; generated by different forms of coloniality **86**; and patterns of exposure and risk 197–200; policy recommendations and ideas for reform 201–203; strategies of resistance against 8–10; and uneven impacts of urbanization 194–196; and urbanization 193–203
environmentalism: re-envisioning **228**; *see also* nativist environmentalism; social environmentalism
environmental justice 64–65, 75, 107–109, 115, 176–180, 187–188, 299–300; academic debates on 11–12; and capabilities 65–74; and degrowth 94–103, **102**; ecofeminist 243–244; environmental injustices generated by different forms of coloniality **86**; and food justice 180–184, 186–187; and food sovereignty 184–187; gender in 234–245; and gender inequality 234–238; history of 6–15; Indigenous environmental justice advocacy 275; and labour unions 249–259; Latin American mobilizations **81–82**; and material concerns 96–99; mobilizing intersectionality for 321; naming and framing problems 111–112; and pluralistic justice 310–312; and the politics of intersectionality 319–321; and the politics of the social metabolism **97–98**; and post-growth 95–96; a prehistory 253–255; prisons and critical environmental justice **295**; and recognition 52–62; reimagining 225–232; romanticizing women's activism 242–243; and strategies of resistance 8–10; and sustainability 109–111, 112–115; and sustainable materialism 303–310, 312–314; and toxic legacies 121–128; and urbanization 193–203; US context of 7–8; why gender matters 238–243; *see also* anti-colonial environmental justice; critical environmental justice; decolonial environmental justice; distributive environmental justice; environmental justice action; environmental justice movement; environmental justice research; environmental reproductive justice; global environmental justice; indigenous environmental justice; multi-species environmental justice; procedural environmental justice; recognition-based environmental justice
environmental justice action 316–317; coalition and the commons 321–325; decolonizing our common future 325–328; and food and agriculture 179; and the politics of intersectionality 319–321; and the term intersectionality 317–319
environmental justice movement *see* environmental justice action

environmental justice research 316–317; coalition and the commons 321–325; decolonizing our common future 325–328; and the politics of intersectionality 319–321; and the term intersectionality 317–319
environmental privilege 13
Environmental Protection Agency 7
environmental racism 7–8, 59, 108, 197, 225, 299–300
environmental regulations 26–27
environmental reproductive justice 296
environmental sexism 237
epistemic violence 87
equality 22–24, 214; equality of respect **237**; income equality **237**; leisure-time equality **237**
equalization 214
ethics: and climate justice 149–152; environmental 270; *see also* kinship ethics
everyday needs **167–168**
evictions 229–230
expendability 294–295
exploitation: anti-exploitation principle **237**
export *see* agro-export
exposure, patterns of 197–200
extinction crisis 134–136
extractivism 15

fairness 150
fair water worlds *see* water justice
farming *see* Black farmers; farmworker livelihoods; urban agriculture
farmworker livelihoods **177–178**
feminism *see* Black feminist activism; ecofeminist perspective
First National People of Color Environmental Leadership Summit 41
Flint water crisis **28**
food 176–180; food, agriculture, and environmental justice 177–180; food justice 180–184, 186–187; food sovereignty 184–187;in the environmental justice movement 179; *see also* food desert; food justice; food security; food sovereignty; food systems
food desert **182–183**, 200
food justice 180–184, 306; evolving areas of focus in 186–187
food security 181, **210**; *see also* community food security
food sovereignty 100, 184–186; evolving areas of focus in 186–187
food systems **136–137**, 176
foreignness 227
forestry **114**
fortress conservation 139
framing 111–112
France 139

Fraser, Nancy 12, 79, 154, 162; and gender 236–237, 244; and recognition 54, 56–59
Fraser–Honneth debate **57**
free prior informed consent (FPIC) 45–46, **46**
frontline communities 22, 37
frontlines 10; *see also* frontline communities
fuel insecurity 199
fuel poverty 167
functionings **65**
future generations 170

gender 10–11, 235; in environmental justice 234–245; why gender matters 238–243
gender balance 238
gender inequality 235; and environmental justice 234–238
gender justice 235, **236–237**
genocide 300
Germany 134, 203, 257; *Ende Gelände* ('Here and no further') protests 103
global climate justice 241–242
global environmental justice 13–15, *see also* global justice
globalization: of just transition 255–256
global justice 87–88
global poor 101
Global South: and energy poverty 169; renewable energy production in 166–167
global union organizations 255
grabbing *see* land-grabbing; water-grabbing
Green Belt Movement **240–241**
growth: and poverty **101**

harm: distribution of 30–31
health *see* human health
Hegel, Georg Wilhelm Friedrich: inter-subjectivity 53–54; master–slave dialectic **53–54**
heritages *see* environmental heritages
heteropatriarchy 294
historical process 79–80
histories: the emergence of just transition 253–255; the globalization of just transition 255–256; a prehistory of labour unions and environmental justice 253–255; the proliferation of just transition 256–258; *see also* environmental histories; historical process
Honneth, Axel 55–56
human agency 284
human health 239–240
humanity: justice beyond 279–288
hydro-social territory 215

identity 54; *see also* environmental identities
inclusion 317
income equality **237**
Indigenous environmental justice 266–275

Index

Indigenous environmental justice advocacy 275
Indigenous peoples 266
Indigenous territorial rights 100
indispensability 109
inequality 135; drivers of urban environmental inequality **198–199**; transit inequality 203; *see also* gender inequality
infrastructure 27–28
injustice *see* conservation injustices; distributive injustice; environmental injustices; participatory injustice; procedural injustice; social injustice; toxic injustice; water injustice
integrated water resources management (IWRM) 213
interactive practices: of water injustice 209–215
interculturality 88
intercultural challenge 88–89
intergenerational justice 109, 133, 148–149, 158
intergovernmental organizations (IGOs) 256
internalization 58
internal pluralism 311
internment, Japanese-American 228–229
intersectionality 109, 316–317, 317; coalition and the commons 321–325; constant struggle of 318–319; decolonizing our common future 325–328; politics of 319–321; the term 317–319; and the undercommons 324–325
intersections 184–186
inter-subjectivity 53–54
intra-generational justice 133

Japanese-American internment 228–229
joined-up approaches 170–171
just deserts 23
justice 132–133, 143–144; and capabilities 308–310; concept of 283–285; and conservation injustices 137–139; and energy consumption 167–171; energy production 162–267; and the extinction crisis 134–137; beyond humanity 279–288; towards just conservation 141–143; non-anthropocentric theories of 285–287; and participation 44–46; and power 307–308; and protected-area conservation 139–141; beyond recognition 85–87; in renewable energy production 163–165; and the social contract 288; and solar panel production and disposal **165**; in sustainable materialist movements 306–310; *see also* climate justice; cognitive justice; conservation justice; distributive justice; ecological justice; energy justice; environmental justice; environmental reproductive justice; food justice; gender justice; global justice; intergenerational justice; intra-generational justice; just transition; landscape justice; pluralistic justice; procedural justice; retributive justice; social justice; water justice
just sustainability 112–115, 304
just transition 4, 249–265, 249–250, 258–259, 304, 322; the emergence of 253–255; the globalization of 255–256; the proliferation of 256–258; voice, choice and ambition in 251–253

kinship 266–275, 267; and anti-colonial action 270–275; Indigenous environmental justice advocacy through 275; kinship ethics 4, 60, 275
kinship ethics 4, 60, 275
knowledge: coloniality of 85

labour unions 249–253, 258–259; and the emergence of just transition 253–255; and the globalization of just transition 255–256; global union organizations 255; a prehistory 253; and the proliferation of just transition 256–258
land: defending land 240; land-grabbing 100, 102, 104n8, 187; land use patterns 25–26
land-grabbing 100, 102, 104n8, 187
landscape justice 164
land use patterns 25–26
Latin America: decolonial environmental justice 78–79; environmental justice mobilizations **81–82**; and roots of decolonial environmental justice 79–84; and theory of decolonial environmental justice 84–89
layers 216–218
legitimacy 216
leisure-time equality **237**
lived experience: and distributive justice 25–32
livelihoods: defending livelihoods 240; *see also* farmworker livelihoods
Love Canal (New York) 7, **123**, 239

management: integrated water resources management **213**
marginalization: anti-marginalization principle **237**
Martínez-Alier, Joan xviii, 3, 95–98, 100, 102
master–slave dialectic **53–54**
material concerns 96–99
materialism *see* sustainable materialism
material participation 306–307
matrix of domination 318
measuring 29–30
Menominee tribal lands **114**, 274
meta-capability 71
methodology 306
migration 225–232
minimum-capabilities approach 24–25, 33

minorities *see* minority move-in hypothesis; racial minorities
minority move-in hypothesis 122, 299
misrecognition *see* recognition
Mississippi Chemical Corridor (Cancer Alley) 124, **124**
mitigation 148
mobility 317–319
mobilizing 317–319; and development 80–81; intersectionality for environmental justice 321
modernity 79; transformation of 79–84; transgressing 82–84
moral agents **281**
moral patients **281**
moral status: of non-human animals 280–281
more-than-human 11, 112
motivation 305
movement-generated theory 306
multiculturalism 54–55
multi-species environmental justice 320

naming 111–112
National People of Color Environmental Leadership Summit 122
nativist environmentalism 4, 226–231
natural resource extraction 230
nature 109–111
natureculture 320
need 23
need-based theory 23
needs *see* everyday needs; need; need-based theory
neocolonialism 79–80
new social movements 54
New York *see* Love Canal; New York City
New York City 32
New Zealand 85, 138, 167, 257, 271, **287**
NIMBY (not in my backyard) 96, 163
non-anthropocentric theories of justice 285–287
non-human animals 4, 60, 74, 171, 283, **287**; and biodiversity 133, 137–138, 143; and gender 239; and Indigenous environmental justice 268, 271–272, 274–275; and intersectionality 320, 322, 325; and justice beyond humanity 279, 282, 284–288; moral status of 280–281; and utilitarianism **280–281**
North Carolina *see* Warren County
not in my backyard *see* NIMBY
Nussbaum, Martha 66–68, 71–74, 79, 162, 307; list of central human capabilities **67**

oppression, systems of 313, 318
otherness 53, 214; construction of 85–87

parameters 250, 252
Paris Agreement 249
parity of participation 56–57
participation 42, 58, **178**, 252; advocacy, Indigenous environmental justice 275; in environmental governance 37–48, **43**; and justice 44–46; moving towards procedural justice 46–48; politics of 212–213; and procedural justice 40–43; tools for **43**; tyranny of **44**; when participation is not justice 44–46; why procedural justice matters 39–40, 48; *see also* material participation; parity of participation; political participation
participatory democracy 202
participatory injustice 24
patients, moral **281**
patriarchy 243; heteropatriarchy 294
patterns of exposure 197–200
patterns of risk 197–200
payment-for-environmental services 215
Pellow, David N. xix, 5, 78, 244, 312; and sustainability 109, 115; and toxic legacies 122, 125, 127–128
Pemon struggle **81–82**
People of Color Environmental Justice Leadership Summit 154
Philadelphia: undercommoning in 325–328
pluralism, internal 311
pluralistic justice 310–312
plurinational state 82
pluriverse 82–84
policy: and food sovereignty 184–186; and urbanization 201–203
policy recommendations: and ideas for reform 201–203
political participation 306–307
polluter pays principle 99, 150
population control 230–231
postcolonial perspective 114
post-growth 95–96
post-materialism 97, 305
poverty: anti-poverty principle **236**; and growth 101; and rural to urban water transfers 211–212; *see also* global poor
power 39, 321; coloniality of 85; in environmental governance 37–48; and justice 307–308; moving towards procedural justice 46–48; and participation 40–43, **43**; state power 294; when participation is not justice 44–46; why procedural justice matters 39–40, 48; *see also* discursive power; energy poverty; fuel poverty
preferences *see* differing preferences
prehistory 253–255
prisoners 299–300
prisons 299–300; and critical environmental justice **295**

privilege *see* environmental privilege
problems: naming and framing 111–112
procedural environmental justice 162; and capabilities 72–73
procedural injustice 22
procedural justice 11, 37–39, 202, 235, 306–307; in environmental governance 39–40, 48; key concepts in **38**; moving towards 46–48; and participation 40–43; *see also* procedural environmental justice
procedure 138; *see also* procedural environmental justice; procedural injustice; procedural justice
proportionality 47
protected-area conservation 139–141
protected areas 133; towards justice in 142–143
protests/protesters 6–7, 60; and critical environmental justice studies **295**, 296–298; and decolonial environmental justice 80–81; and degrowth 96; and gender 239–240; and labour unions 253; and race and migration 226; and sustainability 122
pyramid *see* climate justice

race 7, 225–232
racial expendability 127–128
racial minorities 225–232
racial segregation **26**
racism *see* environmental racism
Rawls, John 22–23
reasons 282–283
reciprocity 268, 268–270
recognition 12, 52, 73, 138, 154, 162, 186, 202, 251; and biodiversity conservation **60–61**; critiques of 58; and environmental justice 52–62; injustices **136–137**; justice beyond recognition 85–87; misrecognizing water normative diversity 212–213; politics of 212–213; subtle recognition politics 209, 214–216; theories of 53–58; versus distribution **57**; *see also* recognition-based environmental justice
recognition-based environmental justice 73–74
redistributive justice 202; *see also* distributive justice
reform: and urbanization 201–203
regulations *see* environmental regulations
reimagining 225–232
relational understanding: of water justice 216–218
relationships 269, 310
renewable energy 166; *see also* renewable energy production
renewable energy production: in the Global South 166–167; justice in 163–165
representation 39; in environmental governance 37–48; and moving towards procedural justice 46–48; and participation 40–43, **43**; when participation is not justice 44–46; why procedural justice matters 39–40, 48
reproductive justice *see* environmental reproductive justice
reproductive rights 242; *see also* environmental reproductive justice
residential differentiation 195
resistance: and development 80–81; *see also* strategies of resistance
resource management: integrated water resources management **213**
respect, equality of **237**
responsibilities 150
retributive justice 284
Richmond (California) **26**
rights 282–283; *see also specific rights*
risk: distribution of 30–31; patterns of 197–200
romanticizing 242–243

sacrifice zone 128
scale 294
scales 216–218; *see also* temporal scales
Schlosberg, David xix, 5, 59, 74, 75n6, 78–79, 286
scholarship 176–177, 187–188; food, agriculture, and environmental justice 177–180; food justice 180–184, 186–187; food sovereignty 184–187
self-determination 187, 271
sentience 280
sentientist cosmopolitan democracy 286
settler colonialism 8
sexism *see* environmental sexism
significant others 55
siting 31–33, 164, 202–204, 253; decisions 25–26, **26**, 31–32, 122, 203; practices 7–8, 14
situated sustainability 112–115
slow Bhopals 126–128
slow violence 126–127
social contract 288
social environmentalism 253
social injustice 24
social justice: interconnected 143
social metabolism: the politics of **97–98**
solar panels: and justice concerns **165**
solidarity activism **177–178**
spheres of interaction 55–56
Standing Rock Indian Reservation 60, 111, 323
state power 294
steady-state economics 95
strategies of resistance: against environmental justice 8–10
sustainability 3, 107–115, 163, 317
sustainable development 109–111, 140
Sustainable Development Institute **114**
sustainable forestry **114**

sustainable materialism 5, 313–314; introduction to 303–306; limits of sustainable material action 312–313; and pluralistic justice 310–312; three notions of justice in 306–310
sustainable materialist action 303–306; limits of 312–313; and motivation 305; notions of justice in 306–310
systems of oppression 313, 318

Taylor, Charles: multiculturalism 54–55
temporal scales 296–297
terra nullius 12
tools for participation **43**
toxic disasters 3
toxic injustice 122
toxic legacies 121–128
transformation 253
transit inequality 203
tribal lands *see* Menominee tribal lands; Standing Rock Indian Reservation
truth regimes 208
tyranny of participation **44**

UCC *see* United Church of Christ (UCC)
undercommons 324–328
uneven impacts, urbanization 194–196
unions *see* labour unions
United Church of Christ (UCC): Commission for Racial Justice 8–10
United States (US/USA): Black farmers in **180**; prisons and environmental justice in **295**; racial minorities in 225–232; US context of environmental justice 7–8; *see also specific states and cities*
urban agriculture: as activism **185**
urban environmental inequality: drivers of **198–199**

urbanization 4, 193–194, 203; in Asia and Africa **200–201**; and drivers of urban environmental inequality **198–199**; environmental impacts of **196–197**; and patterns of exposure and risk 197–200; policy recommendations and ideas for reform 201–203; uneven impacts of 194–196
utilitarianism 22, 283; and non-human animals **280–281**

value: of the environment 281–282
Venezuela **81–82**
Vivir Bien see Buen Vivir
voice 250–251, 251–253
vote-with-your-fork **178**

Warren County (North Carolina) 6, 122
water 4; integrated water resources management 213; rural to urban water transfers **211–212**; water allocation 209, **210**, **211–212**, 212–214; water distributive inequities 210–211; water-grabbing 208 210, 214–217; water normative diversity 212–213; water quality in Baltimore **110**; Water War in Bolivia **82**; *see also* Flint water crisis; water injustice; water justice
water allocation 209, **210**, **211–212**, 212–214
water distributive inequities 210–211
water-grabbing 208–210, 214–217
water injustice 209–215
water justice 207–209; a relational and engaged understanding of 216–218
water rights 212
well-being 24–25, 64, 69–72, 75, 239; capabilities as 65–68
women: romanticizing activism of 242–243; *see also* gender
worldviews: towards alternative worldviews 143